Ecosystem Management
of Forested Landscapes

Edited by Robert G. D'Eon,
Jacklyn F. Johnson, and
E. Alex Ferguson

Ecosystem Management of Forested Landscapes: Directions and Implementation

Proceedings of a conference held in Nelson,
British Columbia, Canada, 26-28 October 1998

KEYNOTE ADDRESS BY J.P. (HAMISH) KIMMINS

Ecosystem Management of Forested Landscapes
Organizing Committee

© 2000 Ecosystem Management of Forested Landscapes Organizing Committee

ISBN 0-9686541-0-X

Canadian Cataloguing in Publication Data

Main entry under title:
Ecosystem management of forested landscapes

Papers from the conference, Ecosystem Management of Forested Landscapes,
held in Nelson, B.C. on Oct. 26-28, 1998.
Includes bibliographical references.
ISBN 0-9686541-0-X

1. Forest ecology – Congresses. 2. Forest management – Congresses. I. D'Eon,
Robert G. (Robert George), 1964- II. Johnson, Jacklyn F. (Jacklyn Flett), 1939-
III. Ferguson, E. Alex. IV. Ecosystem Management of Forested Landscapes
Organizing Committee. V. Slocan Forest Products Ltd.

QH541.5.F6E36 2000   333.75   C00-910022-9

Distributed by UBC Press, Vancouver, BC. 352 p.

Copies available through:
Slocan Forest Products
705 Delaney Avenue
Slocan, BC, V0G 2C0, Canada
tel: 250-355-2100
fax: 250-355-2168
www.ecosystem-management.com

and

UBC Press
University of British Columbia
2029 West Mall
Vancouver, BC, V6T 1Z2, Canada
tel: 604-822-5959
fax: 604-822-6083
www.ubcpress.ca

Contents

Foreword

E. Alex Ferguson

The forest industry in North America has been methodically evolving toward the goal of a sustainable forest management framework. Recognition of social and ecological values in our ecosystems has increased substantially in recent years. Many in the industry are trying to find new ways of addressing these values while providing for economic stability. Within our planning initiatives at Slocan Forest Products (Slocan Division) we have embraced the concepts and methods of ecosystem management. While there is still much to be learned, we have realized success beyond what was thought possible a short time ago. I strongly believe that developing and implementing an ecosystem management framework can provide solutions to many challenges facing the industry today.

Building on a peer review of our planning process at Slocan Forest Products in the spring of 1998, we envisioned a gathering of leading industry, government, and academic representatives to bring state of the art concepts in ecosystem management to interior British Columbia. In the fall of 1998, this vision became a reality when 56 presenters from across North America and the United Kingdom delivered three days of workshops, posters, and presentations to over 240 symposium participants. This book is a lasting record of this achievement. All reasonable effort was expended to include all presenters in these proceedings. Most, but regrettably not all, presenters at the symposium were able to contribute a manuscript for inclusion. While this collection of papers does not contain all the answers, it is a significant step forward in our continuing effort towards better management of our forest resource.

Slocan Forest Products Ltd. was the principle sponsor of the symposium and I would like to acknowledge their financial support. I am also grateful for corporate sponsorship or support in kind received from the British Columbia Association of Registered Professional Biologists, the city of Castlegar, Forest Renewal BC, Kalesnikoff Lumber Company, Hugh Hamilton Ltd., the Interior Lumber Manufacturers' Association, Jerome Timber Company,

Kokanee Forests Consulting, PRT-Harrop Nursery, Riverside Forest Products Ltd., Selkirk College, Summit Environmental Consultants, Thompson Forest Management, Timberland Consultants Ltd., the University of British Columbia Centre for Applied Conservation Biology, and Vista Design Ltd.

Many individuals worked long and hard to make this event possible. I would like to take this opportunity to especially thank our organizing committee: Kandy Akelson, Reiner Augustin, Peter Barss, Rob D'Eon, Doug Ellis, Pat Field, Angus Graeme, Kathy Howard, Roger Ennis, and Paul Jeakins. The first day of the symposium was occupied with a series of well-attended workshops and field tours. I wish to express my gratitude to workshop and field tour providers: Reiner Augustin, Peter Barss, Rob D'Eon, Linnea Ferguson, Sue Glenn, Paul Jeakins, Guoliang Lu, John Nelson, Chris Niziolomski, Grant Thompson, Ralph Wells, and Tim Yanni. I thank very much Garry Merkel, Don Wright, and Gordon Weetman for their opening remarks as well as Janna Kummi and Daryll Hebert for their lunch hour addresses. Students from Selkirk College provided much appreciated volunteer services during the symposium. A special thanks goes to Karyn Zuidinga for providing assistance with preparation of manuscripts. Finally, I am especially grateful to all of our speakers and poster presenters for their wisdom and willingness to share.

Alex Ferguson
Symposium Chair

Keynote Address

Respect for Nature:
An Essential Foundation for
Sustainable Forest Management

J.P. (Hamish) Kimmins

Abstract

There are two major meanings of the word "respect": (1) to notice with special attention, or to take due account of the object of respect and (2) to treat with reverence, or to show honor or esteem for the object. When used in the context of "respect for nature," the first meaning implies that our relationship with nature should be based on an accurate understanding of what nature *is* rather than what we might wish it to be. It is science-based, analytical, and as independent of value systems as science can be. The second meaning implies a value-based interpretation. Respect is defined by belief systems about nature that do not necessarily conform to what nature actually is. Sustainable development and sustainable management of forests will not be successful unless we respect nature in both senses. Reverence, honor, and esteem for nature and the values and services it supplies is an important component of setting management objectives; it can help to define the set of values and services that one would like to sustain in forests. If unfettered by the requirement to carefully consider the ecological characteristics of the environmental and social values we wish to sustain, however, this second definition can lead to non-sustainable policies and practices and to undesirable forest conditions. The analytical definition of respect is vital in identifying the management objectives that are possible in a particular forest ecosystem and in defining how to achieve them.

In this chapter, I give examples of the need to respect nature in the first sense, that of understanding and respecting the ecological role of disturbance in forest ecosystems and recognizing ecological diversity. I reach the conclusion that sustainable development in forestry will not occur unless, amongst other things:

- The forest is subject to ecological site classification to stratify the ecological diversity that is the framework within which biological diversity develops.
- The historical range in frequency, scale, severity, and ecological role of natural disturbance are clearly identified for each type of forest ecosystem.

- The sum of the ecological effects of continuing natural disturbance and management-induced disturbance remains within the historical range of the ecological effects of natural disturbance if it is the objective of management to sustain forests within their historical range of conditions.

Respect for nature requires not only that we have a reverence for nature but that we express that reverence through understanding the ecological characteristics of particular forests and forest values/conditions and that we manage those forests accordingly.

Introduction

With the world's population now (June 1999) at six billion, with the annual increase in global wood demand estimated to be equivalent to or greater than the sustainable annual allowable timber harvest in British Columbia (Sutton 1993), and with the potential for the total world population to be double the present level by the time a seedling planted in Canada's boreal forest in 1999 is a mature tree, it is clear that forests must be managed sustainably. Couple this concern over the future supply of wood products (the current downturn in the global economy and the glut of wood products notwithstanding) with the need to expand the world's area of park, ecological, and wildlife reserves, and the need to modify forest management within critical watersheds to achieve water supply and water quality objectives, and one can only conclude that forestry has never faced greater challenges and responsibilities.

There is nothing new about the loss of forests, their mismanagement, and public concern about them. Major deforestation has been going on for many thousands of years in the ancient cultures of Asia, several thousand years in the Mediterranean region, nearly a millennium in Europe, and over the past two centuries in central and southern Africa, South America, and North America. The beautiful English countryside, beloved of British poets and artists for centuries, is mostly the result of deforestation as land use shifted from forest to food production. Citizens protested large-scale clearcutting and deforestation in Europe as early as the 13th Century, especially in mountainous areas. More recently, concern over deforestation and unregulated forest exploitation late in the last century and early in this century led to the development of the parks system and forest management in North America and to the growth of environmental and wilderness advocacy groups. And finally, the rapid acceleration in loss of tropical forests over the past 50 years led to the international meeting in Rio de Janeiro that resulted in the international convention on biodiversity, a major factor in the current debate over forestry.

The rate of global deforestation increased exponentially over the past few centuries, driven ultimately by the growth of the human population. It accelerated every time the power of military technology and scale of human conflict increased, every time new forest harvesting technologies were invented, and every time new markets for wood products developed. As a result, the two separate activities of forest harvesting and deforestation outstripped the growth of the population. The news is not all bad, of course. Areas of land in Scandinavia, Europe, and North America that were denuded of forest cover for agriculture in the last century and early part of this century have been returned to forest, either naturally as marginal farmland was abandoned, or by reforestation. In some areas, fire control has resulted in increased forest area and forest biomass. These temperate and northern recoveries have been overshadowed at the global scale, however, by continued loss or degradation of forest at tropical latitudes and in developing countries.

Human ability to mismanage forests increased enormously as the power of technology increased. Technological forestry, in which natural patterns and processes are increasingly modified to fit an industrial paradigm (Maser 1988), can be very successful in the short run in providing high rates of timber production with high economic efficacy. We experience increasing concern, however, that high productivity may not be sustainable unless we modify management to respect the ecological processes that support this production. We find equal concern that the industrial paradigm frequently fails to conserve non-timber forest values. An outcome of this concern has been the public demand for a greater respect for nature and for redesigning forestry practices based on this respect. It is hard for anyone to argue against this demand. But a difficulty arises concerning exactly what is meant by the word "respect." Until this difficulty is resolved, we cannot design forest management that will indeed respect nature.

Webster's New Twentieth Century Dictionary gives several meanings for both the verb, *to respect*, and the noun, *respect*. Generally, the meanings can be summarized into two groups. The first is "to notice with special attention, or to take due account of the object of respect." In other words, respect for nature means to carefully observe what nature is and to take this into account in human relationships with nature. The second general meaning is "to venerate, to show reverence towards, to show honor and esteem for." In other words, to regard "nature" as we would a deity, a religious icon, a spiritual place, object, or idea. This meaning can and does, for some people, lead to the conclusion that humans are separate and different from nature, that nature "knows best," and that it is ethically wrong for humans to cause any alterations in ecosystem structure and function from the condition that is deemed to be right for nature. It can also lead to the notion that nature needs to be protected from human-caused disturbance or, in the most extreme ideas, to be protected from all disturbance: the only good nature is that which has not been disturbed at all for a very long time.

The conclusion that we must respect nature if we are to achieve sustainability seems unassailable. The challenge is to agree on what we mean by the word "respect." With the rapid increase in support for the certification of forest management as sustainable, we are similarly challenged to define what we mean by "sustainable forest management that respects nature." Unless we can agree on what is meant by these concepts, it is unlikely that we will be successful in designing sustainable management strategies, and we will have continuing difficulties in defining criteria and indicators of sustainability for forestry.

Sustainable forest management should embrace both meanings of the word "respect." The second of the definitions is important as it relates to the values and environmental services that we want to have sustained; it relates to establishing the objective of forest management. The first definition is equally important as it relates to how we design different sustainable management practices for different kinds of ecosystems and for different values, based on the ecology of those different ecosystems and values.

The take-home message from this paper is that, yes, we must respect nature if we are to achieve sustainability in forestry, but respect means far more than simply revering nature and minimizing human impacts on forest ecosystems. It means gaining an understanding of the ecology of the forest values we want to sustain and designing combined management-induced and natural disturbance regimes that are consistent with, and therefore respect, that ecology.

Respect for Nature: Disturbance and the Role of Humans

Discordant Harmonies: Is Nature Like Mozart or Stravinsky?
Botkin (1990) warned that we will not succeed in working sustainably with nature unless we respect what nature is and how it operates. He noted that the 19th century romantic vision of nature as a perfect machine, or the religious view of undisturbed nature as the perfect temple existing in perpetual balance except when disturbed by human activity, is inconsistent with our current knowledge about forest ecosystems. Rather than being balanced systems in which everything is in equilibrium, harmonious, and well ordered, most forest ecosystems are constantly changing; temporal and spatial mosaics of sometimes violent and frequently large-scale, disruptive change interrupting periods of more orderly development and small-scale change.

The romantic view of nature reminds one of the music of Mozart or Johann Strauss: harmonious, and soothing to the human senses. In contrast, nature is often more like the compositions of Stravinsky or Richard Strauss, with periods of sharp discord and changes of mood and pace woven into a whole that has structure, pattern, and an overall beauty and harmony despite periodic discontinuities and discords.

Sustainable forest management requires the recognition that some forests have a character reminiscent of the music of Mozart while other forests remind one more of Stravinsky. Like some music of Mozart, some forests are dominated by small-scale, infrequent disturbance that gives short-lived humans the impression of balance and harmony. Other forests, of a character reminiscent of Stravinsky, are periodically and violently disturbed – by fire, wind, insects, and/or diseases, events some human observers might find damaging, destructive, and having an ecosystem fabric that has been rent, even when these natural forests owe their ecological character to such disturbance. Still other forests have an intermediate character and level of disturbance corresponding to a wide range of classical and contemporary composers in the world of music. As Botkin (1990) and Attiwill (1994a) point out, failure to respect (as in "notice" and "take due account of") the variable role of disturbance in ecosystems will generally result in failure to achieve sustainable resource management and one's conservation/ biodiversity goals.

Ecosystem Disturbance: Disrespect for Nature or Part of Nature?

A sustainable relationship between humans and forests clearly requires respect for nature in the sense of having due regard for its ecological characteristics. In contrast to the reverential definition of respect, due regard for ecological characteristics suggests that where disturbance is a natural part of the ecology of a forest ecosystem, or of one of its values, we must maintain disturbance regimes that are appropriate for the specific conditions and values of the ecosystem and for its conservation (see Attiwill 1994a,b; Kimmins 1996).

A frequent feature of the reverential interpretation of respect is the idea that forests are fragile and delicate and that they require protection from disturbance. While natural disturbance in forests is accepted by some people as being consistent with this second definition of respect, human-caused disturbance generally is not. How realistic is this fragility concept of nature when applied to forest ecosystems? How much does it respect the ecological character of forests? Is this truly "respect for nature" or is it simply a romanticized view held by people who want nature and forests to be the way they would like them to be or think they should be? Is this philosophy partly a reflection of our spiritual, aesthetic, and emotional evaluations of forests (Kimmins 1999), an indulgence of our feelings about forests rather than a recognition of what forests are and how they really operate?

Consider, for example, the disturbance created by clearcut harvesting of timber. Clearcutting in the hot, dry forests of the southern interior of British

Columbia is inconsistent with sustainable forest management because the microclimatic conditions it creates render forest regeneration very difficult. Clearcutting in such forests is more appropriate for creating rangeland, golf courses, or subdivisions than for sustainable forestry. Clearcutting is equally inappropriate on steep, unstable slopes in coastal areas that experience high rainfall and strong winds, in mountainous areas with high snow instability, and in areas where growing season radiation frosts prevent or severely damage regeneration. In contrast, the disturbance created by appropriate patterns and methods of clearcutting is consistent with the ecology of naturally wind-disturbed or fire-disturbed coastal forests and with the ecology of fire- or insect-disturbed interior forests. Applied with a sensitivity for local conditions and values, clearcutting is appropriate for managing naturally even-aged forests of shade-intolerant, light-demanding species and for forests in which the balances among diseases, insects, tree parasites, and host trees have historically been regulated by periodic natural stand-replacing disturbance events.[1]

The disturbance created by clearcutting is clearly inappropriate for some forests and for maintaining some forest values. But when done with sensitivity for local site conditions, it is clearly appropriate and indeed sometimes necessary in some forests and/or for some forest values. This does not mean that clearcutting can or does exactly mimic natural disturbances. For example, wildfire in black spruce boreal forests in Quebec and Ontario creates appropriate seedbeds, warms the soil, reduces surface acidity, and reduces competition from shrubs and herbs. It also leaves standing dead spruce trees that release abundant seed. "Careful" clearcutting in these forests does not disturb the soil or the shrub layer and leaves the spruce cones on the ground where they get wet and rot, killing the seeds. The consequences of some clearcutting are thus different from the consequences of wildfire in a variety of ways. By the same token, partial harvesting may be no more natural than clearcutting, and with careful planning and implementation, clearcutting in black spruce stands can be as similar to stand-replacing natural wildfire disturbance as any other harvesting system.

Humans: Natural or Unnatural?

For many people, nature is everything except humans; it is ecosystems unaltered by human impacts (see the excellent discussion in Peterken 1996). The effects of beaver that dam and flood valleys and kill trees, the vast areas of forest burned by naturally-caused wildfire, the effect of epidemics of insects that kill trees over large areas, the loss of soil and the sedimentation of fish streams by natural landslides, and the areas without trees that result from grazing by unmanaged populations of wild herbivores are all "natural" and part of "nature." In contrast, harvesting trees, maintaining agricultural land free of trees, soil damage, and forest fire caused by *Homo sapiens* are considered by many people to be "unnatural" and outside of "nature." There are good reasons for this attitude, of course. Although dam-building beavers, nest-building birds, and various other animals use "technology," none do so on a scale comparable with the technology humans use, and no other species uses external, non-metabolic sources of energy to alter their environment the way humans have done. But if we accept that humans equipped with modern technology are different, then we must decide at what point in recent history humans became "unnatural."

For millennia humans have harvested and cleared forests using only human energy, relatively primitive tools, fire, or grazing animals. Both deforestation and human alteration of the structure and function of forests are ancient phenomena that predate modern technology, and they are analogous to the deforestation and alteration of ecosystems caused by high populations of insects or grazing mammals. Much of the world's tropical rain forest has for millennia been subject to shifting cultivation, and temperate forests in various parts of the world are believed to have been burned, for equally long periods, by indigenous peoples. The eucalypt (*Eucalyptus spp.*) forests of southern Australia are believed to have one of the longest histories of burning by early humans.

This relatively benign, chronic level of human impact has been replaced over the past century by much more widespread and sometimes more acute and dramatic alteration of ecosystem structure and function. Two world wars, the Chinese civil war, and more local conflicts like the Vietnam war have done what war has always done to forests – serious disruption of structure and function – but on a much larger scale than before. The powerful technology (e.g., logging trucks and power saws) developed during these conflicts was subsequently used to harvest and manage forests. Technology-addiction combined with the human population growth of the 1950s and 1960s led to a dramatic increase in the area of forests that were significantly altered by humans. This increased human impact on global forests reflected as much the application of an industrial production philosophy to forestry and a failure to respect ecological processes as it reflected the increased power of technology and the population of humans. People simply did not understand the ecology of the values for which they managed nor did they respect them in the practices they used. The unnaturalness of modern *Homo sapiens* is therefore as much a consequence of the failure to take due account of – to respect – forest ecosystem function in using technology as it is in using technology itself.

An unfortunate consequence of the notion that humans are "unnatural" is that we are outside of, not a part of, nature. This notion can lead to a reverence for nature – green religion – that concludes that humans have no ethical right to change or even use nature. It can also lead to the conclusion that nature was made for humans to use and that we are entitled to use and alter nature in any way we want. The notion that humans have no right to change or use nature ignores the reality of a world population of six billion, headed for about 11 billion. The notion that humans are entitled to use nature any way they want ignores the ecological limitations on how much we can alter ecosystems before they fail to sustain the values, functions, and environmental services we expect and need from them.

A more effective philosophy, one that has a greater chance of encouraging sustainability, is to consider humans to be as natural a component of nature as any other species, and equally as dependent on their environment. This philosophy forces us to recognize that, because we are a part of the environment and are dependent on it, failure to respect the ecological processes that sustain the values we want and depend on is unacceptable and that failure to respect ecological processes is ultimately one of the major threats to our continued existence. It leads to the idea that the disturbance to forests caused by humans is simply one of a complex of disturbances that collectively determine the ecological character of our forest environment, but it is also one of a complex of disturbances that can cause undesirable changes in ecosystems if it is not the appropriate disturbance.

Examples of the Ecological Role of Disturbance in Sustaining Forests

Our failure to respect the important ecological role of disturbance can have many unexpected and undesirable consequences. Botkin (1990, 1995) warned that inappropriate paradigms of nature cannot provide a successful foundation for sustainable resource management. Attiwill (1994a, b) reviewed the role of disturbance in conservation and the conservative management of Australian eucalypt forests and summarized the extensive literature that attests to the role of disturbance in forests. He argued that failure to understand this role and to sustain appropriate disturbance regimes is as much a disrespect for nature as is the application of disturbances like clearcutting in environments where it is not appropriate. The following are examples from several biogeoclimatic zones in British Columbia and one example from Australia where disturbance is not only tolerated by nature, but is required if nature is to remain within its historical range of conditions.

The Coastal Western Hemlock Zone

Few useful generalizations apply to all forest site types in any of British Columbia's 14 biogeoclimatic zones (Meidinger and Pojar 1991). Variations in climate, in disturbance type and its severity and frequency, and in geology and topography within any zone combine to define a variety of different local ecological conditions and circumstances. Consequently, the two examples given here from the Coastal Western Hemlock zone are location specific and can be extrapolated only to ecologically equivalent sites.

(1) In the drier portions of the Coastal Western Hemlock (CWH) zone – those in the rain shadow of Vancouver Island mountains and the Olympic Mountains and at drier, lower elevations on the British Columbia mainland – there is usually a significant summer moisture deficit (the demand for water by plants is greater than the ability of the climate and soil to supply water). These forests are therefore not true temperate rain forests, and historically they appear to have been affected by large, stand-replacing fires with an average periodicity of about 300 to 400 years (Green et al. 1998; see also Eis 1962, Schmidt 1970). This disturbance history has maintained a forest dominated by Douglas-fir on all but very moist and wet sites, with a secondary canopy of the shade-tolerant, climax western hemlock (*Tsuga heterophylla*) and western redcedar (*Thuja plicata*) developing as time increases since stand-replacing disturbance.

Western hemlock in coastal forests is the host for a parasitic vascular plant, the hemlock mistletoe (*Arceuthobium tsugense*) (Bloomberg and Smith 1982; Unger 1992). This parasite infects branches and causes abnormal swellings and fan or hand-like growths. Where hemlock is growing rapidly in an even-aged, closed canopy such as in a large gap or a clearcut, the infected branches will be shaded and die before the infection reaches the stem. In contrast, hemlocks growing slowly in small openings or in the lower canopy layers of an uneven-aged, multi-layered, mistletoe-infected hemlock stand frequently retain the infected branches long enough for the infection to reach the stem. Stem infection causes distortion in the development of stem wood and allows the entry of fungal pathogens. This weakens the stem, renders the tree susceptible to stem break during wind storms, and causes the tree to lose much of its value for lumber and pulp. Such stem-infected trees do have value for certain wildlife species. Cavity-nesting birds or mammals may be able to use the decayed and deformed stem sections of live infected trees or the standing snags created when wind causes stem breakage, and birds feed on the insects living in the dead stems. Other species may use as nest sites the deformed fan-like branches or the moss mats that may develop thereon. Heavily infected hemlock forests result, however, in significantly reduced economic and employment values, and they pose safety hazards for power lines, homes, recreationists, and forest workers.

If these relatively summer-dry coastal forests are protected from stand-replacing disturbance, the process of succession steadily depletes the early-seral, shade-intolerant species (e.g., alder [*Alnus rubra*], bitter cherry [*Prunus emarginata*], dogwood [*Cornus nuttallii*], cottonwood [*Populus balsamifera trichocarpa*] on wet sites, Douglas-fir [*Pseudotsuga menziesii*] on medium sites, and lodgepole pine [*Pinus contorta*] on very dry sites), replacing them with shade-tolerant species (western hemlock and western redcedar, with minor components of western yew [*Taxus brevifolia*] and grand fir [*Abies grandis*] on some sites). Where there are relatively even-aged stands with a uniform overstory canopy of Douglas-fir and a relatively even-aged, shade-tolerant lower canopy, the forest will develop into a multi-aged, multi-layered canopy of shade-tolerant species. If the parasite is present in the overstory, this successional alteration in species composition and stand structure generally leads to a progressively more hemlock-dominated forest with increasing mistletoe infection, a condition that is not unnatural but is relatively uncommon in this part of the CWH zone because of the historical role of disturbance in controlling the mistletoe and maintaining higher tree species diversity and lower stand structural diversity.

(2) In the cool and humid portions of the Coastal Western Hemlock zone where summer moisture deficits are uncommon (a true temperate rain forest), wind has been the major natural disturbance factor. Fire has been rare in these areas. These forests are dominated by shade-tolerant western hemlock, western redcedar, and Pacific silver fir [*Abies amabilis*], and are frequently in a self-replacing, climax condition with small-scale disturbance. Wind damage in these forests (windthrow, stem break, crown damage) generally does not push the ecosystem back to an earlier seral stage the way fire does. It simply creates younger, late-successional or climax forests. Nevertheless, it has a major effect on ecosystem form and function. Forests that have experienced complete stand-replacing wind damage in the last one or two centuries may be dominated by western hemlock and Pacific silver fir. They are generally characterized by high levels of net primary production, high tree biomass, and very little understory. Forests that have not experienced complete stand-replacing disturbance by wind for a long time (perhaps a thousand years or more) may be dominated by western redcedar with an understory of ericaceous species of the genera *Vaccinium* and *Gaultheria*. This condition, where it occurs, is associated with a progressive slowing of nutrient cycling, a deep, slowly decomposing forest floor dominated by large decomposing redcedar logs (coarse woody debris), and intense competition for soil nitrogen by the shrubs. The ericaceous shrubs may also reduce the availability of nitrogen by releasing tannins that complex with this important nutrient. Under these conditions, the growth of hemlock and Pacific silver fir decline dramatically. As they grow older, the long-lived redcedars progressively lose their crown, developing a candelabra growth form but remaining alive. This allows more light to penetrate the canopy, promotes the growth of the ericaceous understory, and restricts tree regeneration to the redcedar coarse woody debris. Redcedar appears to be the most successful species at regenerating on these logs. Given sufficient time without disturbance, the forests in this area can, under the right circumstances, change slowly from the productive condition to the unproductive condition and eventually to an ericaceous shrub "heath" with scattered old cedar and moribund hemlock and Pacific silver fir. A full explanation of the

differences between these two ecosystem conditions and their temporal and successional relationships is not yet available. It is more complex than originally thought, almost certainly involving as yet incompletely understood microbial relationships, and the two stand types do not appear to be part of a simple linear successional sequence. Prescott and Weetman (1994) and Prescott (1996) summarize much of our understanding of these forests. It is clear, however, that disturbance in this area is related to the presence of closed-canopy forest and high forest productivity, and the long-term absence of stand-replacing disturbance is associated with a very different forest condition.

The Sub-Boreal Pine-Spruce Zone

In the rain shadow of the largest mountain mass in the British Columbia coastal range, Mount Waddington and its associated ice fields, is a rolling plateau that is warm and relatively dry in the summer and very cold in the winter. The winter cold is partly due to arctic air masses that periodically invade the British Columbia interior from the prairie provinces over the low spot in the Canadian Rockies in the area of Chetwynd and Lake Williston. It is also due to cold-air drainage from the extensive ice fields in the coastal mountains to the southwest. The area is characterized by frequent summer lightning storms. This combination has resulted in a long history of frequent wildfire (both crown and surface fires) and, except in poorly drained sites, the landscape is dominated by a mosaic of monoculture, even-aged, fire-origin lodgepole pine stands. Most of the fires have been small to medium in size, but most of the area burned has been the result of a small number of very large fires (D. Andison, Bandaloop Landscape-Ecosystems Services, Coal Creek, Colorado, personal communication).

Where fire results in complete stand replacement, the regenerating lodgepole pine trees are free of the lodgepole pine mistletoe (*Arceuthobium americanum*) (Hawksworth and Johnson 1989), a species closely related to the hemlock mistletoe described above. Where the fire results in only partial kill (analogous to partial harvesting), the pine stands develop several age-classes and canopy levels. If the surviving overstory trees are infected with mistletoe, most of the trees in this mixed-age stand will eventually become infected and significant growth reduction will result. Intense, stand-replacing fire has been the natural agent that has kept most of this forest relatively mistletoe-free.

The Engelmann Spruce/Subalpine Fir Zone

Climatically one of the most variable of British Columbia's ecological zones, the Engelmann Spruce/Subalpine Fir zone (ESSF), an interior subalpine zone, supports a relatively open-canopied climax forest of Engelmann spruce

(*Picea engelmannii*) and subalpine fir (*Abies lasiocarpa*). Following a stand-replacing, lightning-caused wildfire, the forest regenerates to either shade-intolerant lodgepole pine (in the lower elevations of the zone and where there is a pine seed source), or to spruce (in the absence of a pine seed source), or to a mixture of pine and spruce. Less commonly, it may regenerate to subalpine fir if seed of the other species is not available and fir seed is available. Pure pine stands are invaded and subsequently replaced by the moderately shade-tolerant spruce, and as time passes an understory of subalpine fir develops. The forest at this stage is relatively closed-canopied. The pine can live up to 400 years (although growth declines and mortality becomes important after about 200 years) and the spruce up to about 500 years (dominant spruce are often 250 to 450 years old, but trees up to 600 years are not uncommon). The subalpine fir, being more prone to fungal pathogens, is shorter lived than the spruce (trees older than 250 years are not uncommon but most die before this age, mainly due to root and stem decay, insects, or windthrow [Burns and Honkala 1990]).

In the wetter, deep winter snowpack subzones of the Engelmann spruce subalpine fir zone, the old-growth forest is characterized by patches of ericaceous shrubs growing in canopy gaps on better drained sites and microsites, and a complex of herbs in gaps on the moister sites and microsites. These herb and shrub patches appear to be rather resistant to invasion by tree seedlings, especially pine and spruce, unless disturbed physically or by fire. As mature pine and spruce trees die, the canopy gaps that are created may be invaded by these herbs and shrubs or by subalpine fir seedlings; where trees at the edge of existing gaps die, the gap vegetation may expand to take over the newly released space. Conifer regeneration in unmanaged forests may be limited to decaying logs, a nutritionally poor growth environment but one with reduced light competition and possibly higher winter seed survival (B. van der Kamp, Univ. British Columbia, personal communication).

If these wetter Engelmann spruce subalpine fir forests remain free of stand-replacing disturbance (hot fires) for more than about 400 years, the moderately closed-canopied forests that developed following the previous fire disturbance begin to develop into a rather open, subalpine-fir dominated woodland in a matrix of shrub and herb patches, a process that may be promoted by greater snow accumulation and delayed snowmelt in the expanding canopy gaps. The scarcity of this type of open forest, especially in the drier Engelmann spruce subalpine fir subzones, is a reflection of the historical frequency and severity of stand-replacing disturbance by fire, which has sustained a relatively closed-canopy forest over much of the landscape. Severe disturbance facilitates the reinvasion of the herb and shrub-dominated areas by trees and the re-establishment of pine and/or spruce.

The Boreal White and Black Spruce Zone
If the fire interval is long enough in the humid portions of this zone in northeastern British Columbia, the forest undergoes a post-fire succession from light-demanding early seral hardwoods (various combinations of aspen [*Populus tremuloides*], birch [*Betula papyrifera*], willows [*Salix spp.*], alder [*Alnus spp.*], and poplar [*Populus balsamifera balsamifera*], depending on climate, soil, topography, seed source, and the pre-fire presence of vegetatively reproducing hardwood species) and/or pine, first to white spruce (*Picea glauca*) and then, depending on climate, soil and seed source, to black spruce (*Picea Mariana*) or a combination of white spruce and subalpine fir (or black spruce). This may not be the final climax vegetation, however. In flat or depressional areas or on north slopes, the very cool microclimate and thick, acidic forest floor under the black spruce promotes the growth of feathermoss. As the soil becomes colder under the insulating blanket of litter and moss and the canopy begins to open up, the feathermoss may be replaced by *Sphagnum*. As dead *Sphagnum* accumulates, the mineral soil becomes progressively colder until permafrost is formed, isolating tree roots first from the mineral soil and later from the buried forest floor. Provided with only the nutrient-poor *Sphagnum* peat, tree growth declines, further promoting the growth of *Sphagnum*. If a long time passes without disturbance, what was a closed forest can become a treed muskeg and finally an expanding treeless muskeg. This process can be reversed by a stand replacing wildfire or a summer fire burning through the dry surface layer of *Sphagnum*. Because fire is less common on cold north slopes and in wet depressional areas, the black spruce-muskeg tends to persist in these locations, spreading slowly to surrounding areas in the long-term absence of fire. The climax vegetation in these areas is, therefore, not forest; forest persists over the landscape as a consequence of infrequent but relatively severe disturbance (Heilman 1966; van Cleve et al. 1986; Shugart et al. 1992).

In all these examples from British Columbia, the ultimate climax (i.e., self-replacing vegetation in the long-term absence of stand-replacing disturbance) is probably not a closed-canopy, productive forest. It may be an open community of scattered trees and either ericaceous shrubs, shrubs and herbs or, in cold and humid areas or wet sites in the north, *Sphagnum* bog or treed muskeg. Periodic disturbance appears to be a necessary prerequisite for maintaining a closed-canopy, productive forest on at least some sites in these ecological zones. If the interval between disturbance is long, a high degree of ecosystem disturbance may be required. The more frequent the disturbance, the lower the severity of disturbance required to maintain productive closed forest.

The Fire-Dependent Wet Sclerophyl Forests of Australia
Perhaps the classic example of the necessity for ecosystem disturbance to

sustain a particular forest type is the wet sclerophyl eucalypt forest found in parts of Australia. In Western Tasmania infrequent severe summer wildfires kill the overstory eucalypts, which either resprout from the stump or lignotuber or reproduce from seed released after the fire from serotinous cones. The eucalypts require the exposed mineral soil seedbeds and free-dom from competition that are provided by fire to successfully establish. The result is generally an even-age stand of similar composition to the original fire origin stand. Many of the eucalypts grow to giant sizes, living 300 to 400 years. After this, in the absence of wildfire, they are replaced by the much more structurally diverse and tree-species rich, climax temperate rain forest. Visually beautiful and more biologically diverse, this climax forest generally lacks the cathedral-like grandeur of the eucalypts, which require a periodic massive fire disturbance to perpetuate them as a closed eucalypt forest (Attiwill 1994a,b).

These examples all suggest that periodic disturbance is required to main-tain historical ranges of ecosystem form and function in forests that have a history of periodic, stand-replacing natural disturbance. Maintaining the character of forests that have a history of smaller-scale disturbance will re-quire a less severe disturbance regime. Except in very dry forests, distur-bance regimes tend to become smaller scale, lower severity, but more frequent as one moves from higher to lower latitude, from colder to warmer cli-mates, leading to more of a "gap dynamics" forest mosaic. Even some tropi-cal rain forests, however, are naturally subject to periodic stand-replacing fire and/or wind damage in patches of up to 3,000 ha (e.g., Nelson et al. 1994).

Ecological Diversity, Biological Diversity, and the Great Mistake

Ecological Diversity and the Great Mistake
The Great Mistake that characterizes both exploitive pre-forestry and the early, administrative stage of forestry (i.e., regulations that ignore the ecol-ogy of the values that are to be sustained, as discussed below) is to ignore the spatial variability in the ecological characteristics of forests. This eco-logical diversity is defined by variations in climate and soils, which are re-lated to latitude, longitude, elevation, aspect, slope, and distance to large water bodies, and to geology, topography, slope, climate, and vegetation type. Ecological diversity provides the physical-chemical framework within which biological diversity develops as the combined result of evolution, species migrations, ecosystem disturbance, and the processes of ecological succession. Ecological diversity defines the ecological stage and setting, which in turn defines the ecological actors (species) and the ecological play (the ecosystem dynamics over time – disturbance and succession).

Each biogeoclimatic forest zone (which reflects broad patterns of variation in climate) and each forest site type (which reflects local variations in soil moisture, fertility, and aeration) within each zone in Canada has its own ecological personality and its own characteristic history of type, scale, severity, and frequency of disturbance. This individuality of ecological personalities requires that different management practices and silvicultural systems be applied in different zones and site types and that, where natural disturbance regimes have been altered, their ecological effects be emulated as closely as is possible by management-induced disturbance if it is the management objective to maintain the ecosystem conditions within the historic range characteristic of the particular ecosystem. Management and conservation objectives cannot be achieved if one single approach to forest management and one single silvicultural system is applied everywhere, whether the approach is clearcutting or minimal-disturbance selection harvesting. What is needed is variable-retention harvesting in which the degree of retention of canopy and soil cover, and the degree of ecosystem disturbance, is varied according to the ecology of the site and the ecology of the values that are to be sustained.

In British Columbia, the westernmost province of Canada, ecological diversity is especially high. Arguably one of the most ecologically diverse landscapes of comparable size anywhere in the world, the need for ecological sensitivity in the design of management strategies is even greater in British Columbia than in the rest of Canada and is as great as in any other forest region of comparable size. This ecological diversity is recognized in the biogeoclimatic classification of the province (Meidinger and Pojar 1991), an ecological classification system that recognizes a hierarchy of regional units based on climate (zones, subzones, and variants), and a local hierarchy of ecosystem units within these climatically defined regions based on variations in soils and vegetation along local topographically induced gradients of soil moisture, nutrient availability, and aeration.

Respect for nature in British Columbia requires resource managers to recognize the role that this ecological variability plays in defining the multitude of British Columbia forest types, their structure and function, and their productivity and various measures of biological diversity. It also requires that resource managers respect the role of past disturbance in defining these ecological and biological characteristics. This is equally true for environmental groups and conservationists concerned with sustaining our biological and ecological inheritance.

Ecological Role of Disturbance
As already discussed, natural disturbance in Canadian forests is dominated by fire and insects, with wind and diseases being important in some areas

and landslides important in some mountainous regions. Small-scale disturbance occurs as a part of autogenic successional processes operating between the major disturbance events (allogenic or large-scale biogenic successional processes), but for many of Canada's forests, ecological characteristics are more closely related to these generally infrequent large-scale allogenic (physical) and biogenic (biological) disturbances than to the more frequent, smaller-scale autogenic processes. This is in contrast to many forest types at lower latitudes or in heavily human-modified environments (e.g., Europe) where historical disturbance regimes no longer occur or are reduced in frequency. In such areas, small-scale autogenic (community-based successional) processes play a more important role than large-scale, stand-replacing allogenic or biogenic disturbance events.

Although periodic, large-scale, severe disturbance is a characteristic feature of many Canadian forests and may be necessary to maintain their historical ecological characteristics, these disturbance events are often socially undesirable. Fire can kill people, destroy homes, businesses, and communities, cause the loss of socially important wood supplies with attendant economic impacts, and, in very severe fires, can cause damage to soils, some aspects of wildlife habitat, and fish and streams. As a result, efforts are made to limit forest fire and, often less successfully, to limit insect, disease, and wind damage.

Although it may be socially desirable in the short term, eliminating or reducing ecosystem disturbance causes significant long-term alterations to the ecology, productivity, and biological diversity of forests that have a history of periodic severe natural disturbance. Forest management objectives often include sustaining the historical character of the forest, including the historical temporal and spatial patterns of variation in that character. Where this is the case, resource managers must ensure that the combined effect of the frequency, spatial scale, landscape pattern, and severity of both management-induced and continuing natural disturbance approximate the combined effect of historical disturbance regimes. In forests managed for a sustained supply of wood products, it is harvest-related disturbance that emulates the ecological effects of natural disturbances that no longer occur, and this harvest-related disturbance must be designed with natural disturbance patterns in mind.

Harvest-related disturbance can vary from minimal such as single-tree selection harvesting with little soil and vegetation disturbance (e.g., winter logging) to maximum disturbance such as the clearcutting silvicultural system followed by mechanical site preparation or slash burning. The entire range of disturbance types, covering the entire range of tree retention and a wide range in soil disturbance, is applicable in British Columbia because of the great ecological diversity of the province. Disturbance severity on its own, however, is not sufficient to define a disturbance regime; one must also know the scale and frequency of disturbance, and ecosystem resilience. The combination of these defines the ecological rotation: the time taken for a particular ecosystem or ecosystem value to recover from a particular disturbance. Defining sustainable forestry requires that we define ecological rotations (Kimmins 1974).

The Great Mistake and the Evolution of Forestry

Forestry, the first organized attempt to conserve and sustain desired forest values, has always developed in response to the unacceptable consequences of unregulated forest exploitation. The first stage of forestry, the "administrative" stage, is characterized by laws, rules, regulations, and policies that largely ignore the ecology of the values that are to be sustained; this stage of forestry generally makes the Great Mistake. As a result, this stage is often partially or completely unsuccessful. The next stage is characterized by ecologically based rules, regulations, and policies, where forestry sensitive to the values that have usually characterized this stage is practiced: silvicultural values, game wildlife species, soil fertility, and watershed values. Sustainable in terms of these values, this stage of forestry may nevertheless fail to sustain a variety of other values such as aesthetics, spiritual values, and habitat for certain non-game animal species. As public demand for these other values grows, forestry evolves to the next stage, social forestry, in which a much wider range of values is sustained at either the stand or the landscape level or both (Kimmins 1997).

In many parts of the world, forestry is still in the administrative stage or has only recently evolved to the "ecologically based" stage. In some areas it is still in the exploitive, pre-forestry stage. In North America, public opinion, having recently overtaken forestry, is demanding that forestry practices be selected according to the dictates of the "social" stage. While such an evolution is inevitable and ultimately desirable, its speed has, in many cases, resulted in by-passing the ecologically based stage. In the absence of experience gained from the ecologically based stage, some of the demands for changes in forestry are being based more on belief systems about nature than on experience and on knowledge of the ecology of the values that are to be sustained. Regrettably, in some cases this is leading to a repetition of the Great Mistake: a demand by some people that all forestry must cause minimal disturbance and that the same methods of soft-touch, "respectful" harvesting and the same low-impact silviculture system be applied to all forests.

The evolution of a "green religion" about forests and forestry has made a positive contribution to the development of forestry. It has created political pressures that have made change possible. It has helped to define the broader range of values the public wants in forests. This contribution is balanced, however, by the danger that the same political pressures may lead to undesirable change. Some of the demands based on a "green reli-gion" view of nature would repeat the Great Mistake if they were permitted to become the basis for forest policy and practice. They would return both conservation and forestry to the administrative stage rather than advanc-ing it to a truly sustainable "social forestry" stage (Kimmins 1992).

Design of Appropriate Harvesting Practices

Although many forests require periodic, stand-replacing disturbance if their historical ecological character is to be sustained, the disturbance created by clearcutting has not always adequately emulated natural disturbance events. While large-scale clearcutting may resemble the overall scale of large fire, wind, or insect events and may initiate similar successional sequences, natural disturbances often follow natural boundaries and leave a mosaic of small patches (islands) of live vegetation. In contrast, the shape and landscape pattern of clearcuts has often ignored natural boundaries, creating an un-natural checkerboard pattern of geometric openings across the landscape. Fires, windthrow, or landslides create seedbeds and eliminate or reduce com-peting herbs and shrubs. These disturbances often leave individual seed trees or islands of live forest scattered across the disturbed area, ensuring a seed source for natural regeneration. In contrast, some types of clearcutting, es-pecially winter logging, may not disturb the soil sufficiently to create suit-able seedbeds, and past clearcutting has often removed seed sources from such a large area that natural regeneration has been very slow and unreliable. Cones left on the ground may not open to release seed before they get wet and rot; the same cones on standing fire-killed trees may ensure abundant regeneration. During clearcutting, it has often been required by law, for reasons of worker safety and other considerations, to remove all snags and live trees from within the harvest area. In comparison, natural disturbance leaves most of the woody biomass on the site as snags and downed logs, which may alter microclimate (and thereby favor tree regeneration) and provide habitat for wildlife. Any type of timber harvesting, whether clearcutting or partial harvesting, removes this woody biomass. Equally, in any harvest or silvicultural system the choice can be made to leave snags or coarse woody debris and to retain "wildlife" trees.

If the objective of management-related disturbance is to replace or emu-late natural disturbance, we must first understand clearly the ecological effects of the natural disturbance that has defined the character of a par-ticular ecosystem. The spatial scale and pattern, frequency, and severity of management-related disturbance must then be designed so that it emulates the natural disturbance it replaces or complements natural disturbance events that continue to operate. The following two examples illustrate this.

(1) Fire has been the major agent of disturbance in the boreal and the subboreal forests of British Columbia. Fires have occurred at a high frequency,

approximately every 60 to 150 years, and much of the landscape is occupied by even-aged stands that resulted from relatively few very large fires
(> 20,000 ha). Most of the fires have been much smaller than 20,000 ha,
however, and the natural landscape is a mosaic of a few very large patches
and many smaller patches of various sizes (D. Andison, personal communication.). Harvesting in this area in the past was often an exploitive diameter
limit cut that left the forest depleted in spruce and in poor silvicultural
condition. Next came a period characterized by clearcutting, often followed
by planting. The clearcuts were sometimes large (up to 1,000 ha or more)
and, where insect-killed or fire-killed forest was being salvage harvested,
continuously clearcut areas sometimes extended over tens of thousands of
hectares reflecting the extent of the insect- or fire-caused tree mortality.

Public pressure against clearcutting contributed to a recent limit of 60 ha
on clearcut size in this area. The question arises as to which of the three
types of harvest most closely resembles natural disturbance regimes in these
forests, the earlier diameter limit cut, the varying clearcut size, or the recent
60 ha clearcut limit. In terms of size of disturbance patches, the variable
clearcut size came closest to the natural disturbance regime although it generally failed to duplicate the larger and smaller natural disturbance events.
However, in the shape and spatial distribution of these past clear-cuts, in
the removal of all standing snags and live trees greater than three meters
tall, and in the frequent failure to locate cutblock boundaries along natural
boundaries and to reserve forest around wet areas, lakes, and streams, this
type of harvesting failed to closely emulate natural disturbance.

Clearly we needed to change many aspects of timber harvesting in the
boreal and subboreal forests of British Columbia, but the administrative
approach of requiring a fixed 60 ha maximum harvest area limit on all sites
is inadequate. While recent forest legislation requires that we protect streams,
retain tree islands and snags where appropriate, and follow natural boundaries, imposing a fixed maximum opening size that is much smaller than the
historical natural disturbance regime is inappropriate if the objective is to
sustain reasonably natural landscape patterns. A more natural strategy would
be to have a diversity of types of harvesting and sizes of clearcuts including
some very large clearcuts.

(2) In the drier portions of the Coastal Western Hemlock zone described
earlier, strong public pressure exists to abandon clearcutting and move to
shelterwood or small patch harvesting. Clearcutting, it is argued, is too environmentally destructive.

The structure of much of this forest on well-drained sites is an overstory
of fire-origin Douglas-fir and a secondary canopy or understory of western
hemlock. Removing the Douglas-fir overstory by clearcutting with minimal
disturbance to the soil and understory releases the seedling bank of hemlock. Rather than disturbing the ecosystem and setting succession backwards,

such clearcutting actually accelerates successional development towards a hemlock-dominated climax forest. This is one reason why slashburning has historically been used after clearcutting on such sites. It sets succession back, facilitating the establishment of Douglas-fir and other earlier successional species including, if suitable seedbeds are created, early seral deciduous hardwoods.

Partial harvesting with minimal soil disturbance also promotes successional progression towards the hemlock climax and may exacerbate the hemlock mistletoe problem described earlier (clearcutting tends to reduce the problem by promoting rapid growth of even-age hemlock stands). The forest that results from this "kinder, gentler" type of forestry is not, of course, unnatural. It is relatively uncommon in this ecological subzone, however, because of the historical role of stand-replacing disturbance. If the objective is to sustain the historical structure, composition, and productivity of stands in this region, a clearcutting silvicultural system that is designed to match particular site conditions and desired values will generally respect nature's pattern and process more than the "kinder, gentler" forest practices that the public generally supports. This does not mean no change from past harvesting practices. It does mean designing the changes with a respect for the way nature works in these particular forests.

Conclusions

History suggests that societies that persistently ignore the ecological characteristics of the ecosystems they depend upon will alter those ecosystems in ways that result in negative consequences. They will frequently fail to sustain the historical character of those ecosystems.

The major problem with forestry around the world has been that the laws and regulations that have guided it have ignored the ecology of the values that are to be sustained: the Great Mistake of forestry.

Thanks to public pressure, governments have developed a sensitivity to environmental problems in forestry, and this has created the impetus for necessary change. The very pressure that made this possible, however, threatens to disrespect nature because it, like earlier forestry, ignores the ecology of many of the values it desires to sustain. Pressure from "green religion" threatens to repeat the Great Mistake. As Aldo Leopold (1966) noted, "Conservation is paved with good intentions which prove to be futile, or even dangerous, because they are devoid of critical understanding either of the land, or of economic land use."

Respect for nature, one of the essentials for sustainable forestry, requires that management emulates nature to the extent of maintaining critical ecosystem processes. Nature is not a single entity, and there is no such thing as a single natural disturbance. Fire, insects, disease, and wind exhibit large variations in their frequency, pattern, scale, and severity of impact on forest ecosystems. Management-related impacts should be limited to this range of natural disturbance effects if ecosystem processes are to continue to operate within their historical range of rates.

Because management should attempt to emulate nature if the objective is to maintain forests within their historic range of conditions, and because natural disturbance is so variable and does not necessarily conform to what is socially acceptable, management-related disturbance should be evaluated in terms of unique combinations of frequency, scale, pattern, and severity that in combination will emulate the effects of natural disturbance while sustaining desired values.

Forests harvested for timber will never be exactly the same as forests that are subject only to natural disturbance. To imagine so is a delusion. This is true irrespective of the harvesting method or silvicultural system being used. But by appropriately designing and implementing management systems that truly respect the way nature has affected a particular site (which is not necessarily how we might wish it to be) forestry can respect nature and sustain the many different values we want from our forests.

Note

1 In this discussion I do not refer to all clearcutting. Clearcutting that has left undesirable sizes of forest openings and blocks of one age class, undesirable landscape patterns, a lack of wildlife habitat supply at the landscape level, unprotected streambank environment and fish habitat, no appropriate seedbeds and seedling recruitment where natural regeneration is desired, unacceptable slope instability or soil alterations, and unacceptable alterations in hydrology is generally not acceptable disturbance.

Acknowledgments
A review by Dr. Karel Klinka and contributions by David Andison and Daniel Mailly are gratefully acknowledged, as is the typing of Maxine Horner and Patsy Quay.

References
Attiwill, P.J. 1994a. The disturbance of forest ecosystems: The ecological basis for conservative forest management. Forest Ecology and Management 63:247-300.
Attiwill, P.J. 1994b. Ecological disturbance and the conservative management of eucalyptus forests in Australia. Forest Ecology and Management 63:303-348.
Bloomberg, W.J., and R.B. Smith. 1982. Measurement and simulation of dwarf mistletoe infection on second-growth western hemlock on southern Vancouver Island. Canadian Journal of Forest Research 12:280-291.
Botkin, D.B. 1990. Discordant harmonies: A new ecology for the 21st century. Oxford University Press, New York, NY.
Botkin, D.B. 1995. Our natural history: Lessons from Lewis and Clarke. Putnam, New York, NY.
Burns, R.M., and B.H. Honkala. 1990. Silvics of North America. Agriculture Handbook 654, US Department of Agriculture (USDA), Washington, DC.
Eis, S. 1962. Statistical analysis of several methods for estimation of forest habitats and tree growth near Vancouver, BC. Forestry Bulletin #4, Faculty of Forestry, University of British Columbia, Vancouver, BC.
Green, R.N., B.A. Blackwell, K. Klinka, and J. Dobry. 1998. Partial reconstruction of fire history in the Capilano watershed. Draft report. B.A. Blackwell and Associates, North Vancouver, BC.

Hawksworth, F.G., and D. Johnson. 1989. Biology and management of dwarf mistletoe in lodgepole pine in the Rocky Mountains. General Technical Report RM-169, US Department of Agriculture (USDA) Forest Service, Rocky Mountain and Range Experimental Station, Fort Collins, CO.

Heilman, P.E. 1966. Change in distribution and availability of nitrogen with forest succession on northern slopes in interior Alaska. Ecology 47:825-831.

Kimmins, H. 1997. Balancing act: Environmental issues in forestry. 2nd edition. UBC Press, Vancouver, BC.

Kimmins, J.P. 1974. Sustained yield, timber mining, and the concept of ecological rotation: A British Columbian view. Forestry Chronicle 50:27-31.

Kimmins, J.P. 1992. Ecology, environmentalism, and green religion. Forestry Chronicle 69:285-289.

Kimmins, J.P. 1996. Importance of soil and role of ecosystem disturbance for sustained productivity of cool temperate and boreal forests. Soil Science Society of America Journal 60:1643-1654.

Kimmins, J.P. 1999. Biodiversity, beauty, and the beast. Unpublished manuscript. Submitted to Forestry Chronicle.

Leopold, A. 1966. A Sand County almanac. Ballantine Books, New York, NY.

Maser, C. 1988. The redesigned forest. R and E Miles, San Pedro, CA.

Meidinger, D., and J. Pojar (eds.). 1991. Ecosystems of British Columbia. Special Report Series #6, British Columbia Ministry of Forests, Victoria, BC.

Nelson, B.W., V. Kapos, J.B. Adams, W.J. Oliviera, O.P.G. Braun, and Iêda L. do Amaral. 1994. Forest disturbance by large blowdowns in the Brazilian Amazon. Ecology 75:853-858.

Peterken, G. 1996. Natural woodland ecology and conservation in northern temperate regions. Cambridge University Press, Cambridge, UK.

Prescott, C.E. (ed.) 1996. Salal cedar hemlock integrated research program: A synthesis. Faculty of Forestry, University of British Columbia, Vancouver, BC.

Prescott, C.E., and G.F. Weetman (eds.). 1994. Salal cedar hemlock integrated research program: A synthesis. Faculty of Forestry, University of British Columbia, Vancouver, BC.

Schmidt, R.L. 1970. A history of pre-settlement fires on Vancouver Island as determined from Douglas-fir ages. Pp. 107-108 *in* J.H.G. Smith and J. Worrall (eds.). Tree ring analysis with special reference to North America. Proceedings of a conference on biology of tree ring formation, methods of measurement of tree rings, methods of analysis, and uses of tree ring data, 19-20 February 1970, University of British Columbia, Faculty of Forestry, Vancouver, BC.

Shugart, H.H., R. Lemans, and G.B. Bonan. 1992. A systems analysis of the global boreal forest. Cambridge University Press, Cambridge, UK.

Sutton, W.R.J. 1993. For environmental reasons, should we or should we not harvest British Columbia (BC) forests? Unpublished manuscript. Canadian Forest Service, Pacific Forestry Centre, Victoria, BC.

Unger, L. 1992. Dwarf mistletoes. Forest pest leaflet. Forestry Canada. Pacific Forestry Centre, Victoria, BC.

van Cleve, K., F.S. Chapin III, P.W. Flanagan, L.A. Viereck, and C.T. Dyrness (eds.). 1986. Forest ecosystems in the Alaskan Taiga: A synthesis of structure and function. Ecological Studies 57. Springer-Verlag, New York, NY.

Symposium Presentations

Ecological Characteristics and Natural Disturbances in Interior Rainforests of Northern Idaho

Paul Alaback, Michael Krebs, and Paul Rosen

Abstract

Relatively little published information is available on the ecological characteristics of temperate rainforests that occur within interior continental climates in western North America. Western redcedar/western hemlock/grand fir forests meeting most of the climatic criteria for coastal temperate rainforests occur in isolated pockets from northern Idaho north to the west slope of the Rocky Mountains in eastern British Columbia. In this study we wished to examine how forests in these relatively wet and mild climates contrast with coastal temperate rainforests and with conifer forests in more typical regional climates of the northern Rocky Mountain area. In addition, we did a preliminary assessment of natural disturbances in one site, the Aquarius Research Natural Area in northern Idaho, to see how natural disturbances may affect patterns of biodiversity and stand development and to provide a context against which to assess management activities in these unique ecosystems. Our preliminary data suggest that these forests have extraordinary levels of species richness and that ecosystem recovery is rapid following either stand-replacing fire or wind events.

Introduction

The northern Rocky Mountains contain an extraordinarily rich mosaic of forests and associated grassland ecosystems and have provided ecologists with a fertile area in which to learn about the complex environmental factors that govern forest community composition and structure. Early studies focused on the harsh environmental gradients in the mountains of this region and the vegetation response to those gradients. These studies played a seminal role in developing concepts of vegetation classification and zonation that are now used widely throughout the United States (Cooper et al. 1991; Daubenmire 1942, 1943b, 1952, 1968; Habeck 1976, 1978). While most of the emphasis in these studies was to develop a useful plant or site classification for this diverse region or to describe general patterns of community change across these environmental gradients, many key questions remain about the basic ecological characteristics of moist forests in which western redcedar or cypress (*Thuja plicata*)[1] is dominant or co-dominant and about the natural disturbances that may affect these forests. We wish to pose the general question of whether these moist forests should be treated ecologically as notable anomalies to this generally xeric region or as distinctive rainforest ecosystems with a characteristic composition, structure, and dynamics more like those of coastal temperate rainforest ecosystems (e.g., Alaback 1995).

Moist forests in the Rockies, variously known as cedar-hemlock (*T. plicata-Tsuga heterophylla*), interior cedar-hemlock, or cedar-fir (*T. plicata-Abies grandis*) generally occupy a broad belt of isolated pockets of forest associated with mountain ranges to the west of the principal cordillera of the continental divide such as the Caribou Mountains, the Kootenai Mountains, the Selkirk Mountains, and the Bitterroot Mountains, but also occasionally extend into the western foothills of the main Rocky Mountain cordillera (Daubenmire 1969; Habeck 1977). In these geographic areas, located from British Columbia to Idaho and Montana, the regionally low annual rainfall (300 to 600 mm, mostly falling in the dormant season) is amended to 1,000 mm or more, principally because of local orography and the intrusion of maritime air masses through the maze of mountain chains that are generally to the west of this interior region (Daubenmire 1943b, 1969; Reynolds 1992). These areas provide a major anomaly to the forests of the region since their floristic and climatic affinities generally are closer to distant coastal regions than to the drought-tolerant pine-dominated forests of the region as a whole. Indeed, moisture stress itself appears to be of minor significance to these forests as suggested by data on the shallow depth of annual drought in soils (e.g., Daubenmire 1943a, 1968). In fact, these moist forests have been consistently ranked as the most productive in the region, reflecting the mild, wet climates (Cooper et al. 1991; Daubenmire and Daubenmire 1968).

Studies on natural disturbance in the northern Rockies have almost exclusively focused on the role of fire (Habeck and Mutch 1973; Arno 1980; Stickney 1982, 1986, 1990). Because of the consistent drought that occurs in typical xeric pine-dominated low-elevation forests and the spectacular catastrophic fires that characterize mesophytic forests, ecologists have generally viewed disturbance regimes of forests in the region in terms of a mean fire return interval or fire intensity (Brown et al. 1994; Fischer and Bradley 1987). In moist cedar-dominated forests of northern Idaho, fire return intervals are generally estimated to be in the range of 80 to over 200 years (Smith and Fischer 1997).

While fire clearly plays the dominant role in most ecosystems in the Rockies including these moist types, other disturbances are also potentially important in this region. Insect outbreaks, for example, are a key natural disturbance agent in mesic and subalpine forests in the region (e.g., Byler et al. 1997; Swetnam and Lynch 1993). Drought appears to be of greatest importance in that it can amplify other disturbance agents and their interactions, fire and insects, for example. Wind, recognized as an important agent generally in the Rockies and particularly in older mesic or subalpine forests, has been little studied (e.g., Rogers 1996).

In this study we wished to examine, from a perspective of biodiversity, species composition, climate, and disturbance ecology, the general ecological characteristics of Interior cedar-fir forests to determine the extent to which they form a unique ecosystem in the northern Rockies region. By examining these forests from this larger and more functional perspective, perhaps we could clarify how to set priorities on research needs in this ecosystem and develop a strategy for designing more effective conservation measures for these stands and landscapes so as to implement ecosystem management across its national borders.

Methods

We established vegetation transects in mature and recently windthrown sites on a steep, north facing slope within the Aquarius Research Natural Area along the North Fork of the Clearwater River in northern Idaho. In 1994 a microburst event leveled a narrow swath of forest over more than 300 m in elevation in this site, totaling approximately 16 ha. Because this disturbance occurred within a Natural Area it was not, contrary to standard practice, salvage logged, providing a unique opportunity for a successional study. Three and four years after the disturbance, vegetation and stand structural data was collected from three thirty-metre transects representing the lower, middle, and upper portions of the windthrow area.

To characterize the vegetation structure and composition, we established fifty plots of one square metre each in the windthrow area and eight plots of 0.04 hectares each in adjacent forest sites. Stand characterization of mature forests included tallying all trees greater than 5 cm in diameter at breast height (DBH) and quantifying the abundance of snags, logs, regeneration, and heights and ages of strata within the forest. We assessed vegetation cover and trees less than 10 cm in DBH within 0.04 ha plots nested within the 0.08 ha plots. In the windthrown area we used five-metre-wide belt transects to quantify abundance, orientation and distribution of coarse woody debris, and tree regeneration.

Results

Climate
Coastal rainforests can be simply defined in terms of summer temperatures and annual precipitation. In general, coastal rainforests, which have

relatively unique disturbance regimes, structurally distinctive older forests, distinctive landscapes, and distinctive fauna can be delimited by looking at coastal regions with annual precipitation in excess of 1,300 mm and a July isotherm of 16 degrees C or less (Alaback 1991). It is most useful to divide them into perhumid (continually wet), seasonal (a distinctive summer dry season), and subpolar (high latitude forests with significant sea-level winter snowpack) categories since each of these subdivisions have distinctive climatic, disturbance, species, compositional, and functional attributes (Alaback 1995).

Interior rainforests appear to have some significant climatic commonalities with floristically similar coastal rainforests (Figure 1). With annual precipitation generally ranging from about 700 to 1,500 mm, many interior rainforests appear to come within the minimal definition for precipitation in coastal rainforests. Interior rainforests have a more equitably distributed annual rainfall than corresponding coastal rainforests at the same latitude. It is interesting to note that during the growing season, in some interior cedar (*T. plicata*) sites, the 30-year average rainfall for June is actually greater than for coastal stations whose annual rainfall is two or more times that in the Rockies. At the same time, average rainfall in some interior sites in July and August is substantially below even rainshadow-affected sites on the coast (e.g., Williamette Valley or Puget Sound in the United States). Studies of soil moisture suggest that little consistent drought occurs on the interior sites despite the typically high afternoon temperatures (about 30 degrees C)

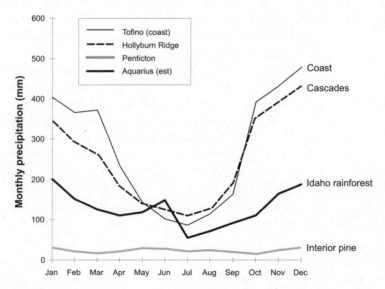

Figure 1. Precipitation patterns for selected locations in British Columbia and Idaho. Data represent 30-year climate normals.

and long periods between significant rainfall during the warmest months (Daubenmire 1968). Clearly one of the key compensating factors for these forests must be the high proportion of snow in annual precipitation (often over 70 percent) and the fact that these forests typically are in valley bottoms or concave slopes at the base of large mountainous terrains. So despite low mid- to late-summer rainfall these forests appear to be quite moist because of snowmelt, groundwater, and cool night-time and morning temperatures.

The second key characteristic that defines coastal temperate rainforests is a small range in annual temperature (a maritime climate) and low growing season temperatures. In coastal forests the cool summer temperatures are significant in increasing water runoff due to low evaporation rates. The mild night-time and winter temperatures are also of key significance in allowing for the full development of bryophytes, lichens, ferns, and other species that thrive in cool, moist habitats.

In interior rainforests the temperature regimes, while distinctive, perhaps provide similar conditions for many plant species because of the generally cool conditions (Figure 2). The Rockies are famous for having large daily and seasonal fluctuations in temperature. Subfreezing temperatures can occur nearly any time of the year. Not surprisingly, interior rainforests have nearly double the annual range in temperatures of coastal forests at a similar latitude. Interior rainforests, however, are still quite cool year-round. Our study site has a mean annual temperature of only 5.4 degrees C putting it in the range of coastal sites 15 degrees or more farther north in latitude but similar to sites in the upper elevational range of temperate rainforests in the Cascade Mountains. It may be argued that the cool annual temperatures are of little consequence to plants that are dormant during much of the winter. But even during the summer, cool temperatures prevail in these interior sites. In coastal forests, for example, the 16 degree C July isotherm appears to be effective in separating coastal rainforests from surrounding mesophytic forest types. In the Rockies, however, most interior rainforest sites can also meet this 16 degree C July isotherm standard, not because of consistently cool daytime temperatures but because of the extraordinarily cool nights. Thus it appears that some climatic basis exists for predicting that interior rainforests may provide environmental conditions similar enough to coastal rainforests to provide habitat for many coastal species. Indeed, many characteristic trees, shrubs, and ferns occur in both regions (Daubenmire 1943b).

Biodiversity
We found our study area to be extraordinarily rich in species diversity both regionally and in general for temperate coniferous forests (Table 1). The lowest diversity regionally (about 20 species) typically occurs in subalpine and xeric low elevation forests, the most extreme environments in the

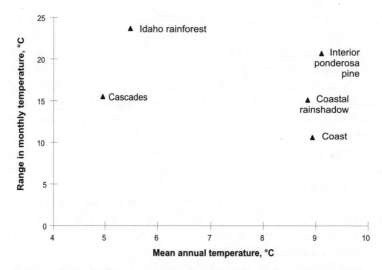

Figure 2. Temperature patterns among moist forest types in the Pacific Northwest of North America and northern Rocky Mountains. Data represent 30-year climate normals.

region. In mesophytic forests, such as western larch (*Larix occidentalis*) and Douglas fir (*Pseudotsuga menziesii*), species richness in the Rockies extends up to 30 species (Elzinga and Shearer 1997; Latham 1996). In previously studied interior "rainforests," variously defined by the presence of redcedar in combination with fir and hemlock, species richness is 30 percent or more than that of the mesophytic Rocky Mountain forests or even that of coastal temperate rainforests (Alaback 1996; Eck 1984; Hawk 1977; Juday 1976). In our study we found species richness patterns similar to those reported by Habeck (1976) in the Selway-Bitterroot Wilderness area south of our study area. These studies collectively represent the highest levels of richness found in this preliminary survey of published data as much as 46 species per 0.1 ha, Table 1. Most of our forest sites were notably lacking in western redcedar, although most sites had old redcedar logs suggesting broadscale representation of redcedar prior to the intense fires of 1910. Dominant trees in our sites were generally grand fir (*Abies grandis*) and red alder (*Alnus rubra*) with scattered individuals of Douglas-fir and western larch on drier microsites. As with so many species, *Alnus rubra* is near its limit of geographic range in our study area and is generally associated with slumps, benches, and other mass wasting phenomena (Steele 1971).

Disturbance

Our preliminary data suggest a rapid ecosystem recovery following windthrow disturbance as compared with other disturbance agents in the

Table 1

Comparison of species richness in interior temperate rainforests of western North America.

Cover type	Location	Zone	Mature richness	Disturbed richness	Source
Interior forests					
Abies lasiocarpa/ Xerophyllum	Idaho	Subalpine	20		Daubenmire (1968)
Pinus ponderosa/ Festuca	Idaho	Semi-arid	21		Daubenmire (1968)
Tsuga heterophylla/ Thuja	Glacier National Park	Interior rainforest	22	30	Habeck (1968)
Pseudotsuga/ Symphoricarpos	Idaho	Mesic forest	26		Daubenmire (1968)
Pseudotsuga/ Symphoricarpos	Montana	Mesic forest	30	32	Latham (1996)
Larix occidentalis	Montana	Mesic forest	30	35	Latham (1996)
Abies grandis/ Pachistema	Idaho	Interior rainforest	36		Daubenmire (1968)
Larix occidentalis	Coram Experimental Forest	Mesic forest	38	18	Elzinga and Shearer (1997)
Tsuga heterophylla/ Thuja	Sundance	Interior rainforest	33	53	Stickney (1986)
Thuja plicata/ Abies	Selway	Interior rainforest	32-45	25-38	Habeck (1974)
Thuja plicata/ Abies	Idaho	Interior rainforest	28-46+	45-52	This study
Coastal forests					
Sequoia sempervirens	California	Warm rainforest	32		Hawk (1977)
Tsuga heterophylla	Oregon	Seasonal rainforest	29		Juday (1976)
Tsuga heterophylla/ Picea	Southeast Alaska	Perhumid rainforest	23	10-20	Alaback (1996)
Tsuga heterophylla/ Picea	South-central Alaska	Subpolar rainforest	27		Eck (1984)

region. Four years after disturbance, species richness is similar to or even greater than that found in the most diverse mature, temperate coniferous forests. We estimate that 25 species were unique to the mature forests, while 17 species only occurred in the recent windthrow site. We found over 100 species of vascular plants in this preliminary sampling. In general we found a strong parallel in vegetation response to that reported for a similar stage in post-fire succession in this forest type (Stickney 1986). Raspberry-related bramble species *(Rubus spp.)* and bedstraw *(Galium aparine)* in particular seemed to consistently be more abundant in our disturbed sites as a result of seed dispersal. Detailed studies of post-fire succession have revealed, interestingly, that even in "holocaustic" stand-replacing fire events the vast majority of plant species persist from the original forest. For example, in the Sundance fire, to the north of our study area, 84 percent of the species in the disturbed sites also occurred in the original forest. Only 16 percent of the species dispersed into the area from other sites (Stickney 1986).

Discussion

It appears that the moist forests on the west slope of the Rocky Mountains are unique from both a regional and continental perspective. They include some of the highest estimates of species richness in temperate coniferous forests. It is unclear at this time what the full implications of this feature of the forest may be for forest conservation and management strategies. Our data and those from studies of post-fire succession and post-logging succession suggest a fairly rapid recovery of species richness following disturbance in these forests. Special attention must be paid to the existence of regional and global endemic species in these forests. Many plant species that characterize these moist forests appear to be relics from coastal forests perhaps dating back to the mid-Miocene before the rising of the Cascade Range plain (Daubenmire 1943b, 1969). Some of these species are quite rare and are associated with older forests and highly specialized wet microsites, especially the ferns (Johnson and Steele 1978). Special attention may be needed to ensure continued viability of these highly disjunct and sometimes small populations.

Most of the surrounding forests in our study area appear to have been nearly completely destroyed by fires in 1910, as deduced from our tree ring samples and surveys of tree-ring scars in the vicinity (Barrett 1982). Amazingly, even the largest trees we encountered, several over 70 cm in diameter, were generally less than 90 years old. This appears to contrast with many coastal sites where, even during years of catastrophic fire, remnant trees are a persistent and key feature of the post-disturbance environment (Spies and Franklin 1991).

The high diversity of these forests would appear to be a product of several key factors. Since these forests are wet microsites in a generally xeric grassland or open woodland region, they provide habitat for the large pool of species that require high soil moisture during the growing season. Also, as noted above, many species appear to be vicarious or disjunct species with coastal affinities following geological and geographic changes in the region both before and during the Holocene. In addition, these forests represent ecotonal forests between more typical xeric Rocky Mountain forests, grasslands, and the boreal forests to the north. Since so many species appear to be near their geographic and presumably their physiological limits of range, these forests may be worthy of special monitoring of climatic changes and anthropogenic impacts. We have not addressed specifically any of the potential genetic issues with these highly fragmented localized species, but this could be another major implication of the unique situation presented by the species composing these moist forested valleys and slopes.

It is not clear what the full implications the unique climate and vegetation have on disturbance ecology of interior rainforests. As in coastal forests, wind may be an important disturbance agent. Trees are generally smaller, so tip-up mounds are less significant and persistent features than they typically are on the coast. The rich mixture of species, from the deciduous western larch (*Larix occidentalis*) to the shallow rooted Englemann spruce (*Picea engelmannii*), make for a complex interaction with wind as contrasted with the typical two- or three-species dominated forests on the coast. It appears that both fire and windthrow result in a large recovery in diversity within five years of disturbance. More detailed analysis will be needed on reproduction and species adaptation to better identify commonalties and contrasts between these disturbance types in the Rockies.

The redcedar, hemlock, and fir dominated forests that characterize interior rainforests are consistently ranked as the most productive forest type in the Rockies. Since this forest type comprises both the most diverse and most productive of forest types in the region, we have compelling reasons to better understand its ecology and dynamics. Implementing ecosystem management strategies will require careful consideration of the unique biological and physical attributes of these forests, the potentially complex interactions between small- and large-scale disturbances, and interactions between disturbances that pervade these unique forests.

Note

1   Nomenclature for vascular plants follows Hitchcock and Cronquist (1973).

References

Alaback, P. 1991. Comparative ecology of temperate rainforests of the Americas along analogous climatic gradients. Revista Chilena de Historia Natural 64:399-412.

Alaback, P.B. 1995. Biodiversity patterns in relation to climate in the temperate rainforests of North America. Pp. 105-133 *in* R. Lawford, P. Alaback, and E.R. Fuentes (eds.). High-latitude rainforests and associated ecosystems of the West Coast of the Americas:

Climate, hydrology, ecology, and conservation. Ecological Studies 113, Springer-Verlag, New York, NY.

Arno, S.F. 1980. Forest fire history in the northern Rockies. Journal of Forestry 78(8):460-465.

Barrett, S.W. 1982. Fire's influence on ecosystems of the Clearwater National Forest: Cook Mountain fire history inventory. US Department of Agriculture (USDA) Forest Service, Clearwater National Forest, ID.

Brown, J.K., S.F. Arno, S.W. Barrett, and J.P. Menakis. 1994. Comparing the prescribed natural fire program with presettlement fires in the Selway-Bitterroot wilderness. International Journal of Wildland Fire 4:157-168.

Byler, J.W., S.K. Hagle, A.C. Zack, S.J. Kegley, and C. Randall. 1997. Successional functions of pathogens, insects and fire in intermountain forests: General discussion and a case history from northern Idaho white pine forests. Pp. 228-232 *in* Proceedings of the 1997 Society of American Foresters National Convention, 4-8 October 1997, Memphis TN.

Cooper, S.V., K.E. Neiman, and D.W. Roberts. 1991. Forest habitat types of northern Idaho: A second approximation. General Technical Report INT-236, US Department of Agriculture (USDA) Forest Service, Intermountain Research Station, Ogden, UT.

Daubenmire, R. 1942. An ecological study of the vegetation of southeastern Washington and adjacent Idaho. Ecological Monographs 12:53-79.

Daubenmire, R. 1943a. Soil temperature versus drought as a factor determining lower latitudinal limits of trees in the Rocky Mountains. Botanical Gazette 105:1-13.

Daubenmire, R. 1943b. Vegetation zonation in the Rocky Mountains. Botanical Review 9:325-393.

Daubenmire, R. 1952. Forest vegetation of northern Idaho and adjacent Washington and its bearing on concepts of vegetation classification. Ecological Monographs 22:301.

Daubenmire, R. 1968. Soil moisture in relation to vegetation distribution in the mountains of northern Idaho. Ecology 49:431-438.

Daubenmire, R. 1969. Ecological plant geography of the Pacific Northwest. Madrono 20:111-128.

Daubenmire, R., and J.B. Daubenmire. 1968. Forest vegetation of eastern Washington and northern Idaho. Technical Bulletin No. 60, Washington Agricultural Experiment Station, Pullman, WA.

Eck, K.C. 1984. Forest characteristics and associated deer habitat values, Prince William Sound Islands. Pp. 235-246 *in* W.R. Meehan, T.R. Merrell, Jr., and T.A. Hanley (eds.). Proceedings of a symposium, Fish and wildlife relationships in old-growth forests, 12-15 April 1982, Juneau, Alaska. American Institute of Fisheries, Alaska District, and US Department of Agriculture (USDA) Forest Service, Pacific Northwest Research Station, Portland, OR.

Elzinga, C.L., and R.C. Shearer. 1997. Vegetation structure in old-growth stands in the Coram Research Natural Area in northwestern Montana. General Technical Report INT-GTR-364, US Department of Agriculture (USDA) Forest Service, Intermountain Research Station, Ogden, UT.

Fischer, W.C., and A.F. Bradley. 1987. Fire ecology of western Montana forest habitat types. General Technical Report INT-GTR-223, US Department of Agriculture (USDA) Forest Service, Intermountain Research Station, Ogden, UT.

Habeck, J.R. 1976. Forests, fuels and fire in the Selway-Bitterroot Wilderness, Idaho. Pp. 305-353 *in* Proceedings of Tall Timbers Fire Ecology Conference, 8 October 1974, Tallahassee, FL.

Habeck, J.R. 1977. Forest succession in the Glacier Park cedar-hemlock forests. Ecology 49:872-880.

Habeck, J.R. 1978. A study of climax western red cedar (*Thuja plicata*) forest communities in the Selway-Bitterroot wilderness, Idaho. Northwest Science 52:67-76.

Habeck, J.R., and R.W. Mutch. 1973. Fire-dependent forests in the Northern Rocky Mountains. Quaternary Research 3:408-424.

Hawk, G.M. 1977. A comparative study of temperate *Chamaecyparis* forests. Ph.D. thesis, Oregon State University, Corvallis, OR.

Hitchcock, C.L., and A. Cronquist. 1973. Flora of the Pacific Northwest. University of Washington Press, Seattle WA.

Johnson, F.D., and R. Steele. 1978. New plant records for Idaho from Pacific coastal refugia. Northwest Science 52:205-211.

Juday, G.P. 1976. The location, composition, and structure of old-growth forests of the Oregon Coast Range. Ph.D. thesis., Oregon State University, Corvallis, OR.

Latham, P. 1996. A process-related quantification of forest structure and understory diversity in inland Northwest forests: From stands to landscapes. Ph.D thesis, University of Montana, Missoula, MT.

Reynolds, G.D. 1992. Climatic data summaries for the biogeoclimatic zones of British Columbia: Version 3. Virga Consulting. British Columbia Ministry of Forests Library, Victoria, BC, Call Number 551.69711R463.

Rogers, P. 1996. Disturbance ecology and forest management: A review of the literature. General Technical Report INT-GTR-336, US Department of Agriculture (USDA) Forest Service, Intermountain Research Station, Ogden, UT.

Smith, J., and W.C. Fischer. 1997. Fire ecology of the forest habitat types of Northern Idaho. General Technical Report INT-GTR-363, US Department of Agriculture (USDA) Forest Service, Intermountain Research Station, Ogden, UT.

Spies, T.A., and J.F. Franklin. 1991. The structure of natural young, mature, and old-growth Douglas-fir forests in Oregon and Washington. Pp. 91-110 *in* L.F. Ruggiero, K.B. Aubry, A.B. Carey, and M.H. Huff (eds.). Wildlife and vegetation of unmanaged Douglas-fir forests. US Department of Agriculture (USDA) Forest Service, Pacific Northwest Research Station, Portland, OR.

Steele, R. 1971. Red alder habitats in Clearwater County, Idaho. M.S. thesis, University of Idaho, Moscow, ID.

Stickney, P.F. 1982. Vegetation response to clearcutting and broadcast burning on north and south slopes at Newman Ridge. Pp. 119-124 *in* D. M. Baumgartner (ed.). Site preparation and fuels management on steep terrain: Proceedings of the symposium 15-17 February 1982, Spokane, WA. Washington State University Cooperative Extension, Pullman, WA.

Stickney, P.F. 1986. First decade plant succession following the Sundance forest fire, Northern Idaho. General Technical Report INT-197, US Department of Agriculture (USDA) Forest Service, Intermountain Research Station, Ogden, UT.

Stickney, P.F. 1990. Early development of vegetation following holocaustic fire in Northern Rocky Mountain forests. Northwest Science 64:243-246.

Swetnam, T.W., and A.M. Lynch. 1993. Multi-century, regional-scale patterns of western spruce budworm history. Ecological Monographs 63:399-424.

The Natural Disturbance Program for the Foothills Model Forest: A Foundation for Growth

David W. Andison

Abstract

For two years the Foothills Model Forest in Alberta supported an extensive natural disturbance research program across more than 2.5 million hectares including Jasper National Park and the Weldwood of Canada Forest Management Agreement in Hinton, Alberta. This long-term integrated program consists of over 20 individual projects, from landscape-level pattern descriptions, to island remnant studies, to understanding fire edge architecture. The complexities of natural disturbance processes are such that the program strategy is allowing us to develop a complete picture of disturbance patterns and processes across scales. Almost half of the projects have begun, and we are beginning to publish, interpret, and implement some of the results both on the Forest Management Agreement and in the Park. We are also using experimentation as a means of testing biological responses to some of our interpretations and testing the potential for reintroduction of fire in some areas. Furthermore, the recently initiated biodiversity monitoring program has already decided to use "natural range of variability" results from the Foothills natural disturbance program as baseline biological indicators.

Introduction

The Foothills Model Forest, in cooperation with Weldwood of Canada, Jasper National Park, Weyerhauser Company, and the Alberta Lands and Forests Service, has been involved in an extensive natural disturbance research and development program for the past three years. The area of study covers approximately 2.5 million hectares of fire-dominated eastern Rocky Mountain foothills and mountain land in Canada. The Foothills Model Forest includes the Weldwood Forest Management Agreement, Jasper National Park, and other Alberta provincial land. To date, we have completed three field seasons, initiated eight projects, and expect to have several more underway in the next two or three years. We have at least preliminary results for four of the projects (Andison 1997; MacLean et al. 1997), have submitted manuscripts for three of them (Andison 1998; Andison et al. 1999; Farr et al. 1999), and are in the process of transferring the results to the partners' area managers and planners.

Expending this much effort to understand natural disturbance may seem to some to be extravagant, or even unnecessary. Like all responsible science, however, the answers must address the questions asked. In this case, a single question posed at the outset precipitated the research: *What are the historical "natural" disturbance patterns and processes on the Foothills Model Forest?*

We are not the first to pose this question, nor are we the first to recognize the tremendous potential of using a natural disturbance paradigm to reach ecosystem management goals (Burton et al. 1992; Franklin 1993; Sapsis and Martin 1994). From the beginning, however, we have recognized that we placed ourselves in a precarious position by selecting a natural disturbance strategy to achieve ecosystem management ideals. Consider that our collective knowledge of disturbance as a process is quite limited. We are only beginning to grasp the complexities of natural disturbance processes in many environments, and in most, we do not have the necessary data. Given that situation, it is ironic that accepting a natural disturbance strategy necessitates assuming that whatever patterns we study and quantify are meaningful in terms of biodiversity. We have no empirical evidence to show that incorporating natural patterns in management strategies results, for more than a handful of species, in higher levels of ecological sustainability. In other words, we are taking a leap of faith that the patterns we measure and subsequently adopt are directly associated with a positive, linear ecological response.

This does not mean that the natural disturbance strategy has no merit, but rather (1) that we are obligated to conduct thorough research, and (2) that we should not take for granted the connection between observed pattern and ecological integrity. It is easy to think we are emulating "natural" patterns and thus being more ecologically sustainable, when in reality we may not be. The key to successful emulation strategies is mainly in the questions we ask and in how we answer them. Three examples from the natural disturbance research program at the Foothills Model Forest demonstrate the importance of asking the right questions and carefully designing research to successfully answer them.

Example 1

A commonly asked question in natural disturbance work is, "What is the frequency and/or cycle of disturbance?" In asking this question, we refer to the frequency with which disturbance affects a given land area or location (Johnson 1992). From frequencies and cycles, we can infer information on the amount of "old-growth" forest to retain, either as a minimum criterion as in the case of the Biodiversity Guide of the Forest Practices Code of British

Columbia (BC Forest Service 1995), or as an average. The question of frequencies or cycles can be answered in simple terms (with the right dataset) using various methods of age reconstruction (Johnson 1992). For the Foothills Model Forest, we found that the fire cycles varied by ecological subregion from 80-90 years in the lower elevation areas of the eastern slopes to over 150 years in the subalpine areas (Andison 1997).

There is no argument that this is useful information, and it answers the question of fire cycles. But it does not fully address the question of natural pattern. The natural pattern of age-class distribution in these and other forests is highly variable and cannot be captured with a single number (Boychuk and Perera 1997; Andison 1996). For the Foothills Model Forest, if we take the last 160 years as a baseline and reconstruct the area consumed by forest fires by 20-year increments, we find average disturbance rates of between 15-25 percent (in 80 to 130 year fire cycles). The 20-year disturbance rate, however, varies from less than one percent to over 50 percent. Thus fire cycles vary depending on the time at which they are measured so that a single fire cycle estimate does not capture the natural pattern of disturbance (Andison 1998).

Abundance of forest in different age classes varies over time as well. For instance, the average proportion of mature forest (> 100 years) in the lower elevation mixed forest on the Weldwood Forest Management Agreement is 5 percent. A simulation exercise that spatially projects the variation in the rates of disturbance for the last 160 years, however, suggests that the amount of mature forest varies over time. Less than one percent mature forest existed 25 percent of the time (or for about 40 of the last 160 years); between two and three percent of mature forest existed another 25 percent of the time; between three and five percent mature forest another 25 percent of the time; and between five and 32 percent of mature forest the other 25 percent of the time. In this way, the natural pattern of mature forest abundance is variable and highly negatively skewed. If we simply had answered the first question about fire cycles and converted the answer to a percentage of mature forest, the management target based on fire cycle estimates would be a constant five percent. As it turns out, managing for a constant level of mature forest is entirely *un*natural.

Example 2

A second question often asked of disturbance analyses is: "What are the ranges of forest patch sizes?" These data are potentially useful for planning disturbances such as harvesting or prescribed burns. As with the fire cycle, this is not a difficult question to answer with the right dataset. Creating summaries of the youngest patch sizes by size-class is a simple process. Once again, this provides a good answer to the question of patch sizes, but taken alone, it is potentially misleading in terms of the natural pattern of patches on a landscape.

Defining a "patch" for the purposes of pattern analysis is not a simple task and many interpretations are possible (Gustafson 1998). For instance, if we look more closely at patch size distributions, we note that as they get larger, they become more complex in shape. This observation is neither new nor surprising. As fires grow, they experience a wider range of climatic and fuel conditions and thus have greater opportunity to change direction and intensity. If we consider the type of edge in each patch, we note that the proportion of edge associated with non-forested areas is higher than expected. For instance, although non-forested patches constitute only ten percent of the landscape on the Weldwood Forest Management Agreement, forest to non-forest edge accounts for almost 25 percent of the total amount of edge in the same area. Furthermore, the shape of non-forested patches is more complex than forested patches of the same size. These findings suggest two important pieces of information with respect to natural patterns of forest patches in the Foothills Model Forest.

(1) Simple patch size and shape metrics cannot be applied to landscapes blindly, or aspatially. Large forest openings tend to be location specific.
(2) Non-forested areas are an important part of the natural landscape pattern. Fortunately for Weldwood, only about ten percent of its landbase is non-forested vegetation. In most other boreal and subalpine landscapes, this percentage is closer to 50 percent. By ignoring the relationship between forested and non-forested patch patterns, we risk not capturing the true pattern of the system.

Example 3
The third and last question also commonly associated with natural disturbance research is: "What is the total area and spatial arrangement of unburnt island remnants (i.e., live vegetation left within the area of a fire)?" (Eberhart and Woodard 1987; DeLong and Tanner 1996). Other researchers have extended this work to look at refugia or island locations associated with permanent landscape features (Camp 1995). The information is potentially useful for harvest block design.

Eberhart and Woodward (1987) and Delong and Tanner (1996) asked and answered important questions that stand alone as scientific contributions. From the perspective of capturing natural patterns, however, better questions may be asked. The main problem is that the concept of "island remnants" is a human-defined construct and, as such, communicates some preconceived ideas. For example, it represents only one aspect of pattern at the scale of an individual fire. We could be considering many other factors.

Suppose we back away from the question of island remnants and look instead at how a fire behaves in relation to local weather patterns, biota, and the land. We could then consider other aspects of pattern at the scale of the individual disturbance. For instance, one could argue that the patch we see today as a fire in forest inventories or satellite imagery is a single time-space snapshot. It is one possible outcome of the ecological processes at work, no different than a time-space snapshot of a landscape mosaic. If we accept that fire is not perfectly random (which is arguable) we can think of fire, as a process, to be either more or less likely to burn certain areas at certain times.

This model of fire is useful, since it introduces two new ideas that may help to better understand the natural pattern at the scale of the individual patch. First, all fire edges are potential islands that have not had time to develop. If we could follow an individual fire through a time series from ignition, we would see bits of edge break off from the fire perimeter. Only at this point is an island formed, yet there is always some edge affected by exactly the same processes of fire behaviour interacting with the land and biota – which is the process we are most interested in. Island remnant research studies only the islands, not the edges.

The second useful idea spawned by the concept of a fire as a time-space snapshot is that some locations have a higher than expected chance of burning while other locations have a lower than expected chance of burning. Island remnant studies focus only on the latter. Where islands and edges tend to form and where they tend *not* to form may be equally important. For example, on topographically complex landscapes such as those found in Jasper National Park, we found a strong *positive* relationship between slope position and the tendency to retain older forest. On the simpler landscapes of the eastern foothills, a *negative* relationship between older forest and certain topographic positions was evident. This suggests that the adoption of a model of positive and negative space for harvest block planning may provide a much clearer picture of the natural pattern than did the old model of island remnants.

Conclusions

The examples discussed here describe only part of the research we are conducting at the Foothills Model Forest (see Andison 1997 for details). They were used to demonstrate some of the risks of conducting natural disturbance research with the ultimate goal of maintaining ecological sustainability: not asking the right questions, not asking enough questions, or not addressing the questions in the best manner. These risks result both from our ignorance and our assumption that natural patterns are connected to ecological sustainability. Until we can provide substantial empirical evidence to support one form of pattern management over another, we must make a leap of faith that all forms of natural pattern are directly linked to ecological function. This assumption means that we tend to accept what is known as ecological truth, when it may not be. For instance, should we worry about harvest block sizes and shapes when we have not looked at how blocks are oriented on the landscape relative to other patches? Similarly, patch size may be irrelevant if we have a limited understanding of what age and structural characteristics exist within those patches or if we assume they are all homogeneous and even-aged.

The success of the natural disturbance strategy will depend on how well we minimize risks. The best way do this is to develop rigorous natural disturbance research programs. The examples given in this paper demonstrate how easy it is to think that one has captured the natural pattern, when this may not be true. Along with research efforts, we should also be working to instill at least a small amount of doubt about what we think we know, thereby gaining respect for what we have yet to learn. We should not assume that all simple questions and answers represent a linear connection to ecological truth. Links to biodiversity monitoring programs are a constructive way of dealing with this problem.

It is clear that the leap of faith we must take when we use the natural disturbance paradigm has potentially serious consequences for our ability to develop truly ecologically sustainable forest management practices. We are obliged to take seriously the research of natural patterns, which extends from the questions asked, to the methods and data sources used, to interpreting results. Finally, given the range of possible interpretations, we should also be very careful about what we label "ecosystem management."

References

Andison, D.W. 1996. Managing for natural patterns in sub-boreal landscapes of northern British Columbia. Ph.D. thesis, University of British Columbia, Vancouver, BC.

Andison, D.W. 1997. Landscape fire behaviour patterns in the Foothills Model Forest. Foothills Model Forest internal report. Hinton, AB.

Andison, D.W. 1998. Temporal patterns of age-class distributions on foothills landscapes in Alberta. Ecography 21:543-550.

Andison, D.W., H. Lougheed, and G. Stenhouse. 1999. Managing for natural levels of age-class variability on fire dominated landscapes. Ecological Applications (forthcoming).

BC Forest Service. 1995. Biodoversity guidebook: Forest Practices Code of British Columbia. British Columbia Ministry of Forests, Victoria, BC.

Boychuk, D., and A. Perera. 1997. Modeling temporal variability of boreal landscape age-classes under different fire disturbance regimes and spatial scales. Canadian Journal of Forest Research 27:1083-1094.

Burton, P.J., A.C. Balisky, L.P. Coward, S.G. Cumming, and D.D. Kneeshaw. 1992. The value of managing for biodiversity. Forestry Chronicle 68:225-237.

Camp, A.E. 1995. Predicting late-successional fire refugia from physiography and topography. Ph.D. thesis, University of Washington, Seattle WA.

DeLong, S.C., and D. Tanner. 1996. Managing the pattern of forest harvest: Lessons from wildfire. Biodiversity and Conservation 5:1191-1205.

Eberhart, K.E., and P.M. Woodard. 1987. Distribution of residual vegetation associated with large fires in Alberta. Vegetatio 4:412-417.

Farr, D.R., C.P. Spytz, and E.G. Mercer. 1999. Structure of forest stands disturbed by wildfire and logging in the Rocky Mountain Foothills. (in preparation)

Franklin, J.F. 1993. Preserving biodiversity: Species, ecosystems, or landscapes? Ecological Applications 3:202-205.

Gustafson, E.J. 1998. Quantifying landscape spatial pattern: What is the state of the art? Ecosystems 1:143-156.

Johnson, E.A. 1992. Fire and vegetation dynamics: Studies from the North American Boreal Forest. Cambridge University Press, New York, NY.

MacLean, K., D. Farr, and D.W. Andison. 1997. Island remnants within fires in the Foothills and Rocky Mountain Natural Regions of Alberta: Part I. Methodology. Internal Foothills Model Forest Report. Hinton, AB.

Sapsis, D.B., and R.E. Martin. 1994. Fire, the landscape, and diversity. A theoretical framework for managing wildlands. Pp. 270-278 *in* Proceedings of the 12th Conference on Fire and Forest Meteorology, 26-28 October 1993, Jekyll Island, GA. [www.wildfiremagazine.com/sapsis.shtml].

Design Your Own Watershed: Top-Down Meets Bottom Up in the Applegate Valley

Dean Apostol, Marcia Sinclair, and Bart Johnson

Abstract

The 28,346-hectare Little Applegate Watershed, located in Oregon's Siskyou Mountains, includes federal and private lands. The diverse landscape of farms, oak savannahs, old growth forests, and subalpine parklands has gained wide recognition for grassroots efforts that support progressive land management. Our planning process built a long-range vision for the Little Applegate watershed using a focused public involvement strategy that integrated local knowledge with landscape ecology, conservation biology, environmental restoration, and landscape architecture.

Introduction

The 1980s and 1990s witnessed several important new conceptual developments in the fields of ecology and biology that have influenced present thinking about natural resource management. Landscape ecology, the study of patterns and processes over large areas, borrows from biogeography, planning, landscape architecture, and other fields. Conservation biology, the study and advocacy of managing landscapes to conserve biological diversity, appeared on the scene at nearly the same moment. Restoration ecology, starting from very small beginnings in Midwestern prairies, has now grown into a significant social and professional movement. Taken together, these three interdisciplinary fields have profoundly altered the way we think about landscapes. Additionally, the range of possible design and management alternatives has been widened considerably.

But managing natural resources has never been exclusively a technical problem, so the adaptation of new technical approaches is not by itself sufficient to create positive changes. In democratic societies, "the people" own public lands. Any shift in management has to be understood and supported by a critical mass of lay citizens. Additionally, it has become clear that conservation cannot stop at public land property boundaries. The coho salmon that spawn in the Little Applegate River don't know or care whose land the waterways go through. The salmon need cool, clean water with silt-free gravel, pools, cover, and insects to eat. Diverse groups of people, often with profound differences, must somehow be brought together to work towards common objectives including ecological and economic health.

To accomplish this, we built a strategic public process that engages the community on its own turf. Complex ecological assessment findings were translated to lay language and presented with graphic images. Local residents and activists who have first-hand, detailed knowledge of the landscape were encouraged to bring their skills to the table in a collaborative effort. The new understanding of the land made possible through landscape ecology, conservation biology, and restoration has been joined to the public involvement process. This marriage allowed us to design a long-term landscape structure and pattern that we hope will retain native biodiversity, restore aquatic conditions, harness the agent of fire, and retain economic benefits associated with farming, logging, and wildcrafting.

The Little Applegate Watershed

The Little Applegate is a 28,346-hectare watershed that flows into the Applegate River near the unincorporated town of Ruch, in southern Oregon. The Applegate in turn flows into the Rogue River, which carves a canyon through the Siskyou Mountains before entering the Pacific near Gold Beach. This is a rugged, twisted landscape with small farms, orchards, vineyards, and ranches hugging the narrow valley floors. South-facing slopes include grassland, chaparral, and pine-oak woodlands. North-facing slopes contain mixed conifer forests including remnant patches of old growth. Subalpine parklands decorate the highest peaks and ridgelines. The federal lands in the Applegate watershed are designated an "Adaptive Management Area" in the Northwest Forest Plan of 1993. This is a regional plan for all the federal lands west of the Cascade Mountains in Washington, Oregon, and Northern California. It is the first regional-scale biodiversity conservation plan in the United States.

Adaptive Management Areas were established to provide for social as well as ecological experimentation. They are to serve as laboratories where federal agencies and local citizens collaborate to plan and carry out land management work. The Applegate is one of nine such areas. Citizens across political lines had formed the Applegate Partnership several years before the Northwest Forest Plan was hatched. The Applegate partnership is a "bottom-up" organization that includes loggers, ranchers, environmentalists, and concerned citizens. They have been advocates for progressive land management including thinning dense forests, protecting rural landscapes, and actively restoring streams.

Public Involvement Process

Most natural resource planners and foresters learn how to manage public involvement, if they learn it at all, in the school of hard knocks (and badly managed public involvement will usually result in a few hard knocks). We mistakenly think we are good at it if we can stand in front of a potentially hostile meeting and maintain our composure. In fact, public involvement should be thought of as a subset of the discipline of public relations. As such, it is worthy of study, testing, and the development of a repeatable process, much like technical silviculture. Public involvement is often done for the wrong reason: THE GOVERNMENT MADE US DO IT! The right reason is to use it as a tool that can and should be harnessed to achieve the desired end – *a plan that is ecologically appropriate, economically feasible, and socially acceptable.*

Public involvement is targeted public dialogue. It is an essential element of a democratic society, yet how many of us think of it in those terms? Democracy is not simply an election every few years. It is an ongoing process of interaction, argument, and hopefully compromise. If one faction has the legal authority to carry out its plans, regardless of the merits of the position of those who oppose it, then democratic processes are replaced by apathy, cynicism, or direct confrontation. The opportunity to build long-term relationships is lost.

The experience of the authors has been that many forest managers, both public and private sector, hold circumscribed views of public involvement. These usually reflect their bias as technocrats who have grown accustomed to setting rather narrow objectives for their projects (i.e., getting timber to the mill in a cost-effective way while ensuring the regeneration of the forest). Most managers also object to open processes that may challenge their plans. Public involvement is viewed as an obstacle to be overcome rather than as an integral component of a successful planning process. So they limit their efforts to rounding up the usual suspects, have the requisite meeting or two, make a few minor tweaks in the plans they already have, and then declare victory. If there is a legal recourse, the unsatisfied citizens, on their own time and expense, can take the project to court. If not, they can use blockades or other adversarial means to stop it. Too often, the goal is to have appeared to involve people without really doing it. To the extent planning and forestry professionals go along with this, we are engaged in a sham.

There is a better way. In the Little Applegate Watershed, we had a client (the Rogue River National Forest in the form of project manager Carol Spinos) who wanted the genuine article: participation. She wanted it done thoroughly and without prejudice. After all, since we were developing a very long range plan (200 years) for a large watershed that included public lands and the private property of several hundred residents, nothing was to be gained by crafting a design in a closet and then attempting to impose it on the people. She needed a critical mass of people to support the concept. If the people were to support it, then they must believe they contributed to its making. They needed to see their own values and ideas reflected in it. To accomplish this, we offered a strategic approach.

One textbook definition of public relations is "the planned effort to influence opinion through socially responsible and acceptable performance, based on mutually satisfactory, two-way communication." This definition formed the foundation for our strategy. We were not initiating the process simply to have an open-ended dialogue. On the contrary, we had an agenda; *crafting a long-range landscape plan for an entire watershed.* We wanted not to sell a preconceived design but rather to influence opinion about the nature of this plan and to help people understand the design process and its potential for solving some of the difficult ecological issues of the area. We did not want to be sneaky or manipulative in our approach. In the end, we wanted citizens to trust the process and take pride in the results. We wanted the process to be repeatable in nearby watersheds. Building local support was essential.

Additionally, our strategy defined the various "publics" concerned with the Little Applegate Watershed. We reject the notion that there is such a thing as a "general public," at least in terms of how a particular project is presented and advanced. We instead defined publics by common characteristics, interests, concerns, or recognizable demographic features. An individual could fit several categories: logger, local neighbor, and environmentalist, for example. Thus we targeted our communication to neighbors, local environmental organizations, elected officials, wildcrafters, the timber industry, and so forth. In all, we identified 21 publics, including personnel of the very agencies that were sponsoring the project. If this seems bizarre, consider that on another project, one of the authors was asked to serve as "a liaison" between two separate departments in the same agency.

For each group, we identified the following:

- Key contacts.
- Members.
- Our objectives for what behavior we wanted from people (i.e., active support, having people feel they were heard).
- Tools for communicating with people. These included articles for their newsletters, attendance at their meetings, and avenues for participation.
- Key messages to get across.
- Measures for success. These are based on emotional response from the varied publics and their expressed willingness to participate in the future.

Figure 1. A three-dimensional image illustrating patch types, complex composition, and structural characteristics of a watershed plan in the Applegate Valley, Oregon.

As a starting point, we were fortunate to have a detailed survey of public values done by a local organization, the Rogue Institute for Ecology and Economy. This survey exposed many common themes that linked conservatives and liberals, young and old. Nearly everyone wanted to restore salmon and steelhead runs. They wanted to retain, at a sustainable level, natural resource-based economic opportunities. They were concerned about wildfire and the health of the surrounding forests. They wanted to maintain both the rural character of the area and a sense of community.

The process for crafting the strategy was long and detailed. We budgeted specifically for this task. We provided more tools than the client could use – a varied palette allowing the greatest flexibility to meet community needs. Thus we implemented less than 75 percent of the overall strategy. But the results have been encouraging. We now have a long-range design for the landscape that has widespread support. The community that is interested in this watershed has a better understanding of the complex ecology of the watershed, and they have some tools for making improvements. The effort from the top met the interest at the bottom.

The Plan

What is a planning process without a plan? We developed a long-range plan for managing the landscape structure across the entire watershed. This plan establishes fourteen distinct landscape patch types, including old growth conifer forest, pine/oak savannah, two-layer forest, and patchy forest, among others. Each of these patch types is linked to three-dimensional illustrations that translate complex composition and structural characteristics to pictorial form (Figure 1). This allows for understanding across technical disciplines at a level most lay people can grasp. The plan merges concepts from landscape ecology, conservation biology, and restoration ecology with time-tested landscape planning principles. Our client hopes to follow up on the work done so far by developing an "owners manual," envisioned as a three-ringed binder that summarizes ecological and design information in a manner that will make everyone who lives in or works in the watershed a more responsible participant in its future development. This "owner's manual" is envisioned as a fluid document that allows for adaptive management in response to new information or changed conditions.

We believe the approach we took is repeatable in complex, large-scale projects. No project is as perfect or smooth as textbooks would have us believe, and ours certainly has had some challenges. But we feel we made real progress in helping to craft a watershed design that is ecologically appropriate, economically feasible, and socially acceptable.

Resource Planning in the Deer Creek Watershed: Historical Approaches to Integrated Resource Management and How They Relate to Ecosystem Management and Strategic Priorities for the Drainage

Reiner Augustin and Greg Rowe

Abstract

Deer Creek is a drainage of about 9,000 ha located approximately 25 kilometers northwest of Castlegar, British Columbia, along the east side of Arrow Lake. The drainage is part of the Arrow Timber Supply Area (TSA). Deer Creek is a community watershed and is a source of water for about 30 homes. The drainage has significant wildlife values, visually sensitive areas, and is an important source of timber for Kalesnikoff Lumber Ltd. A number of planning concepts have been applied here over the last 20 years including the folio system, total resource planning, and Integrated Watershed Management Planning. These plans in conjunction with stand level biodiversity management have led to a variety of forest practices used by the company. The forest management approaches resulted in an award for environmental excellence from the British Columbia Ministry of the Environment in 1986. Forest management in the Deer Creek drainage is presently receiving considerable strategic direction from other plans and processes in the West Kootenays including the Kootenay Boundary Land Use Plan (KBLUP) implementation strategy, the Arrow Timber Supply Area Timber Supply Review, and Landscape Unit Planning in the Arrow Forest District. One of the present challenges facing forest managers in Deer Creek is to use an ecosystem management approach to ensure that stand-level treatments are appropriate for the site, meet overall forest level objectives for the Arrow TSA, and contribute to ecosystem sustainability at the landscape level. This paper reviews what has been done over the last 20 years in Deer Creek, outlines what the present needs are, and discusses some ways of meeting them within the context of the present ecosystem management approach. One focus is maximizing use of existing information and creating a strong link between strategic direction and stand treatments. Some of the concepts discussed include Deer Creek as an operational example of ecological stewardship and the potential for a "stewardship tenure," Geographical Information System impact assessments, potential for the "one plan approval" concept, crop planning, direction for incremental silviculture, and alternative harvest strategies.

Introduction

Individuals involved in forestland management in British Columbia find themselves in an era of continual, sometimes spiralling change. Pressure to change is in large part owing to the public's demand for a more socially conscious land management ethic. Managers are asked to look first at sustainable development rather than the more traditional profit- and risk-oriented approach. It appears that this paradigm shift in forest management philosophy is occurring in British Columbia in a matter of decades compared to jurisdictions such as Europe, where evolution has taken place over centuries. Ecosystem management is one of the forest planning concepts that embodies the demand for sustainable development and reflects the change in forest management philosophy.

Bridging the gap between the theory of ecosystem management and its application is key to its success. From a field practitioner's perspective the pitfalls in application are many but can be summarized in the areas of understanding concepts, adjusting to the speed at which new ideas and direction are expected to be applied, addressing the knowledge gap in understanding how our ecosystems function, blending social and ecological values with economic reality, and having the opportunity to learn through trial and error.

In this paper, presented in two parts, we describe an example of helping bridge the gap between ecosystem management theory and its application. Part one of the paper is a review of resource planning initiatives in the Deer Creek drainage, touching on the products and possible application in the context of ecosystem management. Part two focuses on how the unit fits into the broader picture of both the Arrow Timber Supply Area and the West Kootenay Region in general.

Deer Creek Area Description

Deer Creek is a drainage of approximately 9,000 hectares located in British Columbia about 20 kilometres northwest of Castlegar along the east side of the Lower Arrow Lakes (Figure 1). The drainage is situated in the ecologically diverse Interior Cedar-Hemlock biogeoclimatic zone and contains portions of four Natural Disturbance Types (NDTs 2, 3, 4, and 5) as defined in the Biodiversity Guidebook of the Forest Practices Code of British Columbia (BC Forest Service 1995).

Most of the Deer Creek watershed is within the Arrow Timber Supply Area (TSA). Deer Creek is a community watershed as defined under the Forest Practices Code of British Columbia (Province of British Columbia 1996). The drainage is a source of both domestic and irrigation water to approximately thirty homes. About 44 percent of the total unit is considered timber harvesting landbase as defined in the Arrow Timber Supply Area Timber Supply Review. The remaining area is made up of approximately nine percent

Figure 1. Three-dimensional perspective of the Deer Creek Watershed in southeast British Columbia. Deer Creek Watershed is outlined with white line.

private land and environmentally sensitive areas. Deer Park, located at the delta of Deer Creek, is essentially a summer community. Resettlement along the Lower Arrow Lakes reservoir is causing a slow but continual population increase in the Deer Park community.

The watershed is considered to contain significant wildlife values including avian diversity (Steeger 1995; Steeger and Machmer 1995) and visually sensitive areas and is considered a critical source of timber supply for Kalesnikoff Lumber Company Ltd. (Thrums, BC). Timber harvesting, trapping, and agriculture are the primary economic activities in the drainage today. Construction of the Hugh Keenlyside Dam on the Columbia River and subsequent flooding of the reservoir in the late 1960s resulted in the loss of spawning habitat in the lower reaches of Deer Creek. While not considered a key spawning area, the lower reaches of Deer Creek are today important for various fish species common to the Arrow Lakes.

The Deer Park area appears to have considerable archaeological significance. The extent of aboriginal artifacts that have been and continue to be discovered along the lakefront suggests that First Nations peoples frequented the area.

Planning History

Deer Creek presents an opportunity to educate people about an area where managing for a variety of forestland values has occurred at the sub-landscape unit level (Augustin et al. 1995; Steeger and McLoed 1996). In comparison to other local geographic units, Deer Creek has a comparatively long planning history. The watershed provides an opportunity to review the application of a number of planning concepts including the folio system, Total Chance Planning, Integrated Watershed Management Planning, and the application of the Forest Practices Code of British Columbia requirements.

Resource planning in the Deer Creek area has generally followed the customary planning steps of area identification and description, data assembly and analysis, with the final product consisting of resource objectives and operational guidelines. Goals and objectives were set in consideration of and consistent with the established planning hierarchy. This meant recognizing and incorporating direction and limits set by such documents as the Timber Supply Area Yield Analysis.

Not addressed in these past initiatives were the issues of old growth management, seral stage distribution, landscape connectivity, desired stand structure and composition, and the spatial and temporal distribution of disturbances as currently identified in the Biodiversity Guidebook of the Forest Practices Code of British Columbia (BC Forest Service 1995).

Review of archival BC Ministry of Forests forest cover information indicates that timber harvesting operations in the Deer Park area date back to the early 1900s. Logging in the Deer Creek drainage, limited until the 1980s, saw the first major harvesting by Kalesnikoff Lumber Company Ltd. in the upper Deer Creek drainage in 1981. Incorporating local input in development planning has been a priority for the company. While not without controversy, the operations to date can be considered a success in mixing industrial, social, and environmental goals. The BC Minister of Environment recognized the company's operations in Deer Creek through an environmental award in 1986.

1930s: Reconnaissance Era
A review of past forest surveys and forest cover maps provides a historic perspective of the Deer Creek drainage. The information, gathered through a series of field transects, was plotted on base maps produced by BC Provincial Land Surveys. The topographical information displayed was obtained from the Geological Survey of Canada (surveyed in 1900).

The 1930s inventory divided forest cover information into mature and immature forest (Figure 2). For immature forest polygons the age of stand establishment is provided. Disturbance through windthrow and logging are indicated. All other disturbance associated with stand establishment is assumed to be fire. This reconnaissance mapping provides us with a tool to understand the Deer Creek ecosystems. The mapping provides a temporal and spatial reference of the fire history and/or stand initiating events in this area. A review of the mapping suggests that 11 significant fires occurred between 1847 and 1940. Fire size ranged from 40 to 1,500 hectares with an approximate average fire size of 460 hectares. Based on the mapping it is

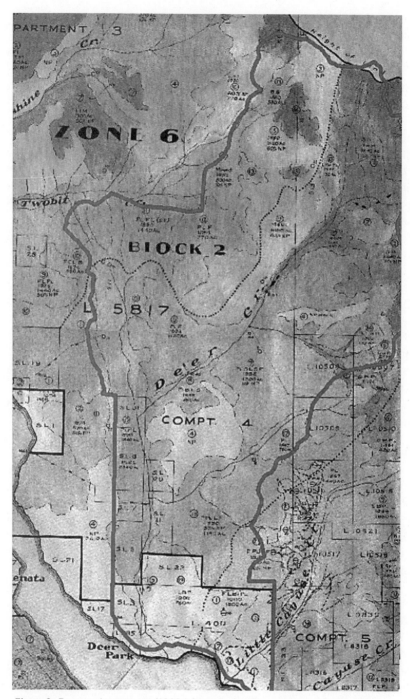

Figure 2. Reconnaissance era (1930s) historic forest cover map for the Deer Creek Watershed in southeast British Columbia.

estimated that up to 55 percent of the drainage may have been impacted by wildfires over this time.

At Deer Park, land use in the early to the middle portion of the twentieth century was predominantly associated with agriculture, mostly fruit growing. Timber harvesting, while active in areas in close proximity to Deer Creek, was relatively limited.

1950s and 1960s: Columbia River Treaty

The 1960s was the era of the Columbia River Treaty. Construction of the Hugh Keenlyside Dam and resulting reservoir caused extensive loss of low elevation lands along the Arrow Lakes. The creation of the reservoir impacted Deer Creek insofar as lower elevation spawning habitat was lost because of fluctuating water levels.

At Deer Park community, like others along the Arrow Lakes, private properties were expropriated. Resettlement of the Deer Park area following dam construction was virtually immediate.

With resettlement came a keen interest in any development plans for Deer Creek. Residents expressed concerns over potential negative impacts on water supplies from timber development that occurred in the late 1960s. Deer Creek was provided watershed reserve status in 1942. This was subsequently rescinded in 1970 and reinstated in 1986. Today Deer Creek is a community watershed as defined under the Forest Practices Code of British Columbia.

Deer Creek has a historical record of water-related information. The creek has been sampled continuously since 1958. Water Surveys Canada undertook the initial sampling. Water information continues to be collected today by the BC Ministry of Environment, Lands and Parks, Kootenay Region.

During this period the status of the Deer Creek drainage with respect to inclusion or exclusion in Tree Farm Licence #23 was a subject of considerable debate. It was not until the early 1970s that a decision was made to exclude Deer Creek from Tree Farm Licence #23. The area became part of the Salmo Public Sustained Yield Unit. It was during this time that Kalesnikoff Lumber Company Ltd. expressed interest in developing the timber opportunities in Deer Creek.

1970s: Deer Creek Folio

The BC Ministry of Forests initiated the resource folio program in the early 1970s. Folio plans were prepared for much of the Arrow Lakes area including Deer Creek. The folios consisted of an assembly of resource layers. Much of the information was based on the Canada Land Inventory Program criteria. Included were map layers identifying fish habitat, ungulate capability, recreation potential, topography, and forest and soil capability.

While the folios primarily identified land capability at a broad scale, they also provided the opportunity to visually combine a variety of resource layers. This information, in combination with BC Ministry of Forests inventory mapping that included the identification of environmentally sensitive areas, represented an important tool in identifying potential resource use compatibility and conflict at a landscape level.

The Deer Creek folio was completed in 1976. Key values identified in the Deer Creek folio were wildlife, fisheries, and timber resources. Not unlike today, values in many instances were described in general terms owing to the lack of specific inventory information.

Timber harvesting priorities in Deer Creek, identified through the folio planning process, were upper elevation Engelmann spruce *(Picea engelmanni)* and subalpine fir *(Abies lasiocarpa)* stands with second priority given to mid-elevation Douglas-fir *(Pseudotsuga menziesii)* and Western larch *(Larix occidentalis)* stands. The folio plan recommended that a variety of block sizes (10 to 25 hectares) and silviculture systems be used in response to various site sensitivities.

Contrary to the priorities outlined in the folio, much of the past timber harvesting has concentrated on mid-elevation Lodgepole pine *(Pinus contorta)* stands in response to Mountain Pine beetle activity. A conclusion that may be drawn from this is that the original plan did not adequately consider the influence of forest health factors.

The folio-based planning process, while considering resource values and compatibility, did not appear to incorporate public input in the decision making. This was significant in the case of Deer Creek given that resettlement of the Deer Park area was occurring. Much of the prevailing public sentiment of the day called for the creation of a Deer Creek wilderness area.

1982/83: Integrated Watershed Management Planning

A watershed plan, building on the 1970s folio, was completed in 1983. The plan was the result of a cooperative effort involving the British Columbia Ministry of Forests, the British Columbia Ministry of Environment, consultant services, and Kalesnikoff Lumber Company Ltd. In response to concerns expressed by residents of Deer Park, the plan focused on protection of the water resource.

Water protection, particularly risks associated with sediment input, was addressed through more detailed terrain stability mapping than was originally available in the folio. The drainage was divided into sub-units for the purpose of monitoring harvest distribution.

Although the plan was submitted for public review that provided opportunity for comment, water-user representatives were not part of the planning team. In the process of plan preparation, value judgements were made that were based largely on land capability and timber goals. A decision-making fallback was the timber targets as set forth in the 1981 yield analysis for the Arrow Timber Supply Area. A product of the process, the identification of engineering corridors, zoned operating units with associated harvesting guidelines.

It may be concluded that the plan had a timber bias. This was a predictable result given the absence of public representatives on the planning team and lack of inventory information for non-timber forest values. The extent of partial cutting regimes undertaken subsequent to the plan, however, lead us to conclude that much of the ecosystem diversity inherent to Deer Creek remains.

1995 Forest Practices Code Application

In 1995 the British Columbia Forest Service and Ministry of Environment ecosystem specialist performed an updating of the 1983 watershed plan for Deer Creek using geographic information system (GIS) technology.

The project started with the timber harvesting landbase as defined in the Timber Supply Review and superimposed the emerging Forest Practices Code of British Columbia requirements as understood at that time. Field work and local knowledge were used in an attempt to meet environmental objectives, as interpreted from draft Forest Practices Code documents, while minimizing the impact on timber supply.

During this planning initiative an assessment of some of the biodiversity components was performed. Items considered in the assessment included old growth requirements, seral stage distribution, temporal and spatial distribution of harvest, connectivity, riparian management, terrain stability, and watershed management through equivalent clearcut area calculations.

The use of GIS, besides providing visually and geographically referenced representation of resource values and their interaction, allowed for a cumulative impact assessment of the various Forest Practices Code of British Columbia requirements. The process provided for both an iterative approach to designing landscape connectivity and old-growth management areas while minimizing timber supply impact and meeting environmental objectives. The application of GIS technology provided a link between strategic and operational planning.

Present operational plans incorporate the results of the various planning efforts. Consideration is currently being given to connectivity corridors, retention of structural and species diversity, riparian protection, and retention of coarse woody debris. Timber harvesting priorities continue to be affected by forest health issues. This is in contrast to the envisioned patterns as described in both the folio and watershed plans.

Historically, harvesting has been predominantly partial cutting systems. What remains is a mix of selection and clear-cutting prescriptions with a wide variety of unit sizes (Figure 3). Areas planned for harvesting incorporate wildlife tree patch needs, retention of snags, veteran trees, and large diameter trees (Figure 3).

Public consultation and input in the operational plans has been strengthened through the practice of continual contact and annual meetings with residents. The forest company has made a commitment to understanding and addressing public concerns.

Deer Creek in the Context of the Arrow Timber Supply Area and the West Kootenays

In the 1980s and early 1990s, prior to the Kootenay Boundary Land Use Plan (Kootenay Inter-Agency Management Committee 1997) and the Forest Practices Code of British Columbia (Province of British Columbia 1996), no formal legal framework for strategic planning existed. Timber supply and harvest flow was and still is managed at the Timber Supply Area level. The allowable annual cut was set for the Arrow Timber Supply Area in 1981 on the basis of a timber supply analysis that indicated a harvest level that, because of an abundance of mature timber, could be maintained for many decades. This initial timber supply analysis also had short-term (20-year) timber supply targets at the inventory compartment (Deer Creek drainage) level of detail. Deer Creek comprises about 2 percent of the total timber harvesting land base in the Arrow Timber Supply Area.

During this period timber management priorities at the Timber Supply Area level were to harvest older stands first and to minimize non-recoverable timber losses to insects, disease, and wind throw. Silviculture objectives included regeneration of harvested stands and reforestation of backlog NSR (not sufficiently regenerated) areas. Non-timber resource values were managed on an integrated resource management basis.

Formal recognition of non-timber resource values in Deer Creek in 1982 included establishing a watershed reserve and developing recommended visual quality objectives. The early 1980s featured a Timber Supply Area that had reasonable flexibility in choosing stands to harvest to meet the annual allowable cut, relatively simple Timber Supply Area level timber and silviculture priorities, drainage level harvest targets, and an informal approach to management of non-timber resource values.

The First Timber Supply Review (TSR1)

The BC Ministry of Forests undertook the first Timber Supply Review for the Arrow Timber Supply Area between 1992 and 1995. This was an aspatial analysis, harvest was not tracked to the landscape unit level, and no geographically located 20-year exercise was associated with it. Although not intended as a source of strategic direction per se, it provided an analysis-based perspective and allowed the BC Chief Forester an opportunity to provide some observations on timber supply management in the Arrow Timber Supply Area through comments in the annual allowable cut rationale document. The BC Chief Forester noted that British Columbia Ministry of Forests staff "should ensure operations are well distributed across the landscape" (Pedersen 1995). In the context of the Arrow Timber Supply Area,

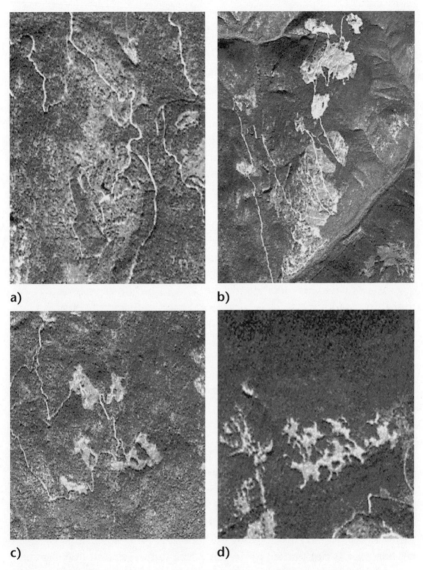

a)

b)

c)

d)

Figure 3. Harvest systems used within the Deer Creek watershed in southeast British Columbia: (a) select harvest, (b) large openings, (c) clear-cut with reserves, (d) group selection.

this means operating in areas with high non-timber resource values including consumptive-use watersheds and visually sensitive areas like Deer Creek. The 1981 timber supply targets proved to be reasonable. By 1992, eleven years later, 55 percent of the 20-year target had been harvested. The performance in Deer Creek with respect to the 1981 timber targets and the BC Chief Forester's comments indicated that, in general, drainage level forest management in Deer Creek has met local priorities and was consistent with the strategic direction of the day. Timber Supply Review 1 resulted in no change to the allowable annual cut.

The Kootenay Boundary Land Use Plan and the Forest Practices Code of British Columbia

The next major initiatives to affect strategic and operational planning in Deer Creek were the formalized structure introduced through the Kootenay Boundary Land Use Plan and the Forest Practices Code of British Columbia. To put this in context, the following recently articulated provincial timber supply objectives from the BC Ministry of Forests should be considered:

- to maintain current harvest levels for as long as possible without creating disruptive shortfalls in future timber supply
- to maintain a future quality profile as good as or better than the current profile.

The Forest Practices Code of British Columbia (Province of British Columbia 1996) initiated a number of additional planning requirements that affected forest management in Deer Creek. These included:

- Deer Creek was designated as a community watershed.
- An Interior Watershed Assessment Procedure was required in community watersheds along with input that included water users.
- Terrain survey requirements were linked to community watershed status.
- Stream classification and riparian management based on stream size, fish presence, and community watershed status were introduced.
- Attention was and must still be given to the Biodiversity Guidebook of the Forest Practices Code of British Columbia (BC Forest Service 1995) and associated recommendations regarding old growth management, seral stages, and wildlife tree patches.

The Kootenay Boundary Land Use Plan, approved by a group of provincial cabinet ministers, contains general resource management direction consisting of regional level objectives, strategies for economic, social, and environmental values, as well as geographically specific resource management guidelines (Kootenay Inter-Agency Management Committee 1997).

For Deer Creek some of the key general direction provided is that timber management is an appropriate land use and water, wildlife, and visuals are important values. The Kootenay Boundary Land Use Plan also identified general objectives relating to economic values (timber resources, energy resources, mineral and coal resources, agriculture, commercial tourism, settlement, access), social values (cultural heritage resources, communities, outdoor recreation, visible areas), conservation values (general ecosystem health, terrestrial ecosystem health, aquatic ecosystem health, air quality, rangeland ecosystems).

The Kootenay Boundary Land Use Plan assigned an intermediate level of biodiversity emphasis to the Cayuse Landscape Unit, which includes Deer Creek. It also identified as important values regional connectivity, ungulate winter range, and visuals and provided a map showing where these values should be managed in Deer Creek. Resource management guidelines were included in the Kootenay Boundary Land Use Plan to provide management direction for these areas.

The overall timber supply and forest management situation is now significantly different than it was in the 1980s in both the Arrow Timber Supply Area as a whole and for Deer Creek. The Arrow Timber Supply Area now has a tightly constrained timber supply. The current annual allowable cut is near the level at which it was first set in 1981 and, because of the presence of large volumes of mature timber, is above the Long Run Sustained Yield. It has proven difficult to access much of that mature volume, however, and actual harvest levels in recent years have generally been below the Annual Allowable Cut. The imposition of Kootenay Boundary Land Use Plan guidelines onto a Timber Supply Area with an uneven harvest distribution and an age class structure resulting from an extensive fire history has resulted in a situation where there is very little spatial or temporal flexibility in timber harvesting. In Deer Creek this has been aggravated by considerable mountain pine beetle activity in recent years.

At present it is difficult to determine how activities in Deer Creek relate to forest-level issues. A multitude of rules and guidelines have been designed to protect a variety of resource values, but clear forest-level objectives for some aspects of forest management (particularly related to timber quantity and quality over time) are lacking, and no formal forest management strategy exists for the Arrow Timber Supply Area where forest management activities (harvesting, silviculture, and protection) are directed at strategic priorities for the management unit.

In the absence of landscape unit objectives we are currently following a manage-by-constraint approach, where we harvest only those few blocks that remain after meeting all of the guidelines.

The current need for strategic direction at the landscape unit/drainage level is clear direction from higher levels that is linked to the condition of the forest at the Timber Supply Area level. Where possible this should be expressed as desired future conditions so that management is directed towards achieving this, rather than meeting static constraints. Landscape unit objectives should be formulated on the basis of careful matching of this higher level direction to the conditions of the landscape unit. This should ensure that the objectives are achievable. Indicators that can measure success with respect to achieving objectives should be developed. It is also important to clearly outline priorities for silviculture investments. The result should be integration and simplification at the drainage level of goals, objectives, guidelines, rules, and plans, which should lead to effective and efficient forest management.

Landscape Unit Planning, Ecosystem Management, and the Innovative Forest Practices Agreement

A number of tools exist that should help meet some of these needs for translating existing strategic direction into something more useful at the drainage level. Three of these are landscape unit planning, ecosystem management, and the Arrow Innovative Forest Practices Agreement (Arrow Forest Licensee Group 1999).

Landscape unit planning is part of the Forest Practices Code of British Columbia planning framework for Crown Land. It can eventually result in objectives for a landscape unit (Deer Creek makes up about half of the Cayuse Landscape Unit) for a number of resource values including old growth retention, wildlife tree patches, connectivity, seral stages, visual quality, timber harvesting, silviculture investments, wildlife habitat, recreation, and water. Landscape objectives will form a key link between the Kootenay Boundary Land Use Plan objectives, Arrow Timber Supply Area management considerations, and the operational level in Deer Creek.

Ecosystem management can be defined as "an approach to land and resource planning and management that emphasizes the recognition and maintenance of ecosystem function and structure" (Kootenay Inter-Agency Management Committee 1997). It can be applied across a series of scales from stand to forest. Sustainability is the underlying premise and goal of ecosystem management. While sustainability is subject to interpretation, a good starting point is the Canadian Council of Forest Ministers criteria (Canadian Standards Association 1996):

- conserving biological diversity
- maintaining and enhancing forest ecosystem condition and productivity
- conserving soil and water resources
- forest ecosystem contributions to global ecological cycles
- multiple benefits to society
- accepting society's responsibility for sustainable development.

Many of the components normally associated with an ecosystem management approach (including a variety of silvicultural systems, species viability and functional roles, and the role of structural complexity in maintaining ecosystem function) are already part of forest management in Deer Creek because of the long history of resource planning in the drainage.

The Arrow Timber Supply Area licensees have recently been awarded an Innovative Forest Practices Agreement. The overall objectives of the Innovative Forest Practices Agreements are:

(1) to test new and innovative forestry practices intended to improve forest productivity

(2) to encourage licensees to carry out new forest practices by offering them the opportunity to apply for an increase to their allocated harvest levels to enhance and maintain employment in the forest industry.

The primary goal of the Arrow Innovative Forest Practices Agreement is to efficiently bridge higher level plans such as the Kootenay Boundary Land Use Plan with operational or stand-level plans and activities. Ecosystem management is a central theme of the Arrow Innovative Forest Practices Agreement.

The Innovative Forest Practices Agreement offers an opportunity to implement the Kootenay Boundary Land Use Plan and ecosystem management concepts in Deer Creek and the rest of the Arrow Timber Supply Area in a manner that will benefit everyone. The three approaches, landscape planning, ecosystem management, and the Arrow Innovative Forest Practices Agreement can be applied together to provide the link between higher level strategic direction and operations in Deer Creek.

How to Use These Approaches to Direct Operations in Deer Creek

A key question about using these approaches with a volume-based tenure system is where to start – at the stand level, drainage level, or forest level? While landscape unit planning in particular and to some degree ecosystem management are focused at the drainage or landscape unit level, given that timber supply is presently being managed at the Timber Supply Area level, some direction from the higher levels is necessary to ensure that forest management in Deer Creek contributes to Timber Supply Area level for timber supply solutions and to regional level objectives for other resource values.

Part of the challenge is to reconcile the ecosystem management principle of outputs that are determined by the capacity of the ecosystem and its sustainability with current legislation, policy, and higher level planning direction.

The following four-step process is one possible approach to reconciling regional, forest, and landscape level issues.

Step 1: Clarify higher level resource management objectives and develop timber crop objectives for the Timber Supply Area for both quantity (in the short-, mid-, and long-term) and quality (species, size, and product).

This can be done by first reviewing the economic, social, and environmental values and objectives in the Kootenay Boundary Land Use Plan and then formulating indicators by which changes in the values can be measured. For indicators to be useful for assessing alternatives and monitoring activities over time, they should be measurable, predictable, relevant, valid, feasible, and as simple as possible. Some work has been done in developing resource indicators during the course of the Kootenay Boundary Land Use Plan. Some valuable work has also been done recently, which could be used to supplement the Kootenay Boundary Land Use Plan indicators, on indicator development associated with the three certification systems (Canadian Standards Association, Forest Stewardship Council, and International Organization for Standardization). To maximize productivity, ensure sustainability, and develop a workable drainage level management approach, the existing Kootenay Boundary Land Use Plan guidelines, while useful on an interim basis as an aid to statutory decision makers when considering approval of operational plans, should eventually be replaced. This should occur when detailed landscape unit analysis has been done so that existing guidelines are replaced by strategies that are specific to the landscape unit.

Step 2: Analyze opportunities and constraints at the forest (Timber Supply Area) level, considering timber supply and indicators for other accounts.

This can be done at the Timber Supply Area level with resolution at the landscape unit scale to deal with Timber Supply Area level age class and harvest flow issues. For instance, partial cutting has been identified as an opportunity for increasing available short-term timber supply in the Arrow Timber Supply Area if it is applied in areas that preclude clear cutting (e.g., some Visual Quality Objective areas and some landscape unit natural disturbance type combinations). This should be tested for its applicability to Deer Creek. We should develop draft silviculture and harvesting strategy at the Timber Supply Area level based on these opportunities and constraints.

Completing the above steps will produce a list of potential forest management activities that, from the perspective of the forest level (Timber Supply Area), could be performed in Deer Creek to meet higher level objectives. This should provide a starting point for detailed landscape unit planning/ecosystem management in Deer Creek. It will specify higher level goals with indicators, preliminary timber targets by period (to be confirmed by the ecosystem management approach), alternate harvest strategies that could be directed at Timber Supply Area level issues (e.g., partial cutting in visual quality objective areas and potential strategic silviculture opportunities). This list is, of course, subject to review at the drainage level through landscape inventory, analysis, and an ecosystem management approach. In some cases the result will be a verification of the list; in other cases improved information for the next Timber Supply Review will result; in yet other cases the outcome could be the advice that higher level goals are not attainable.

Step 3: Do a landscape inventory and analysis for the Deer Creek landscape.

The basic analysis concerns spatial patterns, structure, and functioning of the landscape as an ecosystem. It should consider a number of things including natural disturbance patterns, the Interior Watershed Assessment Procedure, wildlife habitat and use, actual harvest opportunities given development patterns, growth and yield for partially cut stands, location of Old Growth Management Areas (OGMAs), landscape connectivity, forest health, engineering issues, and field investigations of silviculture and innovative harvest opportunities. In the case of Deer Creek, this can be mainly a matter of putting together existing information. While many aspects of the landscape analysis will be common throughout the Arrow Timber Supply Area, some aspects will, in some places, receive more emphasis than in others. In some drainages the emphasis may be on spatial modeling to examine strategies to address patch size, adjacency, and old growth recruitment issues. In Deer Creek the emphasis will likely be on dealing with conflicts between dispersing harvest to meet watershed management objectives and concentrating harvest where there has been mountain pine beetle *(Dendroctonus ponderosae)* activity. At the same time we must maintain habitat for a variety of wildlife species and manage fire to reintroduce a more natural stand structure in this natural disturbance type (NDT 4; BC Forest Service 1995). The choice of silvicultural systems, their effectiveness in meeting resource objectives, and their influence on future site productivity will also be important considerations here. This step will show how the landscape unit can best be managed to meet higher level objectives.

Step 4: Produce draft objectives and management strategies for the drainage and evaluate them against regional and Timber Supply Area objectives using previously developed indicators.

Outputs and desired future conditions should be specified. These can be used for a number of purposes including the following:

- by the district manager as the basis for setting landscape unit objectives
- by Kalesnikoff Lumber Company Ltd. as a development/management strategy from which, under the present system, operational plans should flow easily, or to potentially provide the basis for a simplified approach to operational planning (a "one plan approval" approach)

- by Forest Renewal British Columbia or other program delivery agents, as an aid to setting priorities for investments
- by independent agencies for assessments related to certification, if desired.

Appropriate opportunities for public involvement should be provided. At this point final objectives and strategies should be implemented, monitored, and revised as necessary.

Conclusion and Future Opportunities

Deer Creek is an example of successful evolution of planning and management approaches over a period of several decades. The potential is good for timber production in the Arrow Timber Supply Area since it is located in some of the most productive biogeoclimatic sub-zones in the British Columbia interior. Deer Creek is a drainage with not only a great variety of resource values, but based on past experience, great promise for successful management of them. Many of the components are in place to continue this management success as the focus shifts to ecosystem management and landscape unit planning.

Deer Creek presents an opportunity for ecosystem restoration. A target area may be the Natural Disturbance Type 4 portion of the drainage, impacted by both fire exclusion and human habitation. Restoration in this area could improve the overall drainage diversity.

The move from Timber Supply Review landbase assumptions in the initial stages of planning to an ecosystem management approach (applying our understanding of ecosystem functioning) represents a key change in direction in planning for ecosystem sustainability at the landscape level. Moving to ecosystem management will emphasize production of non-timber forest products. We must improve our planning at spatial and temporal scales to achieve the objectives of sustained productivity and natural diversity. Applying these approaches will forge strong links between strategic and operational planning while maintaining the principles of ecosystem management.

For Deer Creek, we must work further to incorporate such biodiversity considerations as connectivity, patch size, and seral stage distribution. The presence of four Natural Disturbance Types in Deer Creek provides a unique opportunity in ecosystem management. There is concern that the small unit size may be limiting in managing for suggested patch size and distribution.

We can further refine ecosystem management by augmenting it with terrestrial ecosystem mapping. In particular, we need site series mapping, wildlife species inventory, and development of interpretations that address wildlife capability and suitability.

We must monitor the success and failures of the application of ecosystem management concepts. Deer Creek represents an opportunity for such a monitoring program. This could include establishing permanent sampling points and transects that assess ecosystem sustainability and diversity.

In an era of cost cutting, Deer Creek may also present an opportunity to test concepts that may fit into the changing forest management philosophy. In Deer Creek, we could test the "stewardship tenure" concept based on ecological principles and goals. This may lead to "one plan approval" and, with that, provide an incentive for investment.

References

Arrow Forest Licensee Group. 1999. Arrow Innovative Forest Practices Agreement: Forestry Plan. Nelson, BC.

Augustin, R., G.A. Rowe, J. Sherbinin, and P. Field. 1995. Total resource planning in the Deer Creek Watershed. Pp. 577-583 *in* G.I.S. World: Proceedings of the Ninth Annual Symposium on Geographic Information Systems, 27-30 March 1998, Vancouver, BC.

BC Forest Service. 1995. Biodiversity guidebook: Forest Practices Code of British Columbia. British Columbia Ministry of Forests, Victoria, BC.

BC Ministry of Forests. 1998. Guidance towards participation in the Forest Renewal BC planning process and the development of resource management plans. British Columbia Ministry of Forests, Victoria, BC.

Canadian Standards Association. 1996. A sustainable forest management system: Guidance document. Canadian Standards Association, Etobicoke, ON.

Kootenay Inter-Agency Management Committee. 1997. Kootenay Boundary land use plan implementation strategy. Nelson, BC.

Pedersen, L. 1995. Rationale for allowable annual cut determination for the Arrow Timber Supply area. British Columbia Ministry of Forests, Victoria, BC.

Province of British Columbia. 1996. Forest Practices Code of British Columbia Act. Revised Statues of British Columbia, Chapter 159.

Steeger, C. 1995. Winter wildlife diversity in managed and unmanaged stands within the Deer Creek Landscape Unit. Unpublished report. British Columbia Ministry of Forests, Castlegar, BC.

Steeger, C., and M.M. Machmer. 1995. Wildlife trees and their use by cavity nesters in selected stands of the Nelson Forest Region. Technical Report TR-010, BC Ministry of Forests, Nelson, BC.

Steeger, C., and B. McLeod. 1996. Case study: Establishing wildlife tree patches in an operational setting. Research Summary RS-026, BC Ministry of Forests, Nelson, BC.

Ensuring the Social Acceptability of Ecosystem Management

Simon Bell

Abstract

While ecosystem management of forests has become an approach increasingly adopted to enable sustainable development of forests to happen, managers have given little thought to ensuring that ecosystem management is also socially acceptable. Ecosystem management incorporates much of the most recent scientific thinking on dynamic ecosystems, the place of natural disturbances in landscapes, and the maintenance of ecosystem functioning. Applying this recent scientific thinking can involve mimicking large-scale disturbance through forms of forest management that can include prescribed forest fire, electing not to control insect outbreaks, and leaving large amounts of wood on logged areas. To some people this can look like massive clearcutting, forest destruction, or waste of valuable lumber, even though, to experts, it is ecologically sound and even necessary to maintain ecosystem health. Social acceptance of forest practices frequently lags behind the science. Myths and misconceptions about nature abound. They can be used by the unscrupulous to generate misinformation and jeopardize the realization of sustainable management. Experts in any field are subject to mistrust while a number of people willingly prefer theories arising from alternative belief systems that may be treated as contemptible by some mainstream managers and scientists. Problems of knowledge, communication, and philosophical approach must be overcome. Traditional methods such as propaganda, "educating the public," or even some relatively recent attempts at community involvement in land use planning are no longer effective. Proponents of ecosystem management must be philosophers as well as ecologists or foresters; they must develop new methods of communication and directly involve communities. Ethics, aesthetics, and politics must be embraced in the quest for social acceptance, together with fresh approaches to communication and participation.

Introduction

Forestry and forest management are currently undergoing some major and rapid changes. Pressure from both the top down and the bottom up have placed foresters in situations where their professional skills and technical knowledge seem inadequate. Political changes, market changes, and socially driven changes have been institutionalized into concepts of sustainability underpinned by international, national and local treaties, programs, and processes.

Many foresters have accepted that practices must change and have embraced the challenge to develop new models of management that hopefully meet society's concerns while remaining economically viable. Perhaps the most radical and promising of these started out as "New Forestry" and has developed into "Ecosystem Management." The initial idea, presented by Jerry Franklin, the well-known forest ecologist, has been enthusiastically taken up, first by the US Forest Service and later by many private sector lumber companies, so that it has become well established. As will be seen, it conforms to Leopold's Land Ethic and has moved forest management away from Pinchot's utilitarian approach. This shift, an earth-shaking one for foresters, produces visual results very different from both traditional checkerboard clear cuts or from the unchanging climax ecosystem of wilderness mythology.

Put simply, ecosystem management concentrates on the natural functioning of large-scale areas of the forest ecosystem rather than on managing the outputs. The underlying premise is that by maintaining the "functioning" of the ecosystem, typified by the continued presence of natural patterns and processes, the wide range of human values can be obtained sustainably, at a level determined by the capacity of the system to deliver them. Humans manage to emulate natural processes that are excluded from the system, such as large-scale fires; or else natural processes are introduced in a controlled way, such as by prescribed fire. Managers must also allow for some unpredictable natural occurrences. Logging, in the most radical scenario, becomes a management tool to create disturbances in the landscape within the scale and character range for the particular ecosystem. Wood is removed instead of being consumed by fire, for example.

Such an approach is now being practiced in a number of places, not least in British Columbia. If managers lack knowledge about certain aspects of ecosystems, adaptive management is introduced, where plans and designs are modified over time as new information is gathered and the effect of actions taken on the ground is evaluated. This represents a paradigm shift in management approaches and has a sound scientific basis. But is it accepted by society as better than what has gone before? What is being done to ensure that it is socially acceptable?

To build trust and establish ecosystem management as the main approach for the next period of time, it is vital to involve as many sectors of society as possible and to increase the general level of knowledge about ecosystem management. Commensurate with this is the need for foresters to improve their communication methods and raise their own awareness of how and why public attitudes are formed.

Four issues need to be addressed, each of which frequently appears on lists of issues associated with traditional forest management as perceived by members of the public or by special interest groups. The issues that must be addressed if ecosystem management is to be accepted by society may be more powerful in British Columbia, where visible clearcuts continue to be created, than they are in the United States, where the national forests of the Pacific Northwest have been under moratoria in recent years. These issues can be stated as:

- Forest management destroys ecosystems.
- Forest management is unethical.
- Forest management damages the scenery.
- Forest management ignores the wishes of local people.

Common to all of these is the problem of communication between foresters (and other professionals and specialists) and the rest of society. These four issues are examined one by one and methods of overcoming them are suggested.

Forest Management Destroys Ecosystems

One of the main foci for this issue is logging, specifically clearcutting. The problem runs more deeply than that, however, because it also revolves around perceptions of what a forest should be like. No doubt there remain among foresters, let alone the wider public, a number of myths and misconceptions about forest ecology. Daniel Botkin, in his book *Discordant Harmonies* (Botkin 1990), nicely illustrates some of these. One persistent myth is that of the steady state climax forest, where trees grow old and die to be replaced by younger trees leading to a multi-storied forest canopy that remains intact forever unless some outside agent of destruction, such as a fire, hurricane, or insect attack, is introduced. These outside agents are often thought to be negative and unusual elements that should be suppressed, fought, or killed if possible. This picture has been reinforced by the high profile given to coastal old growth forests, which are as close as it is possible to get to this image. Discarded is the fact that coastal old growth forests are actually quite rare when the total amount of forests across the world is considered.

Foresters themselves have helped develop and perpetuate this myth by vigorous fire suppression and insecticide application, as well as by enlisting the help and powerful public relations appeal of Smokey the Bear. The result, taken over 50 or 60 years, is that several generations of people have grown up believing that the sight of endless stretches of pristine mature forest of a healthy green colour is what forests should naturally be like and that any changes to this, whether from natural or man-made causes, is destruction of the forest or the ecosystem. Thus powerful lobbies have thrived on the simple message that forestry practices destroy ecosystems. When supported by images of giant trees, large cuddly animals, and majestic birds of prey, this is very effective propaganda. It leads to the idea that only by preserving areas untouched can the world be saved, thus casting foresters in the role of the bad guys, out to destroy the forest.

Of course, one of the major reasons why managers continue to suppress fires and control insects is to safeguard the wood supply, especially when the output of mills and the well-being of their associated communities depend on continuous timber availability. Thus the forester seems to present a duplicitous figure who, while claiming to be protecting the forest from destruction by fire or insect on one hand, is destroying it by logging on the other. This is exemplified by the sight of clear cuts, especially large ones (now no longer legal in some places), together with associated skid trails, fire sites, slash, and roads.

The problem of out-of-date understanding of ecosystems is often compounded by the development of alternative views based on other traditions and mythologies, such as those embraced by native Americans, or fusions of a range of beliefs selected from a number of spiritual or religious traditions. This opens a wider gulf between some sections of the community and the foresters and ecologists because scientific ecosystem management is open to rejection by people whose belief systems and ethical positions have evolved away from a traditional scientific base. This is a difficult area for managers to address because it may be contrary to their own tradition. The ethical implications of this are considered below.

Political pressures arising from the development of articulate and well-organized protest movements have necessitated legislation in many jurisdictions. Legislation may not have the intended effect of protecting the forest because, while it may result in reducing the size of clear cuts, it may also increase the number of areas cut, so that the end result may be greater fragmentation and negative ecological impact. This suggests that better information is needed all around so that well-meaning but misguided and unscientific practices are not promoted or legislated.

Clearly, what is needed by everyone, foresters included, is a better understanding of the ecology of forests, especially the range of different types of forests and the various processes that are necessary for their survival and that of the functions they fulfil. Generalized models are often applied too widely, and the role of natural disturbance is poorly understood. Destruction of trees is often simplistically equated as being destruction of the forest, while destruction of the forest is often portrayed as irreversible. Since every forest landscape is to some extent different, it stands to reason that an understanding is needed of each landscape and its unique qualities. If this is carried out as part of a community-based participatory process, local communities can gain a better understanding of forest ecology in general and of the local forest in particular. The challenge is to be able to develop silviculture based on natural disturbance regimes that sometimes involve timber harvest in ways that, while modeled on natural patterns, appear to resemble rather closely some forms of clear cutting. This should emerge, however, as an output of the process rather than as a technique imposed on the forest by logging engineers.

The conclusion is that ecosystem management as a form of forestry, far from destroying ecosystems, is the best method yet known for ensuring that ecosystems are preserved. Preservation, however, means attempting to maintain ecosystem functions, not all the individual trees that are present now. Forests must be allowed to change, and where they are already out of balance in terms of their spatial and structural patterns and processes, management should seek to restore them. This might include some logging in areas that appear pristine but are in fact under a degree of stress because of former practices such as fire suppression.

Thus, according to McQuillan (1993), an understanding of forest ecology, especially at a wide range of scales, provides us with the possibilities for obtaining a range of human values. This is particularly important because, while science may inform us of the objectively observed possibilities, and economics and ethics may define the limits of prudence, people generally concentrate on what they want. This hedonism may be at the root of many of the conflicts that develop in some of the public processes recently implemented in forest and land use planning. A more refined process that uses better information to help communities accept what is possible rather than what they want would probably be very helpful.

Forest Management Is Unethical

When Aldo Leopold published *A Sand County Almanac* (1949), he stated his now famous "Land Ethic." This can be considered the starting point for environmental ethics as a branch of moral philosophy. Ethics encompasses the fundamental principles of acceptable behaviour for a person, group, or profession. The main issues here are what is meant by ethics in the forestry context, what position is adopted by foresters, and how this position fits into the range of ethical positions found in society.

Most foresters would probably be offended if they were told that they or their work was unethical. Forestry professional associations incorporate codes of ethics. These may relate in part to business practices but may include references to society and the environment. For example, the Institute of

Chartered Foresters in the United Kingdom includes, in Clause 4 of Regulation 28B Ethics: "Every member shall practice his or her profession with due regard to sound ecological, social, economic and environmental principles to the advantage of future generations."

Laudable though this is, the problem exists that, because ethics is associated with belief systems and because different groups in society adhere to different belief systems, there can be no objectively correct ethical position. Thus the accusation that forestry management is unethical means that the ethics perceived to be followed by foresters is different from that pursued by the accusers. This means that foresters must not assume that the stakeholders of the forest they manage hold the same ethical stance they do. It also means that the practices associated with ecosystem management should be assessed as to their ethical implications.

Environmental ethics as a discipline has undergone extensive development in recent years so that we now have many ways of looking at the subject. The classifications recently proposed by Thompson (1998), summarized below, help clarify different approaches and help people see where they themselves fit. If forest managers are able to explore the ethical positions of different people, perhaps within a stakeholder group, this will help uncover some of the conflicts that exist and open up ways to resolve them.

We can look at different ethical stances in two ways. The first relates ethical theories to the kinds of objects that are deemed to have intrinsic value and be worthy of moral consideration. The second contrasts *ecocentric* and *technocentric* attitudes.

The first way of looking at ethical stances divides theories into two broad categories of human and non-human centred (anthropocentric and non-anthropocentric) theories. Anthropocentric theories place people at the centre of the moral universe where they are the only moral beings and the only creatures to possess intrinsic value. At one extreme is the *egocentric*, or the "me first" position. Such hedonistic theories are associated with *laissez faire* economics, capitalism, and free markets.

Homocentric or people-centred positions are frequently encountered and associated with ideas of welfare and social justice. Utilitarianism, Marxism, and Eco-socialism are political philosophies classified as homocentric. Homocentric philosophies can accommodate respect for other species but do not admit that other species have a moral standing. Social ecologists such as Bookchin (1982) believe that science and technology are necessary but must include recognition that people are dependent on nature. A weaker version of the homocentric position that moves closer to the non-anthropomorphic position holds that other living things can have intrinsic worth, though lower than the worth of people.

Non-anthropomorphic theories are radically different, starting from the view that all living and, in some cases, non-living things have intrinsic moral worth. These are also divided into two variants. *Biocentric* theories include plants and animals as moral beings. Animal rights movements hold this position. Biocentric egalitarianism is a variant in which people are not just part of nature, but an equal part of it. This position attracts a degree of hostility from homocentrists who cannot accept that people should be considered on the same level as slime moulds or mosquitoes.

Ecocentric theories are closely related to biocentric theories but see moral values in the wider ecosystem rather than in individual plants or animals. Many native belief systems can be classified under this heading as can Leopold's land ethic and movements such as deep ecology and Gaianism, although lumping them all together may lead to confusion because they are not all compatible with one another. Deep ecology seeks harmony between people and the cosmos, arguing that to harm nature is to harm ourselves. Some advocates express a view that the real problem is too many people and that they are in some way a blight on the earth. The demotion of people and their rights may pave the way for a form of eco-fascism.

The tensions between the ecocentric and anthropocentric views are apparent when deep ecologists present homocentrically based ethics as shallow ecology in a pejorative sense, i.e., shallow is less aware, deep is more aware. This ignores the fact that homocentric theories have deep philosophical aspects of their own.

The second way of looking at ethical stances is that proposed by O'Riordan (1981) and contrasts *technocentric* with *ecocentric* trends ("ecocentric" is used in a slightly different sense in this case).

Technocentrics believe that science will be able to solve environmental problems in the long term. Their beliefs run a spectrum from those who believe that this will happen in parallel with continued economic growth (the interventionists) to those who put more faith in careful environmental management, assessment of risks, and environmental economics (the accommodators).

Ecocentrics mistrust large-scale technology, although they may be comfortable with "appropriate technology." They are anti-materialist and advocate living in small-scale communities. Their spectrum of beliefs runs from communalists to Gaianists (akin to believers in deep ecology).

O'Riordan (1981) places environmental scientists in the group of technocratic accommodators and the intellectual environmentalists in the group of ecocentric communalists (there is inevitably a somewhat fuzzy boundary between these two).

Forest management has traditionally ranged from crude cutting practices to sophisticated multi-purpose management; this also reflects eras of social development by the countries concerned. Exploitation of forests for purely economic gain with no thought for regeneration, water, wildlife, and native inhabitants would come under the egocentric position. This is not a common position amongst professional foresters but may still be found in certain quarters of the industry in some countries.

In the approach of Gifford Pinchot in wanting the "greatest good for the greatest number," we see a clear adherence to the utilitarian position of homocentric theories. Traditional ideas of the stewardship of nature, supported by some religious points of view, also come under this category. Sustained yield management and even sustainable management generally fall under this category because both envisage the benefits accruing primarily to people. This is perhaps still a very common position among foresters today. It has an honorable place in a tradition of forestry management that goes back to the eighteenth century in Europe. Foresters are also part of the technocentric tradition, both as interventionists (in the sense of the use of plantations, tree breeding, and so on) and as accommodators, which perhaps represents the views of the "greener" end of the forestry profession.

The land ethic of Aldo Leopold has been placed in the ecocentric position as akin to deep ecology. Because Leopold was in some ways a mainstream wildlife manager and forester, although in some ways ahead of his time, his views may be respected by foresters and ecologists who would otherwise be hostile to the deep ecology view. It is interesting that the swing away from forestry planning and management that sought to tidy unruly nature and make forests more productive, a view located well in the homocentric part of the ethical spectrum, should have jumped across to the ecocentric end in a short time. This jump may have left quite a few foresters behind, to say nothing of the public.

Ecosystem management is promoted and practiced by professional environmental scientists including foresters, ecologists, wildlife biologists, hydrologists, and soil scientists. These professions would seem at face value to fall into the category of technocentric accommodators. Many of those who, it is hoped, would support ecosystem management are either intellectual environmentalists in the ecocentric communalist part of the spectrum, or Gaianists. Accommodators and ecocentric communalists may not differ too much, but Gaianists do.

While ecosystem management was identified as having its roots in Leopold's land ethic, it is perhaps less clear than that because, as ecosystem management becomes more accepted, it may develop variants that are constrained by more utilitarian factors, by economics, or by plantation silviculture. Both disadvantages and advantages follow from this. The disadvantage is that if ecosystem management belongs to the homocentric and ecocentric positions simultaneously, it runs the risk of being fought over by adherents of both ethical stances. It can be viewed as weakly homocentric when it uses a range of scientific and sometimes technological tools to manage the forest for a range of values that are defined by and in the interests of people, although recognizing the presence of other values such as that of wildlife. It can also be viewed as ecocentric because, as we have seen, it fits with the land ethic and has an emphasis on the greater ecosystem. The advantage is that it may be possible to accommodate a wide range of ethical positions and combine different perspectives within a common approach without having overtly to dispute the positions held by different stakeholders. To benefit fully from this strength, however, the means of communication and public involvement must be sophisticated.

Thus, as in the case of the need to help all stakeholders better understand forest ecology, it is of prime importance that a thorough exploration of ethical positions is also undertaken. It is particularly relevant to realize that individuals such as scientists, who pride themselves on their scientific objectivity, are not immune from ethics, and that the issue is not objectivity versus subjectivity or facts versus emotions; it is primarily an issue of which ethical position is held. Since no objectively right or wrong ethical position exists, it is incumbent on all parties to desist from attacking those they believe to be wrong and from dressing up ethical differences in the guise of factual or emotional arguments. It is also appropriate to be wary of taking arguments at face value without delving below the surface to uncover the underlying ethical basis. Ecosystem management, uniquely perhaps, offers a solution that combines elements of homocentrism with ecocentrism, permits appropriate technology to be applied, and can reflect the views of both deep ecologists and utilitarianists, as long as communication remains open and entrenched positions are avoided.

Thus the role of an ethical inquiry as part of a participatory planning model is to help stakeholders define, within the ecological possibilities, prudent options. Politics and economics may also be brought to bear at this time. One of the important steps in developing ecosystem management plans is to define potential "desired future conditions" for the forest landscape. Many are ecologically possible; it is a human task to narrow the range and, at this step, ethics helps to determine which are eventually chosen.

Forest Management Damages the Scenery

We hear many complaints and much concern about the visual impact of forestry practices, particularly logging and road construction. Efforts at scenic management (for example, by implementing the US Forest Service's Visual Management System, or the British Columbian version) have either tried to reduce impacts by keeping the scale of activities as small as possible or by screening them from sight. Only recently have there been conscious efforts to design for landscape change. Such approaches concentrate on accepting change as part of the natural condition of landscapes and use natural patterns of vegetation and land form as a cue for the shape and scale of logged blocks. Much of this approach fits very well with the tenets of ecosystem management, so that the two can be worked together, the design process helping to establish the future pattern of the landscape (desired future condition) in a sustainable and ecologically sound way. Because there has not been enough time to implement good design as widely as possible, however, the image remains that forestry damages landscapes. The visual results of disturbance, of active logging, of landscape change to scenes that have looked the same for as long as people can remember, are all part of what continues to cause concern. Thus, to the cynical, forest ecosystem management may appear to spoil landscapes as much as did former management practices. To overcome this problem it is necessary to examine the aesthetic framework that lies at the heart of the concerns to see if a route to a solution can be established.

Aesthetics is a branch of philosophy concerned with the way we perceive things and the emotional reactions stimulated by our surroundings. A long established tradition in the Western world involves appreciating the landscape as picturesque scenery, where the viewer is able to gain a sense of beauty or a sublime experience in a detached way.

Philosophers such as Kant (translated 1982) and Schopenhauer (translated 1969) have attempted to define the qualities of beauty and the sublime. Beauty is found, according to Schopenhauer, when a diversity of objects are found together, clearly separated and distinct but at the same time in a fitting association and succession. The sublime experience occurs when our senses are swamped by the magnitude of a landscape that is difficult to comprehend and suggests limitlessness, where the feeling of danger, potential but not actual, adds an extra sharpness to the emotions.

In the United States this romantic transcendentalist tradition (Gobster 1995) has had a powerful influence, affected by the early landscape painters of the American West and by writers such as Thoreau and Emerson. The vision of John Muir and his sublime experiences at Yosemite add weight to this tradition as do some of the more recent photos of Ansell Adams. This is reflected in the type of scenery chosen for national parks and in the establishment of viewpoints from which to look at it. According to Gobster, this results in three aspects that have affected scenic management in the past, particularly in the United States: an attraction to dramatic and idealized landscapes; a static, visual mode of landscape experience; and an aversion to the death of trees and the mess created by both forestry operations and natural damage such as fire or wind. This is the scenic aesthetic mode of landscape appreciation. Most people, however, have been reluctant to accept change to the landscape and desire to maintain or preserve landscapes as they are.

Recent forest practices, insofar as they have an aesthetic component, are more modernist in outlook (McQuillan 1993). This represents a move away from what are sometimes perceived as elitist tastes to a utilitarian concern for form and function. Louis Sullivan, the architect of skyscrapers, stated that "form follows function" (Connely 1960), and this resulted in simplified Euclidean geometry and the rectangular gridded landscapes of clear cuts or plantations. Originating in Germany, this forestry model was imported to the United States by Bernard Fernow and Carl Schenk at Biltmore, under the influence of Pinchot (Bell 1993). It also has Cartesian origins and relates to the US survey grid and to land allocation to railroad companies. This modernist aesthetic, a failure in architecture and planning, has not yet completely died out in forestry.

Researchers have undertaken many studies on the public's scenic preferences for forest land. These studies have tended to concentrate on the external, static viewpoint. The main findings in North America, however, are that natural or naturalistic landscapes are preferred and that Euclidean geometry is universally disliked. At the same time, people express concern over untidiness and the debris caused, for example, by logging. Instead of concentrating on the actual forest scenes themselves, another set of studies has examined the general properties of landscapes that are universally found to be attractive. According to Kaplan (1983), three components can be found in all scenes that are found to be attractive: coherence, complexity, and mystery. The significance of these may surpass the scenic aesthetic and have a dimension related to human ecology.

Recent developments by aestheticians such as Berleant (1992) emphasize the fact that we are not merely dispassionate, disinterested observers of the scenery but are actively engaged with our environment in all our waking moments. This leads to a different approach to aesthetics, where the quality of our lives depends in part on the quality of our environment. It can be extended to suggest that being actively engaged with our environment is already a kind of ecological approach, where the landscape is an aspect of the human habitat.

Ecosystem management is not about maintaining landscapes unchanged, nor is it about keeping activities tidy. Death to trees, sometimes over large areas by fire, wind, insects, or management activities, is an inherent component. Is there an aesthetic dimension to ecology, or are the two incompatible? If they are incompatible and ecosystem management is perceived to be as scenically damaging as some logging, then we will likely see a negative reaction to many of its practices.

Gobster (1995) has explored the links between ecology and aesthetics in relation to some of Aldo Leopold's theories. In Gobster's assessment, Leopold sees the integrity of the evolutionary heritage and ecological processes as the basis for a positive aesthetic appeal. This eschews the scenic and picturesque qualities and the reliance on vision as the main aesthetic sense. Since the ecological processes are important, Leopold's approach incorporates many of the results of natural disturbances. He overcomes the problem of the acceptability of incorporating the results of natural disturbances by involving experience of the landscape using all the senses and by incorporating knowledge rather than relying on visual perception alone. Gobster argues that the landscape we observe can become more comprehensible as our focus widens to include more complex layers of ecology. In fact this approach also relates to the idea of deeper engagement with the environment as described previously. Moreover, the expressed preferences for coherence, complexity, and mystery relate to landscapes experienced by all the senses in a more engaged way as much as with the scenic view. Natural ecosystems contain these properties in abundance.

The challenge, therefore, is to help people recognize that aesthetics, being more than the beauty of natural scenes, requires a deeper engagement that leads to some understanding of the processes at work and to acceptance of the unavoidable fact of constant landscape change, making a richer experience possible. In fact, the wonder of regeneration after a fire, once it is understood as a natural and necessary event, can engender a rich aesthetic experience when green shoots cover the blackened soil and life emerges from death.

As with the other subjects discussed so far, if ecosystem management with its associated "messiness" is to be accepted, it is necessary to engage all stakeholders in an informed dialogue and to enable them to appreciate the naturalness of nature as being different from the idealized picturesque image traditionally presented. Scenic preferences and the preconceived ideas of all stakeholders should be explored so that the aesthetic implications of management options and the design concepts appropriate to a given landscape can be evaluated openly and fully.

Furthermore, we need to develop or adopt a language that allows better discussion of aesthetics and removes the purely subjective expression of likes and dislikes. The *Visual Landscape Design Training Manual* developed for the British Columbia Forest Service (Bell 1994) helps to achieve this, as does the method of design known as "Total Resource" or Ecosystem Design that facilitates a link between forest activities and the scale and pattern of natural forest processes. This vocabulary of aesthetics must be incorporated into discussions of the forest. It can help articulate the aesthetic results that are likely to occur from different desired future conditions or, more usefully, help develop options that blend together the desired ecological and aesthetic future conditions. This task should not be too difficult if the connections between ecology and aesthetics are properly made.

Forest Management Ignores the Wishes of Local People

Land management in general and forest management in particular attract much attention in terms of the balance between land owners' interests and those of the wider community. In some countries tradition has held that private land is sacrosanct and that the owner can do what he or she pleases, whereas public land should be managed with the interests of the wider community in mind. In practice, however, the scope for private owners to manage as they please is becoming increasingly restricted by international and national legal obligations and by pragmatism in the face of protests over forest management activities. Historically, while public land has been legally managed for the public interest, the institutional culture of forest services has often seemed to many people to mean that managers have followed their own agendas and not been responsive to changing public attitudes.

To many land managers the idea of involving the public in management decisions and planning has been unacceptable for a number of reasons, a primary reason being the perceived loss of power to make decisions. Those managers who do involve local communities in planning and management, however, actually find that the power to manage, which they thought was threatened, is restored. This occurs when sufficient trust has been re-established between managers and other stakeholders.

When a forest is the landscape within which people live, when their water supply comes from it, when their livelihood from tourism depends on the quality of the environment, and when their culture is intimately linked to it, they have a legitimate right to participate in decision making. The question now is not whether communities should be involved, but how far their hedonistic wants can be achieved within the ecological, ethical, economic, and political framework.

A number of participative planning models have been developed that use consensus-building techniques, negotiations, and other methods to extract a community's values and wishes for a unit of management, whether publicly or privately owned. One piece of information needed is to identify the community. In this case we usually refer primarily to the local community identified with a particular place, but possibly subdivided into other communities such as farmers, loggers, recreationists, businesses, and environmentalists. This local community is part of a concentric structure of regional, provincial or state, national, and international communities. The larger entities also need a voice, best expressed perhaps through democratic and governmental institutions, although local concerns may be linked to the wider world via membership in non-governmental organizations such as Greenpeace or the Worldwide Fund for Nature. Thus local communities do not act independently from the wider world and should not be treated as such.

Larger-scale concerns and wants can be accessed by research such as opinion polls, questionnaires, and perception studies. World Heritage Sites, national parks, provincial or state parks, and other protected areas have already been established according to various national and international protocols. A local dimension to this is sometimes not accounted for. Despite the legal channels of elected bodies, however, a clear need exists for more participative planning that reflects the views and aspirations of local communities and puts these in the wider context while at the same time balancing them against the priorities of higher levels of government.

The model of participation must incorporate an understanding of the ecological processes at work in the forest landscape so that the approach to ecosystem management is accepted and the wants expressed as values by the community are defined in terms of what is possible by the ecological "carrying capacity." This can be carried out in a workshop setting where a skilled facilitator leads the community, stakeholders, or their representatives through a detailed analysis of the landscape using experts to explain different aspects, with local knowledge provided by the participants themselves, and with participants working together over maps. Such a model of participation has been carried out in a number of projects. A good example was applied to 40 thousand hectares in the Kispiox District of the British Columbia Ministry of Forests in 1997 (British Columbia Ministry of Forests, Kispiox District 1997). During a thorough analysis of the landscape using principles of landscape ecology and landscape character carried out by a group of about 20 people including experts and local residents, the values placed on the landscape were set in a well worked-out context. From the expression of values it was possible to determine which places in the landscape were able to meet those values and which were not; in other words, setting the values in the realm of what is ecologically possible. The analysis progressed as far as being able to define the desired future condition that would meet those values and fulfill ecosystem functioning and was related to different landscape units within the whole area.

This approach is one of the best yet developed to ensure that human values are incorporated into forest planning in a way that ensures that the ecosystem continues to function within its natural range of variability. The desired aesthetic results are also addressed, and a discussion of the practical and economic issues related to achieving what is ecologically possible helps focus the group on the most prudent results. While ethics are not explicitly discussed, the ecosystem design and management approach is known to be at the ecocentric end, and it seems that accommodating the range of starting points that different people may bring to the project is not a problem. Deeper understanding of the ecosystem helps people to reappraise preconceived ideas as to how forests function and what forests are naturally like. In addition, people come to realize how their existing, perhaps unexpressed, ethical positions fit with those of other stakeholders.

Conclusions

Ecosystem management is here to stay and offers great potential for truly sustainable management. For it to be socially acceptable, practitioners such as forest managers and ecologists should be aware that, to many outsiders, ecosystem management may appear to be just another form of management and a means to continue logging. Public understanding of ecology is limited and often out of date. Mythologies abound and preconceptions of what is natural need to be challenged. Some environmental movements exhibit skepticism, so trust must be developed based on the best science. While science informs, it neither motivates action nor determines what should be done. Motivation is to a great extent driven by what people want; what is actually done is not always to the absolute limit of what is ecologically possible. One factor that defines what is prudent is ethics.

Ethical considerations are often not addressed directly or openly. Many arguments expressed in apparently factual terms turn on ethical points. Land managers need to discuss and explore the ethics of ecosystem management in the context of other approaches, and they must demonstrate that their ethical positions are close to those held by many environmentalists as well as by modern foresters. Arguments among stakeholders about extraction of a resource are likely to contain an ethical dimension that should not be overlooked.

Aesthetics are another important factor that play a role in how people perceive the forest and the effects of management. Ecosystem management enables a close link to be made among advocates of a design philosophy based on natural patterns and processes. Knowledge of what consistently makes a good landscape helps people realize why the traditional forest management approach produced negative results. We must raise awareness of ecology and ecological processes so that we can move the aesthetic from a purely scenic one where the landscape is unchanging to an aesthetic of engagement where change is a necessary part. Stakeholders must be involved in understanding the dynamics of the landscape and the richer aesthetic possibilities that can be achieved.

Following from all of these aspects and running through them as a common thread is the unavoidable need to engage local communities in a meaningful and deep participation in planning for landscape change and ecosystem management. The kind of participatory analysis described at Kispiox is one sophisticated method that produces some excellent results. Communities who understand the way the forest ecosystem works are better able to relate their values to its carrying capacity. Deeper awareness of nature helps to instill a fuller appreciation of ethics and the range of aesthetic possibilities.

References

Bell, S. 1993. At the heart of forestry. Unpublished paper presented to the annual discussion meeting of the Institute of Chartered Foresters, Warwick, Great Britain.

Bell, S. 1994. Visual Landscape Design Training Manual. BC Ministry of Forests, Victoria, BC.

Berleant, A. 1992. The aesthetics of environment. Temple University Press, Philadelphia, PA.

Bookchin, M. 1982. The ecology of freedom. Cheshire Books, Palo Alto, CA.

Botkin, D.B. 1990. Discordant harmonies. Oxford University Press, Oxford, UK.

British Columbia Ministry of Forests, Kispiox District. 1997. Landscape analysis to guide landscape planning, design and management: Proceedings of the McCully Creek Workshop. BC Ministry of Forests, Hazelton, BC.

Connely, W. 1960. Louis Sullivan: The shaping of American Architecture. Horizon Press, New York, NY.

Gobster, P.H. 1995. Aldo Leopold's ecological esthetic: Integrating esthetic and biodiversity values. Journal of Forestry 93(2):6-10.

Kant, I. The critique of judgement. 1982 translation by J.H. Bernard. Oxford University Press, Oxford, UK.

Kaplan, S. 1983. Perception and landscape: Conceptions and misconceptions. *In* J.L. Nasar (ed.). Environmental aesthetics, theory, research and applications. Cambridge University Press, Cambridge, UK.

Leopold, A. 1949. A Sand County almanac and sketches here and there. Oxford University Press, Oxford, UK.

McQuillan, A. 1993. Cabbages and kings: The ethics and aesthetics of New Forestry. Pp. 191-222 *in* Environmental Values 2. Whitehorse Press, Cambridge, MA.

O'Riordan, T. 1981. Environmentalism. 2nd edition. Pion, London.

Schopenhauer, A. The world as will and representation. 1969 translation by E.F.J. Payne. Dover Publications, NY.

Thompson, I.H. 1998. Environmental ethics and the development of landscape architectural theory. Landscape Research 23:175-194.

Applied Landscape Ecology:
A Summary of Approaches

Patrick W. Daigle

Abstract

Conceptual models can help us clarify our approaches in many resource management contexts. I compare four conceptual models developed to apply principles of landscape ecology in the context of an ecosystem-based approach to land use and resource management.

The models compared and contrasted are from four main sources:

- the work of Nancy Diaz, Dean Apostol, and Simon Bell
- Ecosystem Analysis at the Watershed Scale: [United States] Federal Guide for Watershed Analysis
- the United States Forest Service, Northern Region
- the work of the Silva Forest Foundation.

In general, these models have a step-by-step approach, use an array of tools, techniques, and methods, guide practitioners through landscape analysis and design, and result in a product.

The comparison highlights similarities and differences among the landscape approaches. The similarities tend to indicate the importance of the step, tool, technique, or method. The differences provide opportunities for us to learn from others who are working with the complexities of landscape management.

Conceptual Models

Conceptual models can help us learn. Conceptual models can help make sense of the myriad complexities of ecosystem management and landscape ecology. In addition, conceptual models provide a basis for common understanding. With respect to managing landscapes, this can mean enhanced communication among and with members of an interdisciplinary team working on a landscape analysis, senior management, other resource

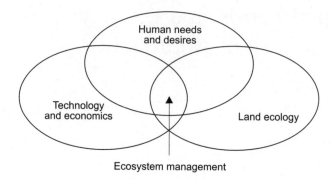

Figure 1. Conceptual framework for ecosystem management
(adapted by Jensen and Everett 1994).

managers who may have to prepare operational plans that must address the requirements of landscape plans, and interested public groups.

A good conceptual model can also improve decision-making. Resource managers working with an understandable and flexible model can more readily exercise professional judgment when exploring viable options designed to achieve ecological and social goals. Of course, a given model will not be perfect – it may not work in every situation.

Figure 1 portrays a simple graphic model of the relationship between ecology, technology and economics, and social aspects of ecosystem management.

Another type of conceptual model uses a step-by-step approach (Table 1). Many forest managers are experienced with stand-level work. When developing a stand prescription, a silviculturist goes through a series of steps, some conscious, some not, en route to completing a stand-level prescription or set of treatments.

A Comparison: Stand Prescriptions and Landscape Designs
Table 1 shows a conceptual model summarizing a silviculturist's or a landscape management team's thinking when preparing a comprehensive stand-level prescription or landscape design. Table 1 illustrates that the process of developing a landscape design is similar to developing a stand prescription; the main difference is scale. I used generic language in the table so that I could describe the steps for both stand and landscapes.

Observe the Context
Think about the lands surrounding the stand or landscape and factors that influence management (e.g., access). A higher-level plan may be zoned to emphasize different objectives across the stand or landscape.

Table 1

Illustration of the similar steps undertaken when preparing either a stand prescription or landscape plan. The primary difference is scale (modified from Daigle 1995).

Concept	Stand level (with some examples)	Landscape level (with some examples)
Observe the context	Adjacent areas Higher-level plans	Adjacent areas Higher-level plans
Analysis stage		
Gather information	Biological Physical Human Other	Biological Physical Human Other
Analysis tools	Office reconnaissance tools Field tools Office analysis tools	Office reconnaissance tools Field tools Office analysis tools
Analysis techniques	Stratify Opportunities within stand Opportunities within stand strata	Stratify Opportunities within landscape Opportunities within landscape strata
Design stage		
Generate options	Even-aged Uneven-aged Opening sizes Other options	Seral stage options Patch size options Other options
Analyze options	Economics Physical access Other	Economics Physical access Other
Reports produced	Prescription containing: • Stand structure and composition objectives • Strata structure and composition objectives • Treatments over time • Maps	Landscape design containing: • Landscape Unit objectives or Desired Future Conditions • Strata objectives or Desired Future Conditions • Treatments over time • Maps

Analysis Stage

Next, use office and field tools and techniques to gather information about the stand or the landscape. The tools could include ecosystem field guides, aerial photographs, and maps. To reduce complexity, the forest manager might use ecological criteria to stratify (segment) the stand or landscape. Stratification, an example of one technique, is the process of categorizing areas of a stand or landscape that are ecologically similar. In the case of a stand, walk the ground. In the case of a landscape, fly over, drive through, and walk representative stands or patches. Use office tools and techniques such as Geographic Information Systems and computer modeling to analyze the potential for achieving the management objectives described in higher-level plans. For each stratum, state measurable objectives or Desired Future Conditions that will provide direction for the design stage that follows.

Design Stage
Generate treatment options that, over time, will take the stand or land-scape from the current structure and composition to the Desired Future Conditions. Analyze the options for economic and physical viability and examine how the options might achieve the objectives stated in the higher-level plans. The product of the design stage is the stand prescription or landscape design. It will contain measurable objectives or Desired Future Conditions (for the whole stand or landscape and for the strata within) that can be monitored so that adaptive learning can be maximized.

The description of the steps in Table 1 makes them appear to be linear, from top to bottom. In real life, iterations occur. For example, if uneven-aged management is considered a possibility, the silviculturist may gather different stand information.

Comparing Approaches to Applying Principles of Landscape Ecology
I surveyed over 30 published papers and landscape plans describing land-scape management approaches in the United States Pacific Northwest and Canada's British Columbia. I also contacted several people involved with these landscape approaches (N. Bilodeau, Kispiox Forest District, Hazelton, British Columbia; B. Bollenbacher, United States Forest Service, Missoula, Montana; T. Bradley, Silva Ecosystem Consultants, Winlaw, British Colum-bia; G. Dickerson, United States Forest Service, Libby, Montana; B. McCammon, United States Forest Service, Portland, Oregon). In this paper I compare four selected approaches and highlight main similarities and differences.

How Do These Four Approaches Relate to Landscape Management in British Columbia?
Many complexities of landscape management are addressed in this com-parison. Most of these complex elements are described in the Biodiversity and Riparian Management Guidebooks of the Forest Practices Code of British Columbia (BC Forest Service 1995a, b). The elements in the four contrasted

approaches include: coarse filter/fine filter approach, seral stage distribution, patch size and arrangement, connectivity, disturbance types, riparian and wetland ecosystems, landscape objectives, structural diversity, and Desired Future Conditions across land areas ranging from 5,000 to 100,000 ha.

The Four Approaches Compared

(1) The work of Nancy Diaz, Dean Apostol, and Simon Bell (hereafter referred to as the "Diaz/Apostol/Bell approach"). During the 1990s, Nancy Diaz, Dean Apostol, and Simon Bell evolved a process for analyzing and designing landscapes. While the work of these people has evolved over time, I will focus on their recent efforts (Diaz and Apostol 1992; Pojar et al. 1994; US Forest Service 1995, 1996a; Bell 1996; BC Forest Service 1997).

(2) Ecosystem analysis at the watershed scale: Federal guide for watershed analysis (United States Northwest Plan Federal Guide). The United States Regional Interagency Executive Committee (1995) developed a watershed analysis process for lands in Washington, Oregon, and California covered by the Northwest Forest Plan (US Department of Agriculture 1994). Additional efforts discussed are those of the US Forest Service (1995, 1996a, 1996b, 1998a).

(3) The United States Forest Service, Northern Region approach (United States Forest Service Northern Region). This landscape approach focuses mainly on United States Forest Service lands in Montana and northern Idaho. Using the same basic approach, the United States Forest Service Northern Region is now favouring landscape analysis and design processes that range from 200,000 to 400,000 ha (Bollenbacher 1992; Spores 1993, 1994; Gilbert and Olsen 1994; O'Hara et al. 1994; Leavell et al. 1995; US Forest Service 1992, 1996c, 1996d, 1997, 1998b).

(4) The Silva Forest Foundation approach (Silva Forest Foundation). The Silva Forest Foundation is a non-profit charitable organization based in British Columbia. The foundation focuses on training, research, and projects related to ecosystem-based forest use. During the 1990s, this approach has been developed for use on British Columbia landscapes (Bradley and Hamond 1993; BC Forest Service 1996; Dunster Community Foundation 1997).

For this discussion I created a framework that allows us to compare landscape and design methods used in these four approaches so that we can enhance our understanding of and improve our skills in managing the landscape issues that confront us daily.

What Are the Purposes for Landscape-Scale Analysis and Design?

In all four approaches the stated purposes are:

- to document a science-based understanding of historic and current bio-physical and cultural ecology
- to identify desired conditions and trends
- to develop management recommendations and priorities that will guide subsequent operations planning and management activities.

The Comparison

Comparisons will be clustered under the following five headings:

(1) context for the process
(2) steps in the process
(3) tools, techniques, and methods used during the analysis and design process
(4) ecological and social analysis procedures
(5) design procedures.

(1) Context for the Process

Features Common to All Four Approaches

Hierarchy of geographic scales is a common feature. Each approach considers flows, connections, and management objectives of adjacent landscapes and other lands within the ecoregion and neighboring ecoregions. Each approach also provides guidance for stand and site management.

Additional Context for the United States Northwest Plan Federal Guide Process and Some Examples of the Diaz/Apostol/Bell Approach

All American examples have land and resource management plans. These higher-level plans contain allocation zones called Management Areas. Objectives and standards are described for each Management Area.

(2) Steps in the Process

The Diaz/Apostol/Bell Approach

The steps in the Diaz/Apostol/Bell approach are to:

- describe landscape structures, patterns, disturbances, succession, and interactions
- stratify landscape into Landscape Character Zones
- synthesize
- identify opportunities and constraints
- state objectives for the landscape

- state Desired Future Conditions for each Landscape Character Zone
- develop design concepts
- monitor key elements.

The United States Northwest Plan Federal Guide Approach
The steps in the United States Northwest Plan Federal Guide approach are to:

- characterize the watershed
- identify issues and key questions
- describe current conditions
- describe reference conditions
- synthesize, interpret, and compare current and reference conditions
- recommend.

The United States Forest Service Northern Region
The steps in the United States Forest Service Northern Region approach are to:

- select priority watershed(s) to be analyzed
- describe historic conditions
- describe existing conditions
- compare existing conditions with historical conditions
- consider protection for rare elements
- model future conditions
- identify management opportunities.

The Silva Forest Foundation Approach
The steps in the Silva Forest Foundation approach are to:

- review and synthesize existing information
- analyze the landscape to identify ecologically sensitive terrain
- identify and map a proposed Protected Landscape Network
- establish Human Use Zones.

(3) Tools, Techniques, and Methods Used during Analysis and Design

Common to All Four Approaches
The following are common to all four approaches:

- interdisciplinary team (to lever the value of each individual's knowledge)
- resource and land allocation information (maps and data) from previously developed plans and analyses
- discussion of key issues and questions

- stratification
- coarse filter evaluation: comparison of historic and current conditions of human aspects (including recreation, visual quality, infrastructure, flows, and impacts) and landscape elements, flows, and processes (including landforms, geology, geomorphology, vegetation patterns, disturbance regimes and succession, aquatic and terrestrial systems, special species and habitats)
- text, tables, matrices, charts, maps, aerial photographs, and clear film overlays.

Additional Tools in the Diaz/Apostol/Bell Approach
Additional tools, techniques, and methods used in the Diaz/Apostol/Bell approach include:

- focus group workshops
- aerial photograph mosaics at 1:15,000 scale
- stratification into Landscape Character Zones and subzones (ecological criteria combining similar landform, features, vegetation, and ecological processes)
- fine filter evaluation: identification and evaluation of Critical Natural Capital (elements and cultural capital that cannot be replaced or relocated) and Constant Natural Assets (elements and cultural assets that must be in a landscape but can be located in different places at different times)
- stratification into design units (social and ecological criteria to address visual resources)
- Desired Future Conditions (usually considered to be a description of the resource or land conditions expected to result if goals, objectives, and standards are achieved)
- photograph and sketch perspectives to assist with visual design
- adaptive management in one example
- short glossaries in some examples.

Additional Tools in the United States Northwest Plan Federal Guide Approach
Additional tools, techniques, and methods used in the United States Northwest Plan Federal Guide approach include the following:

- "issues driven" process; set priorities for the issues
- "core questions" focused on "core topics"
- stratification by ecological process or condition (mainly ecological criteria such as subwatersheds, geomorphology, response units, or resources)
- rating system for biophysical and human use conditions
- scientific method

- classification systems
- systems diagrams
- analysis of ecological and social trends
- literature search
- information about "natural" areas in the subject landscape or similar landscapes
- synthesis, interpretation, and integration
- glossary.

Additional Tools in the United States Forest Service Northern Region Approach
Additional tools, techniques, and methods used during the United States Forest Service Northern Region approach include the following:

- orthographic photographs
- Habitat Type Groups (similar vegetation potential)
- Fire Groups (similar postfire succession and fire risk)
- Landtype Groups (similar soils, geology, and landforms)
- stratification into Vegetation Response Units and subunits (ecological criteria combining Habitat Type Groups, Fire Groups, and Landtype Groups)
- fine filter evaluation: describe rare elements
- modeling future landscape conditions (including a "no action" alternative)
- Desired Future Conditions.

Additional Tools in the Silva Forest Foundation Approach
Additional tools, techniques, and methods used in the Silva Forest Foundation approach include:

- aerial photographs at 1:15,000 and 1:70,000 scales
- satellite imagery
- Old Growth Nodes and an Ecological Viability Rating System for Old Growth Nodes; ecological rating criteria include age, forest area, interior stand area, landscape context, and riparian connectivity; social criterion includes human threat
- stratification of a Protected Landscape Network (based on ecological and social criteria); Network consists of a riparian network, protected Old Growth Nodes, cross-valley corridors, Environmentally Sensitive Areas, and representative ecosystem types (includes a fine filter evaluation)
- rating system for Ecological Sensitivity to [human] Disturbance (ecological criteria include riparian, alpine, avalanche, and wetland ecosystems, continuous slopes with over 60 percent gradient, broken/gullied terrain, very dry and shallow soil sites, and transition zones at the edge of non-forested zones)

- stratification into Human Use Zones (social and ecological criteria used in a hierarchical order, i.e., cultural, ecologically sensitive, fish and wildlife, recreation-tourism-wilderness, and timber use).

(4) Social and Ecological Analysis Procedures

Common to All Four Approaches
Common to all four approaches are the following:

- synthesizing public input for the examples in which public input was sought
- coarse filter evaluation: identifying, mapping, describing, and analyzing historic and current human aspects and landscape elements, flows, and processes.

Other Analyses in the Diaz/Apostol/Bell Approach
Other analysis procedures in the Diaz/Apostol/Bell approach included:

- convening focus group workshops that brought together land managers (representing various resource disciplines) and interested members of the public
- using public and technical input to identify key issues, elements, and indicators that form the basis of a monitoring strategy to optimize adaptive learning
- identifying, analyzing, and mapping biophysical and social constraints and opportunities
- doing a fine filter evaluation that included identifying, analyzing, and mapping Critical Natural Capital and Constant Natural Assets
- stratifying the landscape into Landscape Character Zones and subzones
- synthesizing information about landscape character with issues, plans, and policies
- analyzing landscape character in plan and perspective views.

Other Analyses in the United States Northwest Federal Plan Federal Guide Approach
Other analysis procedures used during the United States Northwest Plan Federal Guide approach included:

- focusing the analysis by answering five to seven "core questions" about seven biophysical and social "core topics"
- setting priorities for the issues and key questions
- stratifying the landscape by ecological process or condition
- conducting coarse filter evaluation that included identifying and explaining the natural and human causes of change between historic and

current biophysical and human use conditions; also explaining the rate and magnitude of biophysical and social trends
- linking the change or trend to management objectives, issues, and key questions, and to the interactions among biophysical and social processes
- synthesizing, integrating, and summarizing ecosystem elements and connections and relationships between ecological features and processes.

Other Analysis Procedures Used in the United States Forest Service Northern Region Approach
Other analysis procedures used in the United States Forest Service Northern Region approach included:

- stratifying into Vegetation Response Units and subunits
- identifying, analyzing, and mapping landscape elements, flows, processes, and management implications across the landscape and by Vegetation Response Unit
- performing coarse filter evaluation by comparing historic and current composition, structure, and processes across the landscape and by Vegetation Response Unit
- performing fine filter evaluation by describing rare elements
- modeling future conditions by analyzing trends and probability of change, for example, wildfire risk, effects of insect and disease, channel stability, and exotic species across the landscape and by Vegetation Response Unit.

Other Analyses in the Silva Forest Foundation Approach
Other analysis procedures used in the Silva Forest Foundation approach included the following:

- rating the ecological viability of Old Growth Nodes
- stratifying, mapping, and analyzing the Protected Landscape Network
- determining the impact on various forest users of establishing the Protected Landscape Network
- rating the stands and sites that are Ecologically Sensitive to [human] Disturbance
- identifying and mapping forest areas or forest uses needing protection from resource extraction
- stratifying, mapping, and analyzing Human Use Zones.

(5) Design Procedures

Common to All Four Approaches
Common to all four approaches were the following:

- communicating scientific information

- providing context for subsequent decision-making processes
- summarizing information in text, tables, matrices, charts, and maps.

Other Design Procedures in the Diaz/Apostol/Bell Approach
Other design procedures used in the Diaz/Apostol/Bell approach included:

- maximizing visual elements in the report to help portray information (photos, maps, and sketches)
- describing landscape management objectives
- developing sketch designs in plan and perspective views
- describing Desired Future Conditions and silvicultural strategies for each Landscape Character Zone
- describing and mapping access and travel plans in some examples
- identifying restoration opportunities in some examples
- identifying data gaps in some examples.

Other Design Procedures in the United States Northwest Plan Federal Guide Approach
Other design procedures used in the United States Northwest Plan Federal Guide approach included the following:

- providing recommendations (not output targets) and priorities for timing and location of management activities for restoration, monitoring, and harvest and recreational development
- ensuring that recommendations are responsive to issues and key questions
- recommending research topics in response to issues and key questions
- documenting data gaps and limitations
- summarizing the logic of the six-step process.

Other Design Procedures in the United States Forest Service Northern Region Approach
Other design procedures used in the United States Forest Service Northern Region approach included the following:

- describing Desired Future Conditions and silvicultural strategies for the landscape and each Vegetation Response Unit
- identifying management opportunities and possible practices that will maintain or restore ecosystems while providing goods, services, and amenities.

Other Design Procedures in the Silva Forest Foundation Approach
Other design procedures used in the Silva Forest Foundation approach included:

- identifying, describing, mapping, and recommending watersheds for large reserve status
- describing and mapping a Protected Landscape Network
- describing and mapping Human Use Zones
- if viable alternatives exist, preparing an options report.

Similarities among the Approaches

So far I have pointed out similar and different steps, tools, techniques, methods, and procedures. The common features and activities indicate convergence. I have noted below a few that are important and may be of special use.

Assemble an Interdisciplinary Team to Garner Collective Knowledge

A team approach can enhance the knowledge and skills of each person taking part in the landscape process.

Use a Step-by-Step Approach

Each of the contrasted approaches uses step-by-step procedures. The United States Northwest Plan Federal Guide approach (Regional Interagency Executive Committee 1995) is particularly concise, transparent, and instructive. The 26-page guide explains the six steps in the process. For each step, the guide describes the purpose, core topics and questions, information resources, techniques, and products. The guide also contains a short glossary.

Work with Spatial Scale Appropriately: Global to Site

Most resource management will take place between ecoregion and site, stand, or reach scales. Note that these are nested within even larger contexts.

Work with Temporal Scale Appropriately: Past, Present, and Future

Management over time considers the needs of future generations. Historic Range of Variability is an important conceptual tool for use at both landscape and stand levels; it is now commonly used in western North America (Daigle and Dawson 1996). The Biodiversity Guidebook of the Forest Practices Code of British Columbia (BC Forest Service 1995a) states, "Native species and ecological processes are more likely to be maintained if managed forests are made to resemble those forests created by the activities of natural disturbance agents." One must therefore understand the history of the ecological processes to describe structure, composition, and ecosystem function. Comparing past and present conditions during analysis may reveal the rate and magnitude of ecological and sociocultural trends thereby permitting resource managers to anticipate and describe future conditions to meet ecological and social goals. Parsons and Morgan (this volume) describe Historical Range of Variability in more detail.

Differences among the Approaches

The differences among the approaches indicate opportunities to share knowledge and collective strengths.

Stratify the Landscape

Although all four approaches stratify the landscape, they do so in different ways. In part, stratification can help clarify ecological and social objectives. To reduce potential confusion when you stratify, do not mix social and ecological criteria (even though they are related). Prepare clear film overlays for social strata. Prepare another set of clear film overlays for ecological strata. When both social and ecological strata films are laid over a basemap, you will get a clearer picture about how to set up objectives and Desired Future Conditions.

The United States Forest Service Northern Region has developed relatively sophisticated ecological strata called Vegetation Response Units. For a given national forest, only some of the Vegetation Response Units apply. As an example, for the Kootenai National Forest, eleven Vegetation Response Units are mapped and described.

The description for each Vegetation Response Unit includes details about disturbances, succession, vegetation structure and composition, age class distribution, ecological function, habitat features, ecosystem health, and general trends. These details lead to a description of management options and silvicultural strategies designed to attain Desired Future Conditions for each Vegetation Response Unit.

Although it has taken time, expertise, and money to map and describe the Vegetation Response Units and summarize suitable Desired Future Conditions and silvicultural strategies, this large effort may pay off because the time, costs, and effort are amortized over millions of hectares. All 12 national forests within the Northern Region (10 million hectares) are moving to this system with the result that each will use similar landscape management protocols, tools, and techniques.

The Diaz/Apostol/Bell approach also uses ecological strata called Landscape Character Zones. These are cryptically described using biological and physical characteristics such as landform, land features, vegetation, and major ecological processes. The United States Northwest Plan Federal Guide also stratifies landscapes by ecological processes or conditions.

If you choose to use both social and ecological criteria, make your decision consciously. Be aware that you are blending social values and biophysical attributes of the land. The Diaz/Apostol/Bell process uses social and ecological criteria when stratifying design units. The Silva Forest Foundation approach stratifies Protected Landscape Networks and Human Use Zones using ecological and social criteria.

Who Might You Involve during the Process? And When?

Who has been involved in the landscape analysis process? In terms of knowledge, attitudes, and behavior, what are the effects of involving a team of resource managers in the landscape analysis process? It would probably improve managers' understandings and generate acceptance and confidence to undertake landscape analyses. Similarly, what are the effects of involving citizens or representatives of interest groups in the landscape analysis?

The Diaz/Apostol/Bell approach gathered public input during focus group workshops at the front end of the landscape analysis. These workshops help focus the analysis by bringing out public and professional issues, questions, and concerns. After the workshop, an interdisciplinary team conducts a technical analysis that addresses the issues, questions, and concerns voiced during the initial workshop. A second focus group can be convened to review the products of the analysis phase.

The United States Forest Service seeks public input during Land and Resource Management Planning efforts. They do not seek public input at the landscape level. At the project level, the United States Forest Service again seeks public comment.

Use an Array of Tools, Techniques, and Methods

During landscape analysis and design, the four approaches used many tools, techniques, and methods. Here are a few important examples.

- search on-topic literature
- focus the process (establish priorities, convene focus groups, examine core topics, and ask core questions)
- develop rating systems for biophysical and human use conditions
- conduct a fine filter evaluation
- use orthographic photographs, aerial photograph mosaics, historic photographs, and satellite images
- use Geographic Information Systems
- model future ecological and social conditions
- use adaptive management and scientific methods to maximize learning

Communicate the Results of the Landscape Analysis and Design

Similar to silviculture at the stand level, landscape management is both a science and an art. To address both of these aspects, when you communicate results of the analysis and design phases, it helps to summarize the "logic" of your approach and maximize visual elements (photographs, maps, sketches, flow charts, and graphs) in your reporting. Successful communication (among team members, senior management, other resource managers, and interested public groups) can enhance learning, expand knowledge, and improve attitudes and behaviors.

Conclusion

In the perspective of ecosystem management, I have considered similarities between stand and landscape management and similarities and differences among four approaches to landscape management in northwest North America. No one "player" (individual, agency, industry, community) has the "best answer." Collectively, we need to lever our knowledge and skills. I encourage you to share your understandings with others and to reach out to others to ask for help.

References

BC Forest Service. 1995a. Biodiversity guidebook: Forest Practices Code of British Columbia. British Columbia Ministry of Forests, Victoria, BC.

BC Forest Service. 1995b. Riparian management area guidebook: Forest Practices Code of British Columbia. British Columbia Ministry of Forests and Ministry of Environment, Lands and Parks, Victoria, BC.

BC Forest Service. 1996. Ecosystem based forest use plan for Coldstream Creek headwaters. Initial plan. British Columbia Ministry of Forests, Kamloops Forest Region, Vernon Forest District, Vernon, BC.

BC Forest Service. 1997. Landscape analysis to guide landscape planning, design and management: McCully and Date Creek watersheds landscape design. Unpublished proceedings, McCully Creek Workshops. British Columbia Ministry of Forests, Prince Rupert Forest Region, Kispiox Forest District, Hazelton, BC.

Bell, S. 1996. Vuokatti landscape ecology project. Kainuu Province, Finland.

Bollenbacher, B. (coord.). 1992. Case studies of silviculture in the landscape. Silviculturist revalidation case studies. US Forest Service, Northern Region, Missoula, MT.

Bradley, T., and H. Hammond. 1993. Practical methodology for landscape analysis and zoning. Silva Forest Foundation, Winlaw, BC. On-line: <http://www.silvafor.org/docs/docs.htm>.

Daigle, P.W. 1995. Landscape ecology: Assessment of research extension needs. Unpublished report. British Columbia Ministry of Forests, Research Branch, Victoria, BC.

Daigle, P., and R. Dawson. 1996. Management concepts for landscape ecology: Part 1 of 7. Extension Note 7. British Columbia Ministry of Forests, Research Program, Victoria, BC.

Diaz, N., and D. Apostol. 1992. Forest landscape and design: A process for developing and implementing land management objectives for landscape patterns. R6 ECO-TP-043-92. US Department of Agriculture (USDA) Forest Service, Pacific Northwest Region, Portland, OR.

Diaz, N.M., and S. Bell. 1997. Landscape analysis and design. Pp. 255-269 *in* K.A. Kohm and J.F. Franklin (eds.). Creating a forestry for the 21st century. Island Press, Washington, DC.

Dunster Community Foundation. 1997. Rausch River watershed analysis. Initial report. McBride, BC.

Gilbert, S., and L. Olsen. 1994. Ecosystem management approach of the Helena National Forest. Pp. 216-223 *in* Silviculture: From the cradle of forestry to ecosystem management. Proceedings of the national silviculture workshop, US Forest Service, Southeast Experimental Station, Asheville, NC.

Jenson, M.E., and R. Everett. 1994. An overview of ecosystem management principles. Pp. 6-15 *in* M.E. Jenson and P.S. Bourgeron (technical eds.). Volume II, Ecosystem management: Principles and applications. General Technical Report PNW-GTR-318. US

Department of Agriculture (USDA) Forest Service, Pacific Northwest Research Station, Portland, OR.

Leavell, D.M., J.D. Head, and E.B. Newell. 1995. A model for the implementation of ecosystem management on the Kootenai National Forest. Pp. 158-183 *in* J. Thompson (compiler). Analysis in support of ecosystem management. US Forest Service, Ecosystem Management Analysis Center, Washington, DC.

O'Hara, K.L., M.E. Jensen, L.J. Olsen, and J.W. Joy. 1994. Applying landscape ecology theory to integrated resource planning: Two case studies. Pp. 225-236 *in* M.E. Jenson and P.S. Bourgeron (technical eds.). Volume II, Ecosystem management: Principles and applications. General Technical Report PNW-GTR-318. US Department of Agriculture (USDA) Forest Service, Pacific Northwest Research Station, Portland, OR.

Pojar, J., N. Diaz, D. Steventon, D. Apostol, and K. Mellen. 1994. Biodiversity planning and forest management at the landscape scale. Pp. 55-70 *in* M.H. Huff, L.K. Norris, J.B. Nyberg, and N. Wilkin (compilers). Applications of ecosystem management. Proceedings of the third Habitat Futures Workshop. General Technical Report PNW-GTR-336. US Department of Agriculture (USDA) Forest Service, Pacific Northwest Research Station, Portland, OR.

Regional Interagency Executive Committee. 1995. Ecosystem analysis at the watershed scale: Federal guide for watershed analysis. Version 2.2. Regional Ecosystems Office, Portland, OR.

Spores, D. (coord.). 1993. Case studies of silviculture and landscape structure. Silviculturist revalidation case studies. US Forest Service, Northern Region, Missoula, MT.

Spores, D. (coord.). 1994. Designing landscapes considering disturbance ecology and management objectives. Silviculturist revalidation case studies. US Forest Service, Northern Region, Missoula, MT.

US Department of Agriculture. 1994. Record of Decision for amendments to Forest Service and Bureau of Land Management planning documents within the range of the Northern Spotted Owl: Standards and guidelines for management of habitat for late-successional and old-growth forest related species within the range of the Northern Spotted Owl. United States Department of Agriculture (USDA), Portland, OR.

US Forest Service. 1992. Our approach to sustaining ecological systems. *In* Sustaining ecological systems desk reference. United States Forest Service, Northern Region, Missoula, MT.

US Forest Service. 1995. Collwash/Hot Springs watershed analysis. Final report. United States Forest Service, Mount Hood National Forest, Gresham, OR.

US Forest Service. 1996a. North Fork Clackamas River watershed analysis. Draft report. United States Forest Service, Mount Hood National Forest, Gresham, OR.

US Forest Service. 1996b. Lower Lewis River watershed analysis. United States Forest Service, Gifford Pinchot National Forest, Vancouver, WA.

US Forest Service. 1996c. Two Joe draft environmental impact statement. United States Forest Service, Lolo National Forest, Missoula, MT.

US Forest Service. 1996d. Protocols for implementing Pathway 1 [project-level planning] and pathway 2 [forest plan revision]. Peer Groups. United States Forest Service, Missoula, MT.

US Forest Service. 1997. Kootenai National Forest Vegetation Response Units and habitat characterization. United States Forest Service, Kootenai National Forest, Libby, MT.

US Forest Service. 1998a. Upper Lewis River watershed analysis, second iteration. United States Forest Service, Gifford Pinchot National Forest, Vancouver, WA.

US Forest Service. 1998b. Wood Rat environmental analysis and decision notice. United States Forest Service, Kootenai National Forest, Libby, MT.

Approximating Natural Disturbance: Where Are We in Northern British Columbia?

Craig DeLong

Abstract

A common theme in current forest management policy is that harvesting should be designed to achieve the landscape patterns and habitat conditions that are maintained by natural disturbance regimes. In British Columbia, the Biodiversity Guidebook of the Forest Practices Code of British Columbia (BC Forest Service 1995) recommends landscape and stand management guidelines that attempt to approximate natural disturbance for distinct areas of the province with divergent natural disturbance histories. In this paper I present details of the guidelines applicable to forests of northern British Columbia, a discussion of how the guidelines might be altered in light of current research, and some aspects of management not covered by the guidelines that must be changed if natural disturbance is to serve as a template for future forest management.

Although we find some discrepancies between certain natural disturbance attributes examined by research and details found in the Biodiversity Guidebook, the Guidebook provides good first steps towards approximating natural disturbance. Other current guidelines, however, prevent forest managers from approximating the level of diversity apparent in natural systems, and these also increase the cost of doing business. The challenge for those involved in research on approximating natural disturbance in northern British Columbia is to adequately describe patterns of natural variability, develop prescriptions that will best approximate natural conditions, and educate the public about why this new management strategy is appropriate.

Background

In British Columbia, the Biodiversity Guidebook of the Forest Practices Code of British Columbia (BC Forest Service 1995) uses natural variability as a basis for recommending landscape and stand level practices for replicating natural disturbance types that are meant to maintain biological diversity. The underlying assumption of the Biodiversity Guidebook is that the biota of a forest, having adapted to the conditions created by natural disturbances,

Table 1

Description of natural disturbance types (NDT) defined by the
Biodiversity Guidebook of the Forest Practices Code of British Columbia
(BC Forest Service 1995).

Natural disturbance type	Description
NDT1	Ecosystems with rare stand-initiating events
NDT2	Ecosystems with infrequent stand-initiating events
NDT3	Ecosystems with frequent stand-initiating events
NDT4	Ecosystems with frequent stand-maintaining fires
NDT5	Alpine tundra and subalpine parkland

should cope more readily with the ecological changes caused by timber harvest if the patterns created resemble those of natural disturbances (Hunter 1993; Swanson et al. 1993; Bunnell 1995; DeLong and Tanner 1996). Natural disturbance types as described in the Biodiversity Guidebook recognize different natural disturbance regimes under which ecosystems have developed and were formed by grouping biogeoclimatic units (areas of relatively homogeneous regional climate that are recognized within the Biogeoclimatic Ecosystem Classification system in use in British Columbia [Pojar et al. 1987]) that were assumed to have similar disturbance intensities and frequencies (Table 1). Significant progress on implementing many of the practices recommended in the Guidebook has been made in the Prince George Forest Region (PGFR). As well, significant new information recently available from research projects will help refine and add to the guidelines for approximating natural disturbance as laid out in the Guidebook.

This paper examines the current status of managing forests within the Prince George Forest Region to approximate natural disturbance and presents views on the current concerns and future requirements for attaining this goal.

Current Status
The Biodiversity Guidebook of the Forest Practices Code recommends a certain range of retention of old forest for each natural disturbance type (NDT) based on estimates of the natural stand replacement disturbance cycle. The level of retention within any particular landscape unit within a certain natural disturbance type varies according to the management objective for the landscape unit. For most landscape units, the objective is to balance wood production with conservation (moderate biodiversity emphasis). If the main objective for a landscape unit is conservation (high biodiversity emphasis), the retention level will be higher than it would be for areas with moderate biodiversity emphasis. Table 2 shows old forest retention levels based on

Table 2

Recommended old forest retention level (percent of forest area within the landscape unit) by emphasis option for forests in the Interior Cedar Hemlock Zone within NDT 1- 3 [adapted from data in Biodiversity Guidebook of the Forest Practices Code of British Columbia (BC Forest service1995)].

Natural disturbance type	Low biodiversity	Moderate biodiversity	High biodiversity
1[a]	13	13	19
2[a]	9	9	13
3[b]	14	14	21

[a] Old forests are considered to be those > 250 yrs old.
[b] Old forests are considered to be those > 140 yrs old.

emphasis option for natural disturbance types 1-3. Within the Prince George Forest Region, an evaluation was done to determine whether the desired level of old forest retention could currently be reached. In most cases the target could be reached. Often the target could be reached by relying on areas of constrained timber (e.g., protected areas, environmentally sensitive areas, or areas unfeasible to harvest). We must, however, determine whether these areas are representative of the ecological variability contained within the remainder of the landscape. Most of the areas where the old forest retention target could not currently be met are areas where the magnitude of previous wildfire history means that most of the area is less than 140 years old. Data from current research conducted in the Prince George Forest Region (DeLong 1998) indicates that in some portions of the region it would be relatively easy to retain old forest consistent with historic levels but in others it would be extremely difficult. In many of the difficult areas, constraints due to high caribou values, difficult terrain, and high visual quality reduce the gap between historic amounts of old forest and proposed levels of old forest retention.

Patch size distribution and other attributes of spatial pattern represent some of the major differences between natural disturbance and past harvesting practices (Hunter 1993; DeLong and Tanner 1996; DeLong 1998). Past harvesting practices generally resulted in large openings related to bark beetle salvage or regularly spaced openings of 50-150 ha. The patch size recommendations contained in the Biodiversity Guidebook are intended to better approximate natural patch size distribution. When compared with actual data for natural disturbance from the Prince George Forest Region, the patch size recommendations provide for a lower maximum patch size and more area in 40-250 ha patches (DeLong 1998) (Table 3). The guidelines, however, represent a good first step in reducing the disparity between natural and harvesting patch size distribution. Within the Prince George

Table 3

Estimated proportion of total disturbance area compared to recommended range in the Biodiversity Guidebook (BC Forest Service 1995) for Natural Disturbance Type 3 (NDT3) for different patch size categories (from DeLong 1988).

Topoclimatic unit[a]	Patch size (ha)				
	< 40	40-250	251-1,000	> 1,000	> 250 (max.)
	(proportion of total disturbed area)				
Recommended for NDT3	10-20[b]	10-20[b]	60-80[b]		
Dry warm plateau[c]	9	13.4	20.9	56.7	77.6 (7,693)
Moist cold plateau	4.5	7.4	15.9	72.2	88.1 (19,030)
Moist cool plateau	5.5	7	9.4	78.0	87.4 (41,787)
Moist cool montane	4.7	9.3	12.6	73.4	86.0 (10,458)
Wet cool plateau	10.4	18.3	33.8	37.5	71.3 (2,515)
Wet cool montane[d]	12.6	25.3	23.6	38.5	62.1 (1,931)
Very wet cool montane[d]	14.4	23.5	49.1	13	62.1 (1,082)
Dry cool boreal	9.3	27.5	22.4	40.8	63.2 (2,691)
Moist v. cold subalpine[d]	4.1	10.3	24.4	61.2	85.6 (4,171)

[a] see DeLong 1988 for discussion of topoclimatic units.
[b] recommended range in proportion of total disturbance area within the patch size category for areas within NDT3 without a major component of Douglas-fir according to the Biodiversity Guidebook.
[c] unit assigned to NDT3 with major component of Douglas-fir according to the Biodiversity Guidebook.
[d] unit assigned to NDT2 according to the Biodiversity Guidebook.

Forest Region, current (< 20 years old) patch size distribution that is primarily related to harvesting has been determined for a number of landscape units to identify differences from recommendations in the Biodiversity Guidebook. In general, the findings indicate an excess of mid-sized patches (e.g., 40 to 250 ha for natural disturbance type 3) accompanied by a shortage of small (e.g., 0 to 40 ha for natural disturbance type 3) and large patches (e.g., 250 to 1,000 ha for natural disturbance type 3). The plans to address this situation are to encourage amalgamation of recent harvest blocks into larger, more irregularly shaped harvest openings along with some small openings in areas of high visual quality or small beetle infestations. Many wildlife managers are in favour of allowing some larger harvest openings. Large openings make access control easier, result in lower road densities which are a critical factor affecting species such as wolves and grizzly bears (McLellan 1990; Thurber et al. 1994), and may be part of an effective management strategy for woodland caribou (Racey et al. 1991). Allowing some larger openings should also reduce the costs of road building and maintenance, transportation of equipment and wood, and development plans and prescriptions.

Figure 1. A harvest block designed by participants of a natural disturbance block design course in Prince George Forest Region, British Columbia.

Personnel are currently teaching courses throughout the region on designing blocks that incorporate characteristics of wildfire. An integral part of block design to approximate wildfire is to increase shape complexity and retain patches of forest (wildlife tree patches) within the block. This practice is meant to replicate the characteristic irregular edges and islands of forests with a history of wildfire (DeLong and Tanner 1996). Figure 1 shows a block designed by course participants.

Although larger disturbance patches (> 1,000 ha) are a feature of all landscapes within the Prince George Forest Region, in wetter portions of the region they occur much less frequently (DeLong 1998). The dominant form of disturbance in these wetter portions is likely to be small patch disturbance caused by disease (e.g., tomentosus root rot) or insect (e.g., bark beetle) outbreaks followed by windthrow. Emulating this type of disturbance process requires some form of partial cutting. A number of experimental and operational trials have occurred within the Prince George Forest Region to examine both ground-based and helicopter partial cutting and it is likely that more will occur in the future. The challenge with any partial cutting system is to reduce the impact of the increased number of entries required to remove timber.

Current Concerns

Although some direction is provided in current guidelines on stand level practices that incorporate characteristics of natural forests, some information gaps exist. One such gap is a lack of guidance on ways to incorporate the diversity in vegetation pattern brought about by differences in disturbance intensity (Lertzman et al. 1998). The impact on above-ground biomass and the forest floor varies greatly within any natural disturbance, particularly wildfire. Within most large wildfires we see a gradation from totally unburned remnants to areas where all above-ground vegetation and the forest floor have been removed leaving only fire-hardened snags and woody debris on bare mineral soil. This variability in disturbance intensity leads to differences in vegetation succession according to varied post-fire conditions. Areas most severely impacted often revert to plant communities dominated by deciduous tree species such as paper birch *(Betula papyrifera)* and aspen *(Populus tremuloides)* and/or shrubs such as willow *(Salix spp.)* or alder *(Alnus spp.)*. To maintain the historic abundance and temporal distribution of these plant communities we must examine past and current forest practices with respect to species conversion and variability in intensity of disturbance. Certain policies direct forest managers to convert land areas dominated by deciduous species to conifer. One such policy directs managers to convert areas within an opening from non-commercial species (e.g., alder) to conifer if they are greater than four hectares and located on potentially productive forest land. More recently exceptions are being made so that managers can map out and exclude these areas from the net area to reforest.

In certain portions of the Prince George Forest Region, former site preparation practices led to significant mineral soil exposure such as would occur after intense wildfire. These practices were deemed to cause excessive soil degradation and have since been curtailed. The strict soil degradation rules presently in place tend to restrict site preparation to techniques that limit the displacement of the humus layer. Thus a conflict currently exists between the objective of approximating natural disturbance and that of minimizing soil disturbance. A potential partial solution may be to reclaim areas of known intense disturbance such as landings and skid roads by planting or seeding species such as willow, alder, birch, or aspen.

Differences in fire return and fire intensity have resulted in large differences in stocking levels and stand structure in young natural forests within the Prince George Forest Region. In mesic to drier plateau areas, natural stands are densely stocked (e.g., 3,000 to 50,000 stems per hectare) and even-aged, while in wetter montane areas they are more open (e.g., < 1,000 stems per hectare) and multi-aged (C. DeLong, unpublished data). Current policy allows for a large range in stocking (e.g., 700 to 10,000 stems per hectare). The penalty system and current practices, however, tend to restrict this range, especially at the lower end. Based on available records for young (< 20 years old) managed stands in the Prince George Forest Region, total stocking is 500 to 21,000 stems per hectare (average 3,475 stems per hectare) for an area on the plateau where natural stands are dense, and 600 to 13,640 (average 2,362 stems per hectare) for a wet montane area where natural stands are open. Thus in the wetter montane areas, current stocking is generally outside the range of natural stocking. The implications to animal movement and to the development of understory vegetation (i.e., denser stands) in the wet montane areas could be significant especially when these stands begin to close crown. To emulate natural variability in stand structure, variability in stocking prescriptions needs to be encouraged and any economic implications closely weighed against the ecological implications.

Future Requirements

To move from the theory of approximating natural disturbance to its practice requires a concentrated effort by people from a variety of backgrounds. The following groups must focus on different aspects of the problem to attain the goal.

Researchers

- spend less time examining the temporal pattern of disturbance (e.g., disturbance interval) and more time on spatial pattern at a variety of scales (e.g., patch size and shape, stand density and composition, vegetation community patterns resulting from variability in disturbance intensity).
- explore innovative ways to express variation and pattern in variation (standard statistics are not very useful for examining pattern).

Policy Makers

- move away from rule-based policies
- spend equal time on formulating policies that encourage the implementation of the existing guidelines that make ecological and economic sense (e.g., aggregate harvesting according to some ecologically based guidelines) as on formulating policies that avoid exerting downward pressures on cutting levels.

Forest Managers

- work with researchers to develop ecologically sound plans and prescriptions for various natural disturbance types
- move away from rule-based management

- find ways to manage for variability while minimizing impacts on economic goals
- remember that what makes ecological sense often makes economic sense.

Environmentalists

- engage in healthy debate about the most ecologically appropriate manner to harvest within the variety of forest ecosystems present. Individual tree selection, like traditional clear-cutting, is unlikely to be most ecologically appropriate for all forest types on all sites
- assist in determining where protection would be an appropriate strategy to maintain biological diversity and where other strategies may be more suitable (e.g., areas where disturbance was frequent and species have become adapted to ecological conditions associated with disturbance).

General Public

- read material outlining the principles of ecosystem management and the concept of emulating natural disturbance and express encouragement and/ or concerns to forest managers and policy makers.

References

BC Forest Service. 1995. Biodiversity Guidebook: Forest Practices Code of British Columbia. British Columbia Ministry of Forests, Victoria, BC.

Bunnell, F.L. 1995. Forest-dwelling vertebrate faunas and natural fire regimes in British Columbia. Conservation Biology 9:636-644.

DeLong, S.C., and D. Tanner. 1996. Managing the pattern of forest harvest: Lessons from wildfire. Biodiversity and Conservation 5:1191-1205.

DeLong, S.C. 1998. Natural disturbance rate and patch size distribution of forests in northern British Columbia: Implications for forest management. Northwest Science 72:35-48.

Hunter, M.L., Jr. 1993. Natural fire regimes as spatial models for managing boreal forests. Biological Conservation 65:115-120.

Lertzman, K., J. Fall, and B. Dorner. 1998. Three kinds of heterogeneity in fire regimes: At the crossroads of fire history and landscape ecology. Northwest Science 72:4-23.

McLellan, B.N. 1990. Relationships between human industrial activity and grizzly bears. International Conference on Bear Research and Management 8:57-64.

Pojar, J., K. Klinka, and D.V. Meidinger. 1987. Biogeoclimatic ecosystem classification in British Columbia. Forest Ecology and Management 22:119-154.

Racey, G.D., K. Abraham, W.R. Darby, H.R. Timmermann, and Q. Days. 1991. Can woodland caribou and the forest industry coexist: The Ontario scene. Rangifer, Special Issue No. 7:108-115.

Swanson, F.J., J.A. Jones, D.O. Wallin, and J.H. Cissel. 1993. Natural variability: Implications for ecosystem management. Pp. 89-104 *in* M.E. Jensen and P.S. Bourgeron (eds.). Eastside Forest Ecosystem Health Assessment. Volume 2, Ecosystem management: Principles and applications. US Department of Agriculture (USDA) Forest Service, Pacific Northwest Research Station, Portland, OR.

Thurber, J.M., R.O. Peterson, T.D. Drummer, and S.A. Thomasma. 1994. Grey wolf response to refuge boundaries and roads in Alaska. Wildlife Society Bulletin 22:61-68.

Landscape Patterns in Managed Forests of Southeastern British Columbia: Implications for Old-Growth Fragmentation

Robert G. D'Eon, Daniel Mack, Susan M. Glenn, and E. Alex Ferguson

Abstract

Forest fragmentation has been cited as one of the most serious threats to loss of biodiversity through species extinction. Despite much debate, little empirical data has been presented to test predictions and assumptions on this topic. We address the assumption that forest harvesting leads to fragmentation of old-growth forests in managed forest landscapes in southeastern British Columbia, Canada. We do this by empirically testing predictions relating the proportion of forest harvested within landscapes to patch indices describing old-growth spatial pattern. Results fail to conclusively indicate a trend towards old-growth fragmentation in our study area. We suggest that historic loss of old-growth habitat is a greater concern than fragmentation in this case.

Introduction

Forest fragmentation has been suggested as one of the most important factors in a current species extinction crisis leading to a loss in biological diversity (Wilcox and Murphy 1985). Fragmentation can be defined as the process of subdividing a continuous habitat into smaller pieces (Andrén 1994). Predicted consequences of fragmentation have been widely cited (Saunders et al. 1991). Many researchers have characterized landscapes as fragmented and have concluded that biological diversity is diminished because a particular landscape is fragmented (e.g., Lamberson et al. 1992; Yahner and Mahan 1997; Laurance and Gascon 1997). Commercial forest harvesting has been one of the most commonly associated causes of forest fragmentation (Franklin and Forman 1987). Such assumptions occur in the absence of a standardized quantitative definition of forest fragmentation and often with little empirical evidence to conclude that fragmentation is occurring.

In this analysis, we address the assumption that forest harvesting leads to fragmentation of old-growth forests in managed forest landscapes. We do this by empirically testing predictions that relate the proportion of forest harvested within landscapes to patch indices describing old-growth spatial pattern. Forman (1997) predicted trends in spatial attributes that should occur as landscape fragmentation progresses. Based on his research, we examine the following six predictions of what might occur if fragmentation of old-growth patches is caused by forest harvesting. As the proportion of harvested forest in landscapes increases, we predict that (1) old-growth patch density increases, (2) mean old-growth patch size decreases, (3) the ratio of edge area to patch area of old-growth patches increases, (4) mean core area of old-growth patches decreases, (5) mean nearest neighbor distance between old-growth patches increases, and (6) dispersion patterns of old-growth patches increase.

Study Area

Landscape pattern data were derived from a managed forest landscape of 295,558 hectares in the Slocan Valley (49° N, 117° W) of the Selkirk Mountains of southeast British Columbia, Canada. Terrain within this mountainous area is generally steep and broken with slope gradients often exceeding 80 percent. Elevation ranges from 525 metres along the Slocan River to 2,800 metre peaks.

Seventy-four percent of the land area within the defined study area is forested. Forests of this area are within the Interior Subalpine and Southern Columbia region described by Rowe (1972) and contain three forest biogeoclimatic subzones described by Braumandl and Curran (1992): Interior Cedar Hemlock Dry Warm at low elevations, Interior Cedar Hemlock Moist Warm at mid elevations, and Englemann Spruce Sub-alpine Fir at higher elevations. Alpine parkland predominates in areas above 2,000 metres.

Logging within the Slocan Valley began in the late 1800s but was primarily confined to localized selective harvesting. Large-scale commercial logging began around 1950 when a forest tenure was issued to Passmore Lumber and subsequently transferred to Slocan Forest Products in Slocan, British Columbia. Side drainages of the Slocan Valley have since been managed, to varying degrees, for forest harvesting and road building. Most areas within the main Slocan River corridor and provincial parks, however, have been excluded from large-scale forest harvesting and are essentially intact and continuous forest landscapes. The majority of low elevation areas along the Slocan River are private land holdings and have been partially cleared for agriculture and urban development. Routine forest fire suppression in the local area began in the late 1930s (British Columbia Ministry of Forests records, Research Branch, Victoria, BC).

Methods

All source data were derived from British Columbia provincial forest cover map information in digital format. Patch indices were calculated within an ARCINFO™ geographic information system platform.

The study area was delineated into 37 distinct landscapes (mean area = 7,827 hectares, SE = 484 hectares). Landscape boundaries were based on drainage patterns and were generally drawn along heights of land distinguishing two adjacent watersheds.

Within each landscape we calculated the proportion of the forested landbase in a harvested state using the sum of all polygon areas labelled as harvest blocks younger than 20 years. We identified old-growth forest patches, consistent with BC Forest Service (1995) standards, as polygons labelled as forests 250 years or older or as forests 140 years or older depending on biogeoclimatic subzone. We calculated the following patch indices for old-growth patches within each landscape: patch density (number of patches per 100 hectares), mean patch area, mean edge to patch area ratio (assuming a 50-metre edge effect on either side of the patch perimeter), mean core area (assuming a 50-metre edge effect from patch perimeter into the patch), and mean nearest neighbor distance (average edge-to-edge distance between a patch and the closest other patch). We also calculated a dispersion pattern index to determine spatial arrangement of patches (Clark and Evans 1954). The dispersion index varies from 0 to 2.1491. Values close to 0 indicate an aggregated spatial pattern, values close to 1 indicate a random distribution, and values approaching 2.1491 indicate even spacing among patches, or maximum dispersion.

Data Analysis

Relationships between the amount of habitat and landscape configuration can be highly correlated (Fahrig 1997). To evaluate relationships between amount of harvesting within a landscape and old-growth patch configuration, independent of old-growth area, we used regression techniques to remove area effects (McGarigal and McComb 1995). The amount of old growth within each landscape was regressed against each of the six landscape indices measuring old-growth patch configuration. We analyzed residuals to remove confounding area effects for indices showing a significant relationship with old-growth area. In these cases, residuals were regressed against harvest level to test our predictions. Absolute values of indices were regressed against harvest level where no area effects were detected. We evaluated the use of a zero-intercept model for relationships that should logically pass through the origin. In all cases, however, we used full regression models (as opposed to zero-intercept models) because of significant differences in y-intercepts from zero (Kozak and Kozak 1995).

Results

Three old-growth patch indices had significant relationships with amount

Table 1

Results of six predicted relationships between proportion of landscape harvested and patch index statistics of 37 managed forest landscapes in the Slocan Valley, British Columbia.

		Regression results		
Patch index[a]	Predicted trend[b]	Observed trend	r^2	p
Pdresidual	increase	increase	0.13	0.04
Psresidual	decrease	decrease	< 0.01	0.78
Edge to patch area ratio	increase	decrease	0.01	0.49
Mcresidual	decrease	decrease	< 0.01	0.82
Mean nearest neighbor distance	increase	decrease	0.03	0.31
Dispersion index	increase	none	< 0.01	0.99

[a] Relationship between proportion of forest harvested within landscapes (x-axis) and the listed patch index (y-axis); PDresidual = Patch density residual, PSresidual = Patch size residual, MCresidual = Mean core residual.

[b] "Increase" indicates increase in index as amount of forest harvested within landscapes increases; "decrease" indicates decrease in index as amount of forest harvested increases; "none" indicates no discernable trend.

of old-growth area: patch density ($r^2 = 0.44$, $p < 0.001$), patch size ($r^2 = 0.23$, $p = 0.003$), and mean core area ($r^2 = 0.28$, $p = 0.001$). In these cases, we used residuals (hereafter referred to as PDresidual, PSresidual, and MCresidual, respectively) to test predicted relationships with harvest levels within landscapes. The remaining three indices (mean edge to core area ratio, nearest neighbor, and dispersion) were not significantly related to amount of old-growth area (all $r^2 < 0.06$, all $p > 0.14$). In these cases, we used absolute values of indices (not residuals) for testing predictions.

The observed relationship between the proportion of forest harvested within landscapes and PDresidual was positive (i.e., PDresidual increased with proportion of forest harvested), was significant ($r^2 = 0.13$, $p = 0.04$), and was consistent with a predicted outcome if fragmentation of old-growth patches is occurring (Table 1). The remaining five old-growth patch index relationships produced insignificant regressions (all $r^2 = 0.03$, all $p = 0.31$) thus falsifying these predictions. Of these remaining five relationships, PSresidual and MCresidual displayed similar trends to those predicted (based on slope of line of best fit) if fragmentation of old-growth patches is occurring. Edge to patch area ratio and mean nearest neighbor displayed a trend opposite to that predicted. Dispersion index had no discernible trend (Figure 1).

Discussion
Our tests provide inconclusive evidence that old-growth fragmentation is occurring in this study area. While one of the six tests, consistent with

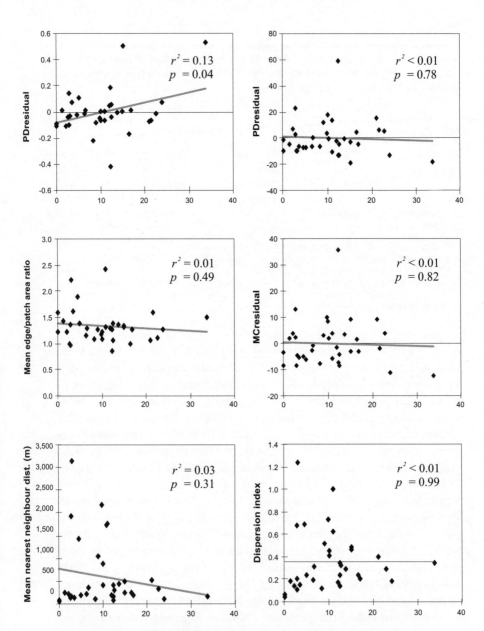

Figure 1. Relationships between the proportion of forest in a harvested state and six old-growth patch indices for 37 landscapes within the Slocan Valley, southeastern British Columbia.

predicted outcomes, provided evidence for the occurrence of old-growth fragmentation, four tests resulted in insignificant results, providing evidence that old-growth fragmentation is not occurring.

The failure to detect a conclusive trend towards old-growth fragmentation in this case can be caused by one or more of the following: (1) harvest levels are not high enough to detect a trend towards old-growth fragmentation, (2) not enough old-growth forest exists to detect a trend, or (3) old-growth fragmentation is not occurring.

Many researchers have predicted critical thresholds in landscape patterns after which significant effects on ecological processes are realized. Those predictions centred on amount of forest harvesting usually cite levels of harvesting that are much higher than those found in our study area (level of harvest for all landscapes combined = 11.4 percent, unpublished data). For example, Franklin and Forman (1987) predict a major loss of obligate interior-forest species in landscapes with 30 to 50 percent of the landscape in a harvested state. O'Neill et al. (1988) predict loss of habitat continuity within landscapes when greater than 40 percent of habitat is removed. Finally, Fahrig (1997) predicts that species survival is ensured in landscapes when up to 80 percent of breeding habitat is removed, regardless of the extent to which this habitat is fragmented. On this basis it is possible that current levels of harvest in our study area are either too low to detect a trend towards old-growth fragmentation or are below a threshold under which fragmentation does not occur.

The amount of old-growth forest in our study area can be considered relatively low (7.4 percent, all landscapes combined, unpublished data). Fires intentionally lit at the beginning of this century by prospecting miners in this area are thought to have contributed heavily to a current seral stage distribution skewed towards mid-successional forests (British Columbia Ministry of Forests data). Fragmentation occurs only when the number of patches increases because contiguous patches break apart (Fahrig 1997). If whole patches are removed, fragmentation, by definition, is actually decreased (Ripple et al. 1991). Fahrig (1997) tested the relative importance of habitat loss and fragmentation and concluded that the effects of habitat loss far outweigh the effects of fragmentation. This prediction was supported by McGarigal and McComb (1995) who found variation in bird abundance among 30 landscapes in Oregon was more strongly related to habitat amount than habitat configuration. On this basis, we suggest that in this case the issue of old-growth habitat loss is likely to be of much greater concern than that of old-growth fragmentation.

Acknowledgments
We are grateful to Slocan Forest Products (Slocan, BC) for partial funding, logistical support, and access to digital map information and forest management records. We thank Kokanee Forests Consulting (Nelson, BC) for providing additional logistical support. We also thank British Columbia Ministry of Forests staff (Arrow District, Castlegar, BC) for providing additional digital map information. Graduate student support for R. D'Eon was provided by the University of British Columbia Department of Forest Sciences and a Canfor Corporation fellowship in forest wildlife management. A critical review of this manuscript was provided by Mark Boyland (University of British Columbia, Centre for Applied Conservation Biology, Vancouver, BC). We are especially thankful to Lenore Fahrig (Carleton University, Ottawa, ON) for her insightful comments on a previous draft and discussions of confounding area effects.

References
Andrén, H. 1994. Effects of habitat fragmentation on birds and mammals in landscapes with different proportions of suitable habitat: A review. Oikos 71:355-366.

BC Forest Service. 1995. Biodiversity guidebook: Forest Practices Code of British Columbia. BC Ministry of Forests, Victoria, BC.

Braumandl, T.F., and M.P. Curran. 1992. A field guide for site identification and interpretation for the Nelson Forest Region. British Columbia Ministry of Forests, Nelson, BC.

Clark, P.J., and F.C. Evans. 1954. Distance to nearest neighbor as a measure of spatial relationships in populations. Ecology 35:445-453.

Fahrig, L. 1997. Relative effects of habitat loss and fragmentation on population extinction. Journal of Wildlife Management 61:603-610.

Forman, R.T.T. 1997. Land mosaics: The ecology of landscapes and regions. Cambridge University Press, New York, NY.

Franklin, J.F., and R.T. Forman. 1987. Creating landscape patterns by forest cutting: Ecological consequences and principles. Landscape Ecology 1:5-18.

Kozak, A., and R.A. Kozak. 1995. Notes on regression through the origin. Forestry Chronicle 71:326-330.

Lamberson, R., R. McKelvey, B. Noon, and C. Voss. 1992. A dynamic analysis of northern spotted owl viability in a fragmented forest landscape. Conservation Biology 6:505-512.

Laurance, W.F., and C. Gascon. 1997. How to creatively fragment a landscape. Conservation Biology 11:577-579.

McGarigal, K., and W.C. McComb. 1995. Relationships between landscape structure and breeding birds in the Oregon coast range. Ecological Monographs 65:235-260.

O'Neill, R.V., B.T. Milne, M.G. Turner, and R.H. Gardner. 1988. Resource utilization scales and landscape pattern. Landscape Ecology 2:63-69.

Ripple, W., G. Bradshaw, and T. Spies. 1991. Measuring forest landscape patterns in the Cascade range of Oregon, U.S.A. Conservation Biology 57:73-88.

Rowe, J.S. 1972. Forest regions of Canada. Canadian Forest Service Publication No. 1300.

Saunders, D., R. Hobbs, and C. Margules. 1991. Biological consequences of ecosystem fragmentation: A review. Conservation Biology 5:18-32.

Wilcox, B.A., and D.D. Murphy. 1985. Conservation strategy: The effects of fragmentation on extinction. American Naturalist 125:879-887.

Yahner, R., and C.G. Mahan. 1997. Behavioral considerations in fragmented landscapes. Conservation Biology 11:569-570.

Predictive Ecosystem Mapping: Applications to the Management of Large Forest Areas

David J. Downing, Keith Jones, and Jan Mulder

Abstract

Ecological land classification is becoming an integral part of forest management in many jurisdictions given the current regulatory and policy environment; the necessity for more accurate estimates of standing timber, long-range sustained yields, and allowable annual cuts; and the benefits of identifying areas that provide the best returns for silvicultural investments. Conventional ecological mapping provides detailed information on ecological conditions within a forest management area, but it takes a long time to complete and is relatively expensive. In the last ten years significant advances have occurred in both spatial and aspatial predictive modeling approaches to mapping. These "predictive ecosystem mapping" methods capitalize on the knowledge of expert ecologists and use digital information normally available to forest managers such as terrain models and forest cover maps. Characteristics of good predictive ecosystem mapping systems include: the flexibility to adapt to changing information sources and ecological knowledge in a timely and economical fashion; the ability to assign estimates of accuracy to predictions; and the ability to generate explicit traces of the allocation path followed for each prediction. A case study of a forest management area in western Canada demonstrates the benefits of using a commercially available predictive ecosystem mapping system, ELDAR, in support of contemporary forest planning requirements including timber supply analysis, silvicultural planning, and non-timber modeling such as biodiversity and wildlife habitat assessments.

Introduction

Complexity, uncertainty, and higher demands on natural resources and the land base characterize current forest management in today's fast-changing world. Plans and business decisions, often determined in an atmosphere of conflicting values, are usually made with limited availability or access to quality land information and with inadequate or inconsistent application of practical forestry experience and scientific knowledge. Forest management has evolved from a timber-based focus toward a broader forestland and forest landscape perspective that is starting to incorporate ecosystem qualities in terms of both timber and non-timber resources. Many forces are changing the way we manage forestland, including the following:

- more comprehensive regulatory policies and planning and reporting requirements
- the need to forecast and plan for future forest ecosystem and landscape conditions, values, natural processes, and management practice that affects spatially and over long timeframes
- global market demands for certification of forest management areas and forest products
- international agreements related to sustainable forest management and associated efforts toward monitoring and reporting criteria and indicators of economic, ecological, and social importance
- acute timber supply shortages in some regions and the need to focus silvicultural investments on the most productive land
- heightened public interest in natural resource management and the importance of having informed stakeholders who understand resource management and decision trade-offs.

In total, these factors create a substantial challenge for management planners and for those who are faced with actual operations on the ground. As regulatory requirements and societal demands increase, the margin for errors narrows. In this regard, the forest sector is at an important transitional stage in its history. Industry executives and government officials in many jurisdictions are scrambling to respond more effectively to pressures such as charges levelled by environmental lobby groups, issues of local employment and its effect on community, and changing societal values. A major challenge for the forest sector is to rebuild public trust in its stewardship of the forestland resource. Action is required in the form of sound management that demonstrates the use of sound forestland practices in a manner that reflects appropriate local and global environmental, social, and economic values.

Resource Inventories Are Changing
This new reality in resource management is having a significant effect on how foresters and natural resource professionals conduct resource inventories and develop land information systems. Increasingly, managers and inventory specialists must fine-tune their ability to predict resource potential and management impacts across a wide range of values. They also have to interpret, forecast, and monitor management effects along with natural

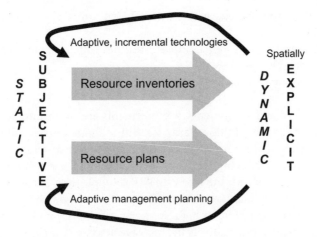

Figure 1. Ecological inventories and resource plans must be flexible, explicit, and able to adapt to rapidly changing conditions and technologies.

processes and then adjust subsequent plans and practices on the basis of well-documented experience (i.e., adaptive management). Integral to these changes is the need for both resource inventories and resource planning to change from relatively static and subjective processes to procedures that are flexible and explicit (Figure 1). Resource inventories need to be incremental, iterative, and adaptive, making full use of the information available at the time and tailored to the specific interpretive requirements of the resource manager. Similarly, the inventory information and interpretive modeling systems must be able to aggregate and decompose information at geographical scales and levels of information resolution that are consistent with the planning and operational questions being asked.

Geographic information systems (GIS), knowledge-based systems, and enhanced data acquisition technologies are playing an increasingly important role in the integration of data, information, and knowledge. The tools and techniques allow resource managers and inventory specialists to capitalize more fully on existing land information (e.g., forest cover maps, soil surveys, digital terrain data, spectral imagery, hydrography), existing ecological classification frameworks, understanding of ecosystem processes, and ecosystem-landscape response experience (Jones 1993). An attractive aspect of these developments is that it enables a more continuous linkage or integration of the following components:

- classification and spatial distribution pattern knowledge of forestland conditions

- empirical, process-based, and heuristic-based knowledge of ecosystem processes
- an understanding of forest landscape features important in controlling these ecosystem processes
- spatial, temporal, and statistical tools and models.

Paralleling these developments, advancements are continuing in the development of data acquisition hardware and more automated integration and interpretation software (Pollock 1998). A number of remote and airborne data collection technologies are being developed and integrated with interpretive software with the intent to either supplement or supplant conventional inventory methods. The latest techniques show the potential to provide data of exceptionally high accuracy, resolution, and information value in an efficient and relatively inexpensive manner. These developments suggest that ecological classification and mapping approaches are soon to be enhanced.

Organizations are becoming increasingly dependent on knowledge of ecological characteristics and its effects on the assessment and valuation of timber and non-timber resources. Ecological land classifications characterize ecological features useful for forest management, but the circumstances under which these classifications are conventionally mapped – intensive ground surveys and photo-interpretation – limit their usefulness in the following ways:

- demand is growing across a wide range of management agencies, companies, and stakeholder groups to support planning and operational decision requirements for ecologically oriented information of high accuracy and resolution
- ecological classification systems in themselves are insufficient; many were not originally designed as mapping systems per se
- most resource inventory methods continue to be expensive and are done in isolation from other inventory activities
- resource inventory programs and contractors are under considerable pressure to reduce costs and identify ways to increase efficiency and add value for the client
- experienced professionals in natural resource inventories such as soils and ecological surveys are harder to find because of organizational downsizing, retirements, and a lack of recruitment in these fields.

These realities create a considerable challenge for future ecological inventories. Some of the ideal features for future inventory systems include:

- clear identification at the outset of the most important interpretive requirements for the inventory including an understanding of what input the inventory will be expected to provide to the different forecasting or assessment models
- approaches that focus on "mappability," pattern description, and limiting features without necessarily having to impose an existing aspatial classification taxonomy on the process
- integral to the inventory process, a means to assess and report map reliability for interpretations and anticipated consequences of incorrect interpretations or decisions being made with the information
- creation of a digital information base that uses an unambiguous data structure and can be shared and linked easily with other thematic information and models
- an ability to organize and portray the information so it can be aggregated and decomposed for several survey and planning intensity levels.

With the above as background, the remainder of this paper advances some definitions and concepts for reference and consideration, outlines a spectrum of predictive ecosystem mapping approaches, describes a case study in applying a predictive ecosystem mapping software modeling system to a large forest management license area in Western Canada, and closes with a discussion of some of the advantages we have experienced or envision with predictive ecosystem mapping systems.

Definitions and Concepts

Ecological classification systems provide a framework to organize and communicate our knowledge about the nature of physical and biotic features of landscapes (Jones 1994). The fundamental premise of ecological classification systems is that future responses of forestland to management activities and natural events can be predicted – i.e., that similar forestland conditions respond in a similar manner to similar practices. Virtually all classifications are built around the concepts of soil formation (Jenny 1941) and vegetation formation (Major 1951). Soil, vegetation, or ecosystem entities integrate the combined effects of climate, relief, parent material, organisms, time, and function as meaningful entities in themselves, suitable for classification and interpretation (see upper portion of Figure 2). As Williams (1988) states, "Classification could be considered the science of defining subsets of the universe in which a given process or processes act in a common manner."

Predictive ecosystem mapping is an approach to ecosystem delineation that stratifies the landscape into ecologically oriented map units on the basis of

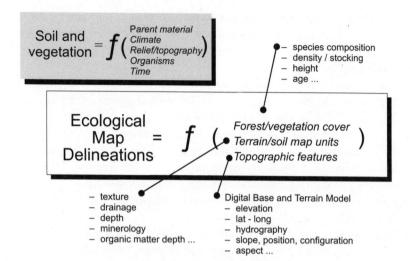

Figure 2. Ecosystem classification and predictive mapping approaches are founded on analogous concepts of integration.

combinations of existing mapped features, point information, and knowledge about ecosystem-landscape relationships and their distribution pattern. In essence, predictive ecosystem mapping applies the concepts of Jenny (1941) and Major (1951) to a mapping context (Figure 2). Just as we can approximate ecosystem types on the basis of relatively homogeneous landform (soil and vegetation) features, we can likewise approximate ecological map delineations on the basis of combined evidence of spatial themes like forest (vegetation cover, terrain) soil map units and derived topographic features from digital terrain models. Predictive ecosystem mapping tools can cover a wide range of techniques from basic geographic information system (GIS) overlays with manual expert interpretation and adjustment of apparent relationships to more automated and parametrizable software systems involving knowledge-based tools in conjunction with GIS.

Predictive Ecosystem Mapping (PEM): Use, Test, and Generate Hypotheses

Existing ecological land classification systems represent a set of hypotheses about different ecosystem properties and how ecosystem entities are organized and related within a particular region. Ecosystem and other resource inventory maps (including legend information) portray a set of hypotheses about how ecosystems are distributed spatially across a map area. Consistent with the concept of dynamic resource inventories, predictive ecosystem mapping processes use, test, and generate these classification (aspatial) and map (spatial) delineation hypotheses. On the spatial side, these processes can work with discrete polygonal input data sources or with more continuous raster- or grid-based input data sources.

Typically, we use an existing ecosystem classification system as an initial start (i.e., set of working hypotheses) for defining the predictive value of any already mapped land resource features (such as species composition from a forest cover map or soil depth from a soil map). When these mapped features are processed by polygon or grid cell through the knowledge base, they are assigned a predicted ecosystem class. Some polygons or grid cells may not be assigned because of shortfalls in either the source input map data or the knowledge base – for example, oversights when the knowledge base is created or true mapped feature combinations that represent "holes" in the existing classification. Whatever the case, predictive ecosystem mapping processes provide a mapped feature-oriented approach to testing and refining ecological land classification systems.

Advantages of Predictive Ecosystem Mapping

Much of the drive behind predictive ecosystem mapping development has been to address the high cost of conventional ecosystem mapping approaches. While considerable cost and time savings have been realized over traditional approaches, we have also found other advantages or envision them with the future use of predictive ecosystem mapping software modeling systems. The other advantages fall into three categories and are summarized as follows.

Improvements in Inventory Cost, Production, and Human Resource Capacity

- Predictive ecosystem mapping approaches provide a more systematic, consistent, and repeatable stratification process when working with existing resource inventory information.
- Predictive ecosystem mapping systems offer opportunities to increase mapping efficiency and the speed of mapping by directing field effort to complex areas and by being able to extend established knowledge bases somewhat automatically to similar geographic areas.
- These systems provide increased production capacity through the transfer of knowledge to junior staff more efficiently, explicitly, and consistently.
- These systems offer increased efficiency with map production and updating, since map-based information quality is reconciled as a part of the mapping process.

Capitalize On, Add Value to, and Protect Investments in Classifications and Resource Inventories

Predictive ecosystem mapping:
- Capitalizes on and adds value to already available digital land information, providing features to predictive ecosystem mapping that are known to control or influence ecosystem conditions and their distribution pattern.
- Capitalizes on and adds value to ecological classification systems and resource inventory "expanded legends" that are already available for many jurisdictions and that provide ecosystem and landscape relationship information.
- Protects the large corporate investments that have been made in ecological classifications and resource inventory information by capturing this knowledge in a structured, easily accessed, and shared manner.

Increase Accessibility, Knowledge Improvement, and Acceleration
- Relationship knowledge captured in predictive ecosystem mapping makes the knowledge more widely accessible and transferable across the range of public and private sector clients and stakeholders which, in turn, allows for more consistent use and understanding of the information value and its limitations.
- More explicit, documented capture of the relationship knowledge in predictive ecosystem mapping will enable faster improvement in our understanding of ecosystem and landscape relationships and how best to map these features in subsequent iterations.

Approaches to Predictive Ecosystem Mapping

We have several approaches to the process of classifying landscapes into useful forest management units. For example, at the stand level, models such as PrognosisBC (BC Ministry of Forests 1998) and MGM - Mixedwood Growth Model (Titus 1998) are used to refine growth and yield estimates and can make use of various stand attributes including ecological characteristics. We can use these models to improve timber supply analyses at the forest level, but the models are not spatially based and are therefore of limited use for understanding the relationship between individual stands and the landscapes within which they are situated.

Until the advent of computer-based spatial analyses and knowledge-based tools, ecologically based landscape models were derived through an intensive process of field data collection and photointerpretation by specialists trained in the art and science of ecological land classification. This is still the primary approach to applying ecological classification knowledge in, for example, British Columbia. But the process has been streamlined to some

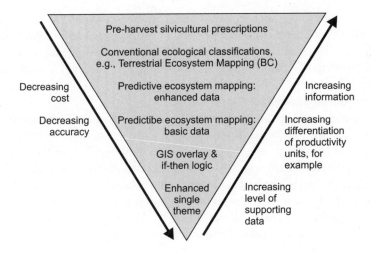

Figure 3. Hierarchy of ecological classification schemes.

extent through improved abilities to generate thematic products using GIS and relational databases.

Figure 3 shows the spectrum of ecosystem mapping approaches ranging from simple reinterpretations or enhancements of a single theme (bottom of the triangle) to highly intensive field and mapping efforts in support of operational silvicultural applications (top end of the triangle). Each of the approaches across the spectrum has strengths and weaknesses. The four core approaches are described below.

Single-Theme Inventories

Single-theme inventory approaches cannot reveal much about the finer details of ecological structure. A pure pine stand on a dry south-facing slope, for example, is classified the same as a pure pine stand on moderately drained undulating till, based on forest cover overstory characteristics alone; but site conditions, and hence stand growth potentials and probable successional pathways are markedly different. While this approach is inexpensive, it may not be cost-effective if it provides ambiguous information.

GIS Overlay and If-Then Logic

Inventories based on a GIS overlay of existing information (such as a forest cover map and a slope/aspect coverage) and the subsequent application of "if-then" logic are practical if the classification rules are simple and the total number of attributes being assessed is small. These approaches rapidly become intractable, however, as complexity increases. For example, consider a conditional logic scheme to identify five different ecosystem types using three different cover classes for five key overstory species and fifteen different slope and aspect combinations. This scheme would require at least 225 if-then statements to cover the range of possible combinations. Moreover, every time the classification changes, the logic statements would have to be modified so that assigning an explicit probability to a prediction is difficult.

Conventional Ecological Classifications

Conventional ecological classifications provide an excellent landscape stratification for both timber and non-timber resources (e,g., the Terrestrial Ecosystem Mapping processes followed in British Columbia). These classifications are based on intensive field checking and photointerpretation, and their quality is highly dependent on the abilities of the ecological classifier. This type of information is relatively expensive and time-consuming to collect; once a map has been drawn, if the model concepts (e.g., the classification system) change, it may be time-consuming to change polygon labels.

Software and Knowledge-Based Predictive Ecosystem Mapping

Broadly speaking, most predictive ecosystem mapping of this type involves aspects of all of the above approaches. In this context, predictive ecosystem mapping might better be thought of as an alternative approach to ecosystem inventory, one that combines some of the elements of conventional ecological land classifications such as field checking, the involvement of expert ecologists, and some of the elements of single-theme approaches or simple GIS overlays.

Since the 1980s researchers around the world have been looking for ways to make better use of existing information to generate ecological inventories more quickly and with less intensive effort. The two main strategies that have been pursued use: (1) the abilities of GIS technology to produce derivative maps and attribute databases through overlays; and/or (2) the organizational and inference capabilities of knowledge-based systems to capture and infer the knowledge of ecologists. In the second case, attributes derived from the GIS overlay process are used by the knowledge base and ecosystem types are assigned polygon by polygon or grid cell by grid cell. These results are portrayed spatially in map form by the GIS. Coughlan and Running (1989) developed an ecological process-oriented grid-based algorithm that created homogeneous biophysical polygons by aggregating cells having similar ecological properties – leaf area index, soil water holding capacity, or climatic efficiency. Skidmore et al. (1996) developed a probability-based expert system for predicting Australian soils with results that were both comparable to conventional soil maps and meaningful to forest managers. Mulder and Corns (1994) employed a similar knowledge-based software modeling system in west central Alberta, achieving validation accuracies of between 85 to 94 percent in several trials. At least seven different predictive ecosystem mapping approaches are currently under consideration in British Columbia, employing variations on the approaches outlined above (Biggs et al. 1997).

Based on our operational experience using predictive ecosystem mapping methods and software modeling systems, we have found the following characteristics to be advantageous:

Flexibility
Flexibility refers to the ability of the system and process to adjust to changing information sources and ecological knowledge. As forest inventories are revised, as better terrain modeling tools and information bases are developed, and as our understanding of ecological relationships improves, it is desirable to have a system that is flexible and dynamic enough to incorporate changes. Conventional ecological inventory products, although rich in information, have less flexibility than software and knowledge-based predictive ecosystem mapping products because typically each polygon is assessed individually and labelled by an interpreter. Thereafter, whenever changes in classification schemes occur or enhanced point or spatial data become available, considerable effort might be needed to update ecological maps on a polygon by polygon basis.

Knowledge-Based Transparency and Prediction Tracing
This refers to the ability to display the knowledge base and the ability to trace the path taken to reach any prediction. Every decision should be explicit and transparent to the user.

Accuracy Assessment
Accuracy assessment refers to the ability to assign some measure of accuracy to a predictive map. It is advisable to define in some quantitative manner the risk involved in applying the ecological inventory to particular management interpretations.

Representation of Complexity and Ambiguities
This refers to the ability to make predictions that are neither oversimplifications of naturally complex units nor overly complicated representations of simple units. A system that predicts many complex ecosystems (e.g., two or more ecosystem types) when few complex ecosystems actually exist is less useful to a forest manager than a system that does not. This is especially true for ecosystem types that have specific potentials or constraints. If such types are combined with other ecosystem types that do not have the same potentials or constraints, it is difficult for the forest manager to assess the spatial extent and distribution of, for example, "good" versus "average" sites for fibre production.

The availability of enhanced landscape information (Figure 3) can exert a strong influence on both accuracy and ambiguity. The availability of a predictive ecosystem mapping system to make unambiguous and accurate predictions is determined in part by the nature of input information and how well matched the knowledge base is to the input information. Benefits can be gained from producing and applying simple landscape characterizations in addition to the usually available data sources (forest cover and terrain models) as the case study below will demonstrate. It is usually not appropriate, however, to combine information sources from radically different scales (e.g., a 1:126,000 scale reconnaissance soils map and a 1:20,000 scale forest cover map). Both uncertainties in boundary placement and the aspatial nature of complex polygon attributes (e.g., reconnaissance soil surveys usually have polygons that combine two or more soil series) preclude the assignment of just one attribute from a small-scale map to a derivative map produced through a GIS overlay and can actually lead to further ambiguity in a classification.

Case Study: Predictive Ecosystem Mapping across a Forest Management Area

In 1998, Timberline Forest Inventory Consultants were contracted by a forest company with long-term land management responsibilities to provide a timber supply analysis. The inclusion of an ecological land classification and inventory as part of this analysis was requested for the following reasons:

- development of site-specific yield curve estimates through ecological classification of permanent sample plots
- more accurate timber supply analyses, long-range sustained yield estimates, and allowable annual cut estimates as a result of site-based yield curve estimates and an understanding of the spatial distribution and area of ecosystem types upon which the yield curves are based
- requirement for a forest-level site productivity classification to facilitate both enhanced forest management (e.g., stand tending) and reforestation planning
- requirement for spatial ecological information to model biodiversity, habitat, and other non-timber values.

Methods

The company had a limited time frame within which to conduct the timber supply analysis and ecological land classification and mapping. It would not have been possible to complete a full ecosystem map using conventional ground and photo-interpretation methods. We therefore used a knowledge based software modeling system (ELDAR).[1] We developed an eight-step process using this predictive ecosystem mapping system.

(1) *Data preparation* involves gathering data sources, overlaying these using GIS, and producing an ASCII dataset in the required format. For this project, a forest cover map provided vegetation attributes (species composition is of primary importance), an ecoregion map provided information on broad climatic characteristics, a digital terrain model provided slope and aspect data, and a modified landform map provided local information on moisture and nutrient variations.

(2) *Evidence base preparation* was necessary because the software system classifies ecosystems by matching characteristics of the attribute data (e.g., slope, aspect, overstory composition) to a set of rules, or probability functions, defined by an expert ecologist. The evidence base consists of a hierarchy framework that reflects the regional ecological classification and local ecological variations within the region; a set of evidence tables that contain the probabilities of attribute occurrence for any ecosystem type within the hierarchy; and a rule definition tool to facilitate the definition of higher-order interactions between evidence tables.

(3) *Initial classification* involves running the input data using starting values for specific attributes such as the proportion of black spruce that is allowed in certain ecosystem types or the slope and aspect ranges most commonly associated with specific ecosystem types. These starting values may be obtained from field guides or from previous projects that used the same software modeling system for the same biogeographic area.

(4) *Field data collection*, an integral part of the modeling process, provides both the information necessary to modify the evidence base and to test the accuracy of predictions. Three essential elements to a fieldwork program are:

- The fieldwork must be conducted by ecologists who are very familiar with the landscapes in the project area. If decisions made in the field are wrong (i.e., if the evidence collected is inadequate or is improperly interpreted by the ecologist), the errors propagate across the entire landscape for all map units with similar attributes.
- Plot locations must be highly accurate. GIS overlays result in smaller polygons in the derivative layer than in any of the contributing layers. If plots are not located correctly, attributes from the wrong polygon could be improperly linked to field data with the result that errors occur for all polygons with that attribute set.
- The initial modeling system classification (point 3 above) can be used to guide the fieldwork to pay attention to areas where the system was unable to classify, areas where the classification is suspect, and areas where complexes occur. All these situations should be targeted for more intensive surveys. Data collected must include all

elements required to confirm the ecological classification assignment, and all elements that are used as attributes (e.g., slope, aspect, and cover type) must be measured so that field values can be compared to the attribute values from the input data sets.

(5) *Evidence base modification* using the field data as further input can be done to adjust the evidence tables. For example, slope and aspect might be more or less influential in the current project area than in a previous project where they may have provided the starting knowledge base for the software system.

(6) *Iterative runs* are done. The classification is run with iterative changes to the evidence base until an acceptable level of accuracy and the lowest possible level of ambiguity (complex predictions of two or more ecosystem types) is achieved. Typically, about three runs are required to determine and correct systematic errors in predictions. The incorrect predictions that remain are almost always caused by either landscape variations too small to be captured by the available coverages or inaccuracies in the terrain models (e.g., overestimates or underestimates of slope, or incorrect aspects). The software modeling system predictions can be traced through all stages of the prediction process, which helps identify problem areas.

(7) *Accuracy assessment* is the next step. We assess validation accuracy (validation accuracy is not the same as map accuracy; only subsets of the map polygons are checked) where field plots considered to be correct are compared to the modeling system predictions. To do this, we convert each ecosystem type, both predicted and verified by field work, to its equivalent position on a moisture and nutrient (edaphic) grid and calculate the Euclidean distance between the two. A hypothetical example is presented in Figure 4. If the distance is no more than 1.414 (the square root of 2), the prediction is considered correct. For this project, if the nutrient prediction differed from the actual plot value between the C and D classes (the part of the moisture-nutrient grid that is probably most significant overall to forest managers), the prediction was also considered incorrect. Predictions across the entire management area were on average 81 percent correct, with values for individual compartments ranging from 79 percent to 84 percent correct.

The importance of unambiguous classifications (one prediction for a given polygon versus two or more possible ecosystem types) was viewed as a positive characteristic for predictive ecosystem mapping. For this area, just under 20 percent of the predictions were ambiguous (two or more ecosystem types). This relatively low level of ambiguity was made possible by including a modified landform theme. Simple air photo

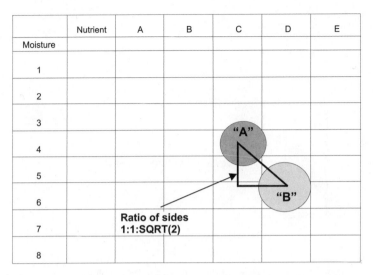

	Nutrient	A	B	C	D	E
Moisture						
1						
2						
3						
4				"A"		
5						
6					"B"	
7		Ratio of sides 1:1:SQRT(2)				
8						

Figure 4. Hypothetical ecosystem units and illustration of Euclidean distance measurement for accuracy assessment.

interpretation of landscape units delineated on the basis of relative moisture and nutrient status accompanied by intensive field checks resulted in a map of edatopic conditions across the entire area. This map contributed input attributes to the modeling system that reflected local moisture and nutrient conditions. The attributes were then related to ecosystem units through known associations.

(8) *Final map and coverage production* is the last step. The predicted ecosystem types are merged back with the original derivative coverage and may, for further analyses or cartographic presentation, be grouped in whatever way is required using GIS database functions.

Results

The predictive ecosystem mapping system resulted in both a spatial ecological model of the management area and a set of aspatial statistics that gave total areas within each ecological type. We linked total area statistics for ecosystem types that had been aggregated into useful management interpretive units to ecologically based yield curves for yield classes that had sufficient ecological data to support this relationship. These yield curves indicated higher volumes at rotation age for "good" sites than for "average" sites (Figure 5).

The resulting timber supply analysis using ecological stratification indicated, after deletions and other net-down activities had been taken into account, an increase of approximately 15 percent in the estimate of standing

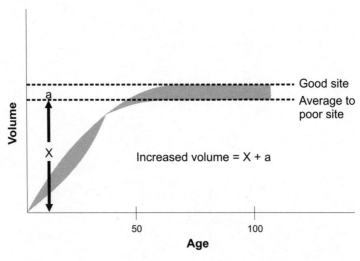

Figure 5. Generalized yield curve incorporating ecological stratification.

timber over an identically structured analysis without stratification. This in turn led to about a 10 percent increase in allowable annual cut with no adjustments for silvicultural enhancements.

In terms of costs and benefits to the company, the value of the resulting incremental cut increase in the first year was approximately twice the cost of the undertaking, which equates to a payback period of approximately six months. In subsequent years, no further costs will be associated with ecological classification unless the model is run again using new knowledge and/or improved information. In fact, being able to do this in the future is a major advantage of this approach. In addition, an ecological inventory will now be available for forest-level landscape assessments of productivity, silvicultural potential, and non-timber values at essentially no cost after the first year.

It is reasonable to consider whether timber supply estimates will always show an improvement if ecological stratification is improved. The degree of improvement will naturally be related to the relative proportions of "good" versus "fair" or "poor" sites in the management area. The advantage of an ecological land classification is that it assists greatly in determining this proportion.

Conclusions

Predictive ecosystem mapping that integrates GIS technology and knowledge-based systems provides excellent opportunities to better use existing information and expert local and regional ecological knowledge. The case study and other projects that have used this type of predictive ecosystem mapping provide evidence for the ability of knowledge-based predictive ecosystem mapping to supply ecological map information in a short time and at relatively low cost in comparison to conventional methods or simple reclassifications of thematic maps.

Future improvements to predictive ecosystem mapping will likely incorporate advancements in both input information and knowledge bases. More accurate and explicit terrain models that represent not only slope and aspect but also slope position are desirable for ecosystem classification, and some investigations have been made in this area (e.g., Macmillan et al. 1993). Forest vegetation inventories must become more sensitive to capture key variables important to ecosystem mapping and, as noted above, improved remote sensing technologies may provide opportunities. Ecological classifications (e.g., field guides) will undoubtedly change over the next decade as forest managers better understand silvicultural responses and habitat requirements, which will drive the further development of classification schemes. Continued advancements in hardware and software capabilities will also facilitate the use of predictive ecosystem mapping by those who wish to focus more on modeling and less on technology, and this will streamline data sharing among applications.

Note

1 ELDAR (Ecological Land Data Acquisition Resource) was developed by the Alberta Research Council in cooperation with the Canadian Forest Service, Weldwood of Canada - Hinton Division, and the Foothills and McGregor Model Forests. Timberline Forest Inventory Consultants are licensees for this product in Western Canada.

References

BC Ministry of Forests. 1998. PrognosisBC. Forest Renewal British Columbia, British Columbia Ministry of Forests. Fact sheet. Victoria, BC.

Biggs, W., P. Eligh, R. Sims, and R. Wiart. 1997. A business approach to terrestrial ecosystem mapping (TEM) and TEM alternatives. Resources Inventory Committee, BC Ministry of Forests and Ministry of Environment, Lands and Parks, Victoria, BC.

Coughlan, J.C., and S.W. Running. 1989. An expert system to aggregate forested landscapes within a geographic information system. Artificial Intelligence Applications in Natural Resource Management 3(4):35-43.

Jenny, H. 1941. Factors of soil formation. McGraw-Hill, New York, NY.

Jones, R.K. 1993. Next generation forest site classification: Ecologically-oriented predictive classification and mapping systems. Pp.143-151 *in* Proceedings of GIS '93 Symposium, 15 - 18 February 1993, Vancouver, BC.

Jones, R.K. 1994. Site classification: Its role in predicting responses to management practices. Chapter 7 *in* W. Dyck, D.W. Cole, and N.B. Comerford (eds.). Impact of forest harvesting on long-term site productivity. Chapman and Hall, London, UK.

MacMillan, R.A., P.A. Furley, and R.G. Healey. 1993. Using hydrological models and geographic information systems to assist with the management of surface water in agricultural landscapes. Pp.181-209 *in* R. Haines-Young, D.R. Green, and S. Cousins (eds.). Landscape ecology and GIS. Taylor and Francis, Basingstoke, UK.

Major, J. 1951. A functional, factorial approach to plant ecology. Ecology 32:392-412.

Mulder, J., and I. Corns. 1994. Knowledge based ecosystem prediction: Field testing and validation. *In* Decision support strategies in Canada's model forests: Reprint of symposium papers. Eighth annual symposium on geographic information systems, 21-24 February 1994, Vancouver BC.

Pollack, R. 1998. Individual tree recognition based on a synthetic tree crown image model. Presented at the International Forum on Automated Interpretation of High Spatial Resolution Digital Imagery for Forestry, 10-12 February 1998, Victoria, BC.

Skidmore, A.K., F. Watford, P. Luckananurug, and P.Ryan. 1996. An operational GIS expert system for mapping forest soils. Photogrammetric Engineering and Remote Sensing 62(5):505-511.

Titus, S. 1998. Mixedwood growth model (MGM), Version 97, May 1997. Fact sheet. Department of Renewable Resources, University of Alberta, Edmonton, AB.

Williams, T.M. 1988. Integration of land classification and modeling on lower coastal plain loblolly pine. Pp. 165-173 *in* T.M. Williams and C.A. Gresham (eds.). Proceedings of the IEA/BE project CPC-10 workshop on Predicting Consequences of Intensive Forest Harvesting on Long-term Productivity by Site Classification, 2-9 October 1987, Baruch Forest Science Institute of Clemson University, Georgetown, SC.

Biodiversity Assessment in Forest Management Planning: What Are Reasonable Expectations?

Peter N. Duinker and Frederik Doyon

Abstract

Biodiversity, one of the six criteria of the Canadian Council of Forest Ministers' Criteria and Indicators (1995), must now be taken seriously in the management of all publicly owned working forests in Canada. This means that we need a whole new set of analyses in forest management planning and redirected actions in implementation. We will focus here on planning for forest biodiversity conservation. As forest planners embrace the challenge, what is reasonable to expect them to be able to accomplish in the first planning efforts where biodiversity is a dominant theme?

Our thoughts on this topic come from a variety of sources: (a) our Biodiversity Assessment Project (BAP), in which we are developing and testing a suite of biodiversity assessment models for use in guiding development of a management plan for an industrial forest in central Alberta; (b) the Sustainable Forest Management Standard of the Canadian Standards Association (1996a, b); and (c) the adaptive management approach to forest biodiversity conservation promoted by Wildlife Habitat Canada (1996) in its Forest Biodiversity Program. Included in our set of reasonable expectations are requirements to: (a) establish values, goals, indicators, and objectives; (b) develop and use explicit quantitative models to forecast reasonably possible futures for biodiversity indicators; and (c) use the biodiversity indicator forecasts to help design and choose a forest management strategy. The array of chosen indicators should be kept small for tractability.

Introduction

Biodiversity, one of the six criteria of the Canadian Council of Forest Ministers' Criteria and Indicators for Sustainable Forest Management (1995), must now be taken seriously in the management of all publicly owned working forests in Canada (Duinker 1996a). This means that we need a whole new set of analyses in forest management planning as well as redirected actions in implementation. Our objective in this paper is to elaborate on a set of reasonable requirements for incorporating biodiversity considerations into forest management planning. In other words, as forest planners embrace the challenge, what is reasonable to expect them to be able to accomplish in the first planning efforts where biodiversity conservation is a dominant theme?

Basis for Setting Expectations

Our inspirations and ideas for the following set of basic expectations come from a wide variety of sources, but the principal ones are these: (a) the Biodiversity Assessment Project for Millar Western Forest Products Ltd.; (b) the sustainable forest management standard of the Canadian Standards Association; and (c) the "Sam Jakes Strategy" of Wildlife Habitat Canada's Forest Biodiversity Program. These initiatives have a high degree of concordance with each other, and each embodies a firm implementation of adaptive management. Each also is dedicated, in whole or in part, to the conservation of biodiversity in Canada's forests.

The Biodiversity Assessment Project for Millar Western Forest Products Ltd.

The Biodiversity Assessment Project (BAP) was initiated through a contractual agreement between Millar Western Industries Ltd. of Whitecourt, Alberta (now Millar Western Forest Products Ltd., or MWFP), and the Chair in Forest Management and Policy at the Faculty of Forestry, Lakehead University (Thunder Bay, Ontario). Millar Western Forest Products Ltd. recently negotiated with the Government of Alberta for a forest management agreement (FMA) area. The firm determined that it would need to prepare a comprehensive forest management plan for the area shortly after signing the forest management agreement. The purpose of the Biodiversity Assessment Project is to create relevant biodiversity assessment models and test them in undertaking a preliminary assessment for the forest management plan for the forest management agreement. More specifically, the project's objectives are to:

(a) develop models appropriate for Millar Western Forest Products Ltd.'s forests of spatial landscape patterns, ecosystem diversity, and wildlife habitat carrying capacity

(b) assess performance of a range of forest development forecasts in terms of landscape patterns, ecosystem diversity, and wildlife habitat carrying capacity

(c) advise Millar Western Forest Products Ltd. on interpretation of the assessment, design of a promising forest management strategy and plan, implementation of a sound research and monitoring program for forest biodiversity, and further operation and development of the assessment models.

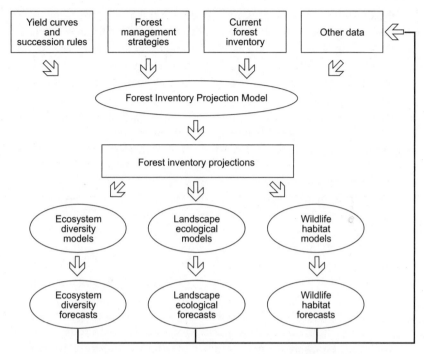

Figure 1. Basic framework for the Biodiversity Assessment Project, Millar Western Forest Products Ltd., Whitecourt, Alberta.

To convey in simple form how the Biodiversity Assessment Project is designed to function, we portray its process in terms of datasets and models (Figure 1). Such a portrayal is consistent with well-established frameworks for natural resource management (e.g., Walters 1986) and environmental assessment (Holling 1978; Munn 1979; Beanlands and Duinker 1983, 1984; Duinker and Baskerville 1986). The framework for the biodiversity assessment is as follows:

(1) We begin with data inputs to a forest projection model (Figure 1). These include:

- yield curves and succession rules
- forest inventory (as complemented by any biogeoclimatic classification information that may be available)
- forest management strategies (including intensive management of common forest ecosystem types).

(2) The forest projection model generates forecasts of the forest inventory under alternative management strategies and assumptions about yield

development and succession. It also determines the long-term sustainability of various cut levels and allocates silvicultural treatments. Depending on model sophistication, it may also permit adjacency constraints in harvest-block allocation and other planning functions. In our case, we need a model that preserves the spatial identity of treated stands.

(3) For each management strategy and set of assumptions to be tested, the forest projection model will generate a spatially explicit forecast of the forest inventory. The forecasts will stretch to 200 years into the future, and the Biodiversity Assessment Project will use no finer than 10-year snapshots of the predicted inventories.

(4) The core of the Biodiversity Assessment Project is its biodiversity assessment models. The models will be produced to interpret the inventory forecasts in terms of landscape pattern metrics, ecosystem diversity (coarse filter), and species-specific wildlife habitat suitability (fine filter). Where possible, population models for specific species (e.g., moose [*Alces alces*]) will be used to explore the interaction of habitat dynamics and wildlife harvest.

(5) Landscape pattern metric forecasts, ecosystem diversity forecasts, and wildlife habitat suitability forecasts for each strategy tested will be analyzed, compared, and evaluated, leading to reformulation and retesting of management strategies. This process continues until an acceptable management strategy is achieved.

We have decided to analyze the forest in terms of future habitat potentials for a small yet diverse array of vertebrate species. We have chosen vertebrates because (a) the public is generally quite concerned about the welfare of vertebrate species in managed forests, (b) some vertebrate species have high economic importance, and (c) approaches for analysing forests in terms of vertebrate habitat potential are relatively well developed. With these factors in mind, the species chosen for initial analyses in the Biodiversity Assessment Project include some twenty mammals, amphibians, and birds (see Doyon and Duinker 1998 for the procedure and outcome of species selection).

In choosing indicators of forest sustainability as related to ecosystem diversity and landscape fragmentation, the first-approximation biodiversity indicators developed for Ontario's boreal forest by Plinte (1995) and Siberian boreal forests by Plinte (see Duinker 1996b) will be used as a framework. This base of indicators will be complemented with spatial measures of ecosystem diversity. The final set will be a handful of practical indicators that can be measured with current forest inventory data.

A unique feature of the Biodiversity Assessment Project is our use of a natural-disturbance simulation model (in our case, the LANDIS model of

Mladenoff et al. 1996). With it we explore the range of natural variation that one might expect to find in the future behaviour of the forest under circumstances of no human intervention. The resulting biodiversity responses will help define a benchmark against which to compare responses under alternative forest management strategies.

The Biodiversity Assessment Project represents just one of several so-called Impact Assessment Groups (IAGs) that are contributing to the preparation of the Millar Western Forest Products Ltd. forest-management plan (Millar Western Industries Ltd. 1997). Additional Impact Assessment Groups cover other major forest values such as soil, water, timber, forest health, non-timber economic values, and socio-cultural values. One intention is to bring all the Impact Assessment Groups together once first-round analyses are done to explore the trade-offs inherent in trying to meet multiple, potentially conflicting objectives at the same time in the same forest.

The Sustainable Forest Management Standard of the Canadian Standards Association

Late in 1993, representatives of the Canadian forest products industry requested the Canadian Standards Association (CSA) to develop a standard for sustainable forest management (SFM). Using a consultative, consensus-seeking process, the Canadian Standards Association developed the required standard (Canadian Standards Association 1996a, b) to be consistent with the ISO 14000 standard series on environmental management systems. The standard is voluntary and focuses on the following four components: commitment, public participation, management systems, and continuous improvement. The management system, firmly rooted in the concepts and principles of adaptive management (Holling 1978; Baskerville 1985; Walters 1986; Lee 1993; Ontario Forest Policy Panel 1993; Maser 1994) has five elements: preparation, planning, implementation, measurement/assessment, and review/improvement (Figure 2).

Sustainable Forest Management, as defined by the Canadian Standards Association standard (Canadian Standards Association 1996a), requires that we must provide explicit planning and monitoring for all indicators chosen to represent the status of the selected values. The indicators are to be chosen locally but need to link to a set of nationally accepted criteria (Canadian Council of Forest Ministers 1995). This means that biodiversity will be represented by a significant portion of the indicators chosen for use in management planning.

The Sam Jakes Strategy of Wildlife Habitat Canada's Forest Biodiversity Program

Wildlife Habitat Canada (WHC), a non-profit foundation dedicated to habitat conservation, has initiated and supported a program to work with forest

---

Preparation:
- Define DFA (Defined Forest Area)
  - forest area and current conditions
- Values, Goals, Indicators, Objectives
  - identify values, define goals, choose indicators, set objectives
- Inventory
  - maps/data/records

Planning:
- Forecast
  - forecast expected changes within the DFA under a range of reasonable management strategies
- Choose
  - select forecast and associated strategy

Implementation:
- Take planned actions to achieve objectives

Measurement and assessment:
- Compare performance to objectives
  - assess actions taken against actions planned
  - measure forest conditions
  - compare the actual and forecast outcomes in terms of the indicators

Review / improvement:
- Understand differences between actual and planned outcomes
  - use learning to improve management

Figure 2. Steps in sustainable forest management according to the Canadian Standards Association standard Z808 (adapted from Canadian Standards Association 1996b).

companies wanting to understand how to manage and conserve biodiversity (Wildlife Habitat Canada 1996). Currently six forest companies are participating in the Forest Biodiversity Program (FBP). During Phase I of the program, Wildlife Habitat Canada convened a workshop with company staff and a group of North American biodiversity experts to develop a framework for building a biodiversity conservation strategy and a process to assess a company's current capacity to plan for a variety of biodiversity conservation objectives. Participants adopted four guiding principles for the Forest Biodiversity Program (Wildlife Habitat Canada 1996):

(1) know where you are going and what you want to achieve
(2) start now and work with what you have
(3) implement so that we can learn by doing
(4) ensure continuous feedback in an iterative process.

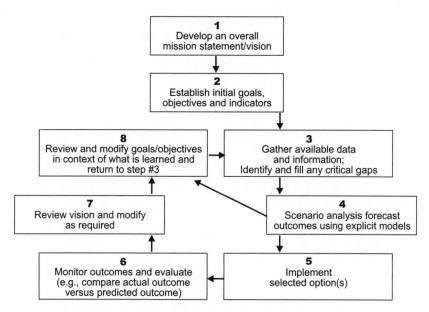

Figure 3. The Sam Jakes Biodiversity Strategy: An eight-step process for a forest company to develop and implement a biodiversity conservation strategy (Wildlife Habitat Canada 1996).

The generic framework, called the Sam Jakes Biodiversity Strategy (Figure 3), is an eight-stage process for a forest company to develop and implement a biodiversity conservation strategy. The approach relies on an adaptive management process for achievement of biodiversity conservation in forest management.

Expectations Elaborated

In our view, all forest management planning exercises should include a minimum of the following elements to address biodiversity conservation.

Establishment of Values, Goals, Indicators, and Objectives

We agree with the early setting of values, goals, indicators, and objectives as outlined in the Canadian Standards Association (1996a) standard (Figure 2). Thus forest stakeholders, interest groups, and other elements of the public must be engaged early with forest managers and analysts to establish values, set goals, choose indicators, and develop objectives. We extend caution, however, on two counts: (a) the Canadian Standards Association definitions are not as crisp as they could be; and (b) objective setting must be seen as tentative until late in a planning exercise. We will look at each of these in turn.

Values, Goals, Indicators, Objectives: Meanings and Relationships
We believe that the following definitions offer helpful consistency:

Value: things for which someone would deem a forest important, for example:

- process (e.g., carbon sequestration, water quality regulation, recreation)
- physical thing (e.g., timber, moose, marten *[Martes americana]* pelts)
- forest condition (e.g., biodiversity, soil bulk density).

(In this context, a criterion is a sort of 'mega-value' or large grouping of values.)

Goal: a directional statement for a value (need not be stated in quantitative terms), for example:

- have forests become long-term net sinks of atmospheric carbon
- produce a continuous non-declining flow of quality wood to meet mill needs
- maintain current levels and types of biodiversity.

Indicator: a measurable variable relating directly to one or more values, for example:

- for carbon sequestration, kg/ha/yr net carbon flux
- for timber, cubic metres per year harvest volume
- for biodiversity, age-class structure of the forest.

Objective: a directional statement for an indicator (must be stated in quantitative terms), for example:

- more than zero kg/ha/yr (i.e., a positive number) for net carbon intake
- at least 500,000 cubic metres per year of softwood pulp
- 10 percent or more of total forest area in any of five development stages at any time.

A hierarchical organization is probably required to work with a wide array of forest values. Under the 'mega-value' called biodiversity, three common sub-divisions or elements are ecosystem diversity, species diversity, and genetic diversity. One can imagine even further divisions of these elements into finer values, whereupon it should be possible to identify appropriate indicators.

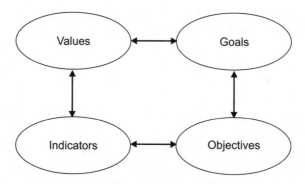

Figure 4. Relationships among values, goals, indicators, and objectives in sustainable forest management.

Using the above definitions, it becomes clear (Figure 4) that each value requires a goal statement and assignment of one or more indicators. Each indicator, in turn, requires an objective statement. Thus if the structure is carefully put together for any value, the goal statement should be fulfilled if the objectives are fulfilled. Adoption of this approach to sorting out criteria and indicators will reduce much confusion at the planning table, especially as planning tables now include a wide assortment of technical specialists, interest and stakeholder groups, and members of the public.

Provisional Objectives
An age-old paradox arises in forest planning when goals and objectives are being discussed (Duinker 1994): How can we know how much we should want until we know how much we can get, and how can we know how much we can get until we know how much we want? The solution is obvious to anyone who uses formal computerized models for wood-supply analysis – we need to set and test first-round, very tentative objectives to explore a range of scenarios of wood supplies and the associated action sets required to achieve them.

We have next the problem of value trade-offs. How can we know how much of two values we should want and can get when the pursuits of the values conflict with each other? Surely it is necessary to undertake trade-off analysis to discover the trade-off functions and also to discover how to relax constraints so that more of each value might be satisfied.

Given these dilemmas, it seems reasonable that forest planners might propose objectives early in a planning process. Those objectives, however, are likely to be revised throughout a planning exercise, and they will be clinched only at the very end when decisions are made.

Development and Use of Explicit Quantitative Models to Forecast Reasonably Possible Futures for Biodiversity Indicators

Forest management planning is an exercise in decision making under huge uncertainties. The decisions are choices about the long-term future and about actions that will help attain a desired long-term forest future. The long-term future cannot be adequately analyzed unless explicit long-term forecasts are made to depict many possible futures. From these possible futures, one might try to choose the most desirable.

In many jurisdictions in Canada, managers of public forest land are now required to forecast wood supply and forest structure for a period of one to two centuries into the future. This is fortunate for analysts who want to assess biodiversity conservation in forest management, because many forest biodiversity conservation issues actually hinge on the future state of the forest structure. It means that, for many biodiversity indicators, analysts must develop indicator models that interpret the forest structure forecasts in biodiversity terms.

Problems arise, of course, when one discovers that forest inventory simulators are not projecting variables that are needed for biodiversity analysis (e.g., abundance of coarse woody debris, or herbaceous vegetation). Under these circumstances, analysts must develop unique relationships so that such variables can be inferred from the forest structure forecasts. Another challenge arises when analysts wish to model species population responses to habitat change, for which special models are needed (e.g., Duinker and Baskerville 1986; Duinker et al. 1996). For the most part, however, biodiversity indicator forecasts can be relatively simple to develop once wood supply and forest structure forecasts are in hand.

Use of the Biodiversity Indicator Forecasts to Help Design and Choose a Forest Management Strategy (Objectives and Associated Actions)

The whole point of making a series of biodiversity indicator forecasts in forest management planning is to check whether each management strategy under consideration has favourable biodiversity outcomes. If the only strategies that are highly desirable for timber production are at the same time rather unfavourable for biodiversity conservation, then the analyst's task is to try to develop strategies that retain the desirable timber production outcomes and also generate favourable biodiversity outcomes. Thus forest management planning becomes a complex design exercise.

The approach we are taking in the Biodiversity Assessment Project includes two rounds of biodiversity indicator assessments. In the first round, we assess some half dozen alternative forest management strategies with an eye to searching for circumstances where specific biodiversity indicators are responding rather unfavourably. These first-round strategies were not designed in any meaningful way to be "biodiversity friendly" but rather to represent different ways of trying to produce timber sustainably. Detailed assessment of the indicator forecasts in relation to the assessed strategies should reveal any biodiversity conservation problems and point the way to redesigning the strategies so that the biodiversity indicators respond favourably (or, at worst, unfavourably to an acceptably small degree). The redesign will be guided by the envelope of natural variation defined through use of the LANDIS model. The second round of analysis checks whether the indicator responses to the redesigned strategies are more biodiversity friendly than those of the first round.

Finally, we expect to find opportunities to implement different management strategies in different parts of the forest in an attempt to accelerate learning about some of the most uncertain yet important relationships between management actions and biodiversity responses. Good design of such experimental management regimes depends on incisive sensitivity analyses using the forecast models (Duinker and Baskerville 1986). Such an approach is consistent with the concept of active adaptive management (Walters 1986; MacDonald et al. 1997).

Conclusions

We have represented just a partial picture of adaptive management for biodiversity conservation. At the end of planning, a forest-management plan is adopted. The plan contains: (a) a set of objectives for all the indicators analyzed; (b) a set of action prescriptions that, in sum, are expected to cause the indicators to respond in such a way that the objectives are met; and (c) a set of monitoring programs designed to track the indicators through time during management implementation. Regarding monitoring, we believe strongly that a forecast modeling exercise is the most sensible basis for designing monitoring studies for biodiversity indicators (Duinker 1989).

Once planning is finished, the work of the biodiversity analyst is hardly over. Much remains to be done in the overall scheme of forest management. The analyst needs to:

(1) monitor the indicators to determine whether they are responding in nature as they were expected to respond (as described in the plan).
(2) design special field studies to determine the validity of the models used to generate the indicator forecasts.
(3) correct erroneous relationships in the models and document the learning process thoroughly.
(4) prepare for another round of biodiversity indicator forecasting in the next planning exercise.

This completes the adaptive management loop. Our intention here was to explicate the major activities for analysts in the planning stage of forest management. We remain firm in our conviction that every management plan for public forest land in Canada must have explicit analysis of biodiversity indicators along the lines we have described before it can be deemed to be professionally competent. If there are no biodiversity indicator forecasts, the plan is incomplete. The expectations we have laid out are not beyond the analytical capabilities of the forest scientific community in Canada and not beyond the financial capabilities of the forest-management companies.

We do not promote the use of hundreds of indicators to be analysed in response to dozens of management strategies. Rather we urge that a small and relevant set of biodiversity indicators be assessed against a few carefully designed alternative management strategies using a rigorous analytical process. Much fieldwork is needed, of course, but it is best directed by attempts to build appropriate models and make much-needed forecasts.

Acknowledgments
The Biodiversity Assessment Project is funded by Millar Western Forest Products Ltd. of Whitecourt, Alberta, and the Sustainable Forest Management Network based at the University of Alberta in Edmonton, Alberta. We heartily thank Jonathan Russell of Millar Western Forest Products Ltd. and Laird Van Damme of KBM Forestry Consultants in Thunder Bay, Ontario, for their inspiration and leadership in the Millar Western Forest Products Ltd. forest management planning process.

References
Baskerville, G.L. 1985. Adaptive management: Wood availability and habitat availability. Forestry Chronicle 61:171-175.

Beanlands, G.E., and P.N. Duinker. 1983. An ecological framework for environmental impact assessment in Canada. Institute for Resource and Environmental Studies, Dalhousie University, Halifax, NS, and Federal Environmental Assessment Review Office, Hull, QC.

Beanlands, G.E., and P.N. Duinker. 1984. An ecological framework for environmental impact assessment. Journal of Environmental Management 18:267-277.

Canadian Council of Forest Ministers. 1995. Defining sustainable forest management: A Canadian approach to criteria and indicators. Canadian Council of Forest Ministers, Canadian Forest Service, Ottawa, ON.

Canadian Standards Association. 1996a. A sustainable forest management system: Guidance document. CAN/CSA-Z808-96, Environmental technology: A national standard of Canada. Canadian Standards Association, Etobicoke, ON.

Canadian Standards Association. 1996b. A sustainable forest management system: Specifications document. CAN/CSA-Z809-96, Environmental technology: A national standard of Canada. Canadian Standards Association, Etobicoke, ON.

Doyon, F., and P.N. Duinker. 1998. Biodiversity assessment at the species level: Species selection for Millar Western Industries Ltd. and Alberta Newsprint Company FMA territories in Central Alberta. BAP Report #2. Biodiversity Assessment Project, Chair in Forest Management and Policy, Faculty of Forestry, Lakehead University, Thunder Bay, ON.

Duinker, P.N. 1989. Ecological effects monitoring in environmental impact assessment: What can it accomplish? Environmental Management 13:797-805.

Duinker, P.N. 1994. On setting goals in forest resources management. Pp 1-8 *in* A.J. Kayll (ed.). Forest planning: The leading edge. Ontario Forest Research Institute, Sault Ste. Marie, ON.

Duinker, P.N. 1996a. Managing biodiversity in Canada's public forests. Pp. 324-340 *in* F. di Castri and T. Younes (eds.). Biodiversity, Science and Development. CAB International, Wallingford, UK.

Duinker, P.N. (ed.) 1996b. Biodiversity of Siberian forests: Concepts, preliminary analyses, and proposed research directions. IIASA Working Paper WP-96-79, International Institute for Applied Systems Analysis, Laxenburg, Austria.

Duinker, P.N., and G.L. Baskerville. 1986. A systematic approach to forecasting in environmental impact assessment. Journal of Environmental Management 23:271-290.

Duinker, P.N., C. Daniel, R. Morash, W. Stafford, R. Plinte, and C. Wedeles. 1996. Integrated modeling of moose habitat and population: Preliminary investigations using an Ontario boreal forest. Final Report prepared under the Northern Ontario Development Agreement for the Canadian Forest Service. Faculty of Forestry, Lakehead University, Thunder Bay, ON.

Holling, C.S. (ed.). 1978. Adaptive environmental assessment and management. No. 3, Wiley International Series on Applied Systems Analysis. John Wiley and Sons, Chichester, UK.

Lee, K.N. 1993. Compass and gyroscope: Integrating science and politics for the environment. Island Press, Washington, DC.

MacDonald, G.B., R. Arnup, and R.K. Jones. 1997. Adaptive forest management in Ontario: A literature review and synthesis. Forest Research Information Paper No. 139, Ontario Forest Research Institute, Ontario Ministry of Natural Resources, Sault Ste. Marie, ON.

Maser, C. 1994. Sustainable forestry: Philosophy, science, and economics. St. Lucie Press, Delray Beach, FL.

Millar Western Industries Ltd. 1997. Detailed forest management plan terms of reference. Millar Western Industries Ltd., Whitecourt, AB.

Mladenoff, D.J., G.E. Host, J. Boeder, and T.C. Crow. 1996. LANDIS: A spatial model of forest landscape disturbance, succession and management. Pp. 175-180 *in* M.F. Goodchild, L.T. Steyaert, and B.O. Parks (eds.). GIS and environmental modelling: Progress and research issues. GIS World Books, Fort Collins, CO.

Munn, R.E. (ed.). 1979. Environmental impact assessment: Principles and procedures. SCOPE 5. John Wiley, Chichester, UK.

Ontario Forest Policy Panel. 1993. Diversity: Forests, peoples, communities. Report of the Ontario Forest Policy Panel. Ontario Ministry of Natural Resources, Toronto, ON.

Plinte, R.M. 1995. Indicators of forest sustainability for Ontario boreal forests: A first approximation. M.Sc.F. thesis, Lakehead University, Thunder Bay, ON.

Walters, C.J. 1986. Adaptive management of renewable resources. MacMillan Publishing, New York, NY.

Wildlife Habitat Canada. 1996. Wildlife Habitat Canada Forest Biodiversity Program: Report on the February 1996 workshop. Wildlife Habitat Canada, Ottawa, ON.

Spatial Silviculture Investment Planning as a Means of Addressing Landscape Performance Objectives: A Case Study

B.G. Dunsworth

Abstract

The coastal Enhanced Forest Management Pilot Project is a combined MacMillan Bloedel/government initiative funded by Forest Renewal British Columbia to develop a silvicultural investment strategy integrated with a total resource harvest plan for a large forested area on northern Vancouver Island, British Columbia. A Working Group, formed to develop the planning process, used the Adam-Eve landscape unit as a starting point. We have completed total chance engineering using computer-assisted planning tools, conducting a variety of spatial and aspatial harvest schedule runs to assess the social and economic impact of various spatial harvest constraints. This helped identify spatial "pinchpoints" that significantly limited our ability to find the assigned annual allowable cut (AAC) on the ground. The two most constraining elements were adjacency and Forest Ecosystem Networks (FENs). We identified silvicultural opportunities to overcome those pinchpoints and ranked them to provide the greatest harvest revenue and maintain Forest Practices Code of British Columbia (Province of British Columbia 1996) compliance. These included an old-growth recruitment strategy, a riparian restoration strategy, and a rapid early establishment strategy. We also established a local advisory group to provide input and direction from local communities. To assist in communicating, an Enhanced Forest Management Pilot Project web site has also been constructed (www.mbltd.com/environ/mb-enfr.html).

Introduction

Biodiversity has become a major issue in forestland management in Canada and British Columbia (Wilson and Peter 1988; Fenger et al. 1993). The original concern about species loss in tropical rainforests has been transformed into a global concern about stewardship and land ethic. This occurred through linking those land use practices that led to fragmentation, degradation, and elimination of habitat with those practices that compromised the provision of free ecological services (clean air, clean and abundant water, and productive soils). In this more abstract form, the term "bio-diversity" became easily transportable to other regions in the world where people's fears of losing forest productivity, future economic opportunities, healthy ecosystems, and species are dominant social concerns (Wilson 1988; Ehrlich and Mooney 1983; Ehrlich 1988).

British Columbia is such a place and, in response to the social concerns reflected in the Convention on Biodiversity (United Nations 1992), has recently ushered in a new regulatory environment by legislating the Forest Practices Code of British Columbia (Province of British Columbia 1996). All of British Columbia's public forests are now managed under the direction of this legislation. While the management objectives defined under previous legislation included provisions for non-timber resources, the Forest Practices Code of British Columbia includes more explicit regulations and guidelines for the practice of commercial forestry. One of the aims of the regulations and guidelines is to maintain a wide range of resource values by limiting the impact of timber production on related non-consumptive uses.

Currently, in the coastal region of British Columbia, considerable debate centres around what stand and landscape-level guidelines are appropriate to conserve biological diversity. The Biodiversity Guidebook for the British Columbia Forest Practices Code (BC Forest Service 1995) has rules for seral stage balance, patch size distribution, old-growth and interior old-growth retention, and forest ecosystem network (FEN) construction. In addition, the guidebook recommends stand-level rules for wildlife tree patches and objectives for stand structure, vegetation and tree species diversity, and maintenance of coarse woody debris. Other non-timber harvest constraints in British Columbia can be grouped into those protecting soil, riparian, and aesthetic values. In short, these rules deal more with forest stewardship than with forest habitat and can significantly reduce permitted timber harvest levels and increase harvest costs (Nelson and Shannon 1994; van Kooten 1994; Haley 1996).

These constraints can have the effect of reducing the commercial landbase, affecting the rate of timber harvest, or dispersing the harvest blocks. For different combinations of constraints, spatial and aspatial harvest models can help find solutions to maximize the net present value or the volume of the harvest. This allows people to calculate the opportunity cost for each set of constraints. Thus it is possible to partition constraints and determine their relative impacts (Dunsworth and Northway 1998).

Stewardship policies that use rule-based approaches are administratively convenient, but compliance can give little assurance that we are working towards species conservation objectives (Daust and Bunnell 1992). Recent work on the Olympic peninsula in Washington State and elsewhere in the Pacific Northwest (Carey and Curtis 1996; Carey 1989; Kohm and Franklin 1997; and Hansen et al. 1991) have suggested ways in which silviculture interventions can be used to create ecological attributes more rapidly than by simply aging a forest. They also suggest that alternatives to achieving the performance objectives exist that can provide greater forest management flexibility and reduced costs.

MacMillan Bloedel has used a spatial and aspatial harvest schedule model to explore the interactions of harvest constraint rules and determine where the greatest economic impact occurs. The explicit objectives of this case study were to:

- assess the cost of regulation for a variety of harvest scheduling rules on a typical Vancouver Island landscape unit
- rank the spatial and aspatial constraints by socio-economic impact
- develop a spatial silviculture investment plan that would target silviculture activities to help mitigate the impact of the harvest scheduling rules.

The Enhanced Forest Management Pilot Project (EFMPP) was initiated in September 1995 as a cooperative effort among industry, the British Columbia Ministries of Forests and Environment, Lands and Parks, Forest Renewal British Columbia, labour, and the academic community. A broad-based provincial Steering Committee, established to guide the process, selected MacMillan Bloedel (Tree Farm Licence 39) as one of three areas for pilot projects.

The underlying premise of the pilot project is that societal values for providing a wide range of forest values (i.e., recreation, aesthetic, hydrologic, and biodiversity values) from landscapes is currently reflected in the Forest Practices Code of British Columbia. The Tree Farm Licence 39 pilot project used the biological growing potential as estimated by the current assigned harvest level and compared this to the current spatial reality of the regulatory environment. The difference defined the disparity between potential forest growth and administratively available harvest rates. The pragmatic objective is to identify the spatial "pinchpoints" in achieving the potential of the forest, to target stand level silvicultural activities that overcome these "pinchpoints" (i.e., enhanced forest management opportunities), and to integrate the activities that overcome "pinchpoints" in a comprehensive landscape management plan.

The landscape management plan rationalizes the timing, location, and intensity of stand level management practices, the selection of harvesting systems, and the focus of inventory and research expenditures. In short, it finds spatial solutions to implementing sustainable harvest rates.

Location and Key Issues
The Adam-Eve landscape unit in north central Vancouver Island covers

approximately 69,000 ha that includes approximately 57,000 ha of productive forest. The headwaters are above 2,200 m in Strathcona Provincial Park and are drained by two major river systems – the Adam and the Eve. These are approximately 60 km long and 5 km to 8 km wide valleys that drain eastward into Johnstone Strait on the east side of Vancouver Island.

These U-shaped and previously glaciated valleys have bedrock geology of mainly basalt and pillow lava. The surface geology largely reflects the retreating glaciers of 10,000 years ago that left large amounts of morainal material; subsequent weathering has resulted in some colluvial and fluvial deposits.

The climate of the valley is classified as West Coast marine. Eighty percent of the more than 400 cm of annual rainfall occurs between October and April. The landscape unit is broken into three biogeoclimatic variants: CWHvm1 (Submontane Very Wet Maritime), CWHvm2 (Montane Very Wet Maritime), Coastal Western Hemlock Variants, and Mhmm1 (Windward Moist Maritime) Mountain Hemlock Variant (Green and Klinka 1994).

The recent Timber Supply Review for Tree Farm Licence 39 determined an appropriate harvest level of 1.335 million m^3 per year for the study area consisting of four landscape units. The area not included within the timber harvesting landbase includes environmentally sensitive areas such as sensitive soils, streamside buffers, deer and elk winter ranges, and important recreation features.

Native mammals in the landscape unit include important large species such as wolf *(Canis lupus)*, beaver *(Castor canadensis)*, marten *(Martes americana)*, wolverine *(Gulo luscus)*, black-tail deer *(Odocoileus hemionus)*, Roosevelt elk *(Cervus elaphus rooseveltii)*, and riparian small mammals such as the Pacific water shrew *(Sorex palustris navigator)*, Townsend's mole *(Scapanus townsendii)*, Trowbridge's shrew *(Sorex trowbridgii)*, river otter *(Lutra canadensis)*, mink *(Mustela vison)*, and northern bog lemming *(Synaptomys borealis)*. Native and migratory birds in the forests of the area include species of management concern such as marbled murrelets *(Brachyrampus marmoratus)*, northern goshawks *(Accipiter gentilis)*, great blue herons *(Ardea herodias)*, and pileated woodpeckers *(Dryocopus pileatus)*. Adjacent marine and estuary habitats support populations of notable species such as Peale's peregrine falcons *(Falco peregrinus pealei)*, bald eagles *(Haliaeetus leucocephalus)*, trumpeter swans *(Cygnus buccinator)*, and harlequin ducks *(Histrionicus histrionicus)* (Meidinger and Pojar 1991).

Tourism and recreation are an important aspect of this forest area. The forest provides easily accessible recreation opportunities to local communities, populations of the middle and south Vancouver Island, and the lower mainland of British Columbia. The scenic qualities of the forest landscapes are also important, especially along the major highway and the marine corridors.

Recent changes in forest planning, introduced through the Forest Practices Code of British Columbia, reflect a desire for a higher level and more persistent provision of a wider range of non-timber resource values than in the recent past. In coastal British Columbia and in Tree Farm Licence 39 specifically, implementing the Code has led people to anticipate a large gap between the current state of the forest, which reflects historic social values (i.e., providing fewer non-timber values at lower levels), and the desired future condition. The primary concern is how to bridge this transition with the least impact on local communities and the provincial economy.

The key technical issues are:

- minimizing the difference between the biological potential and what is spatially achievable on the ground
- adjacency constraints (free-to-grow, adjacency, visual greenup, hydrological greenup); access to adjacent blocks contingent on reaching height targets
- rate-of-harvest (biodiversity, hydrology, and visual quality objectives); limits on proportions of forest areas in various age classes
- management opportunities in areas with high non-timber values (e.g., partial harvest in areas of restricted clearcutting)
- impact of silviculture activities on fibre flow and log value.

Methodology and Results

Block Design and Harvest Scheduling

Our spatial analysis required that all of the productive forest be assigned harvest blocks regardless of whether it was an excluded or reserved area. This blocking framework allowed the flexibility to turn spatial constraints on and off and assess individual and combined impacts of constraint layers on harvest levels and costs. A contract engineer using a computer-assisted spatial planning tool (TYPO) originally designed the harvest blocks. He used the existing five-year development plan and the proposed road network provided by the logging division as a starting point and placed blocks in productive forest where he felt they would be most appropriate. He was asked to use, as much as possible, a 40-ha block size. Harvest system choices were based on proximity to roads and were assigned programatically as follows: if a road was a within the block it was a conventional clearcut; if a road was next to the block it was a long-line system; and any others were assigned as helicopter blocks. Over 2,000 blocks were created in this landscape unit.

Once the blocks were created the engineer was asked to schedule the harvest manually in a fashion that would keep the harvest as high as possible for as long as possible. He was asked to do this while maintaining immediate adjacency rules, using a 3 m green-up rule, and respecting all spatial constraints. The intent here was to provide a basis to compare with computer-optimized block scheduling. The computer-optimized schedule maximized volume in all periods and used 400 m adjacency and 3 m green-up. In both cases forest cover polygons were eligible for harvest at 300 m³/ha.

Assessing the Gap and Pinchpoints

We assessed in three stages the impact of spatial constraints on achieving desired harvest levels. The first stage was to look at the assignment of a target harvest level for the landscape unit. We used the British Columbia Chief Forester's Annual Allowable Cut assignment for Tree Farm Licence 39, Block 2 from Management Plan 7. The harvest level apportioned to the Adam-Eve landscape unit was proportional to its contribution to the productive forest in the study area. This landscape unit has approximately one-third of the productive forest, old growth, and second growth in the study area, and the aspatial harvest level could be easily approximated with a one-third factor.

The second step was to assess the spatially available harvest and the gap between the aspatial assignment and the spatially available harvest. To assess this we used the computer-optimized harvest schedule with the best schedule of blocks that maximized volume harvested and complied with all spatial constraints. The final step was to apportion the gaps to spatial constraints and to rank these constraints. To do this we first assessed the

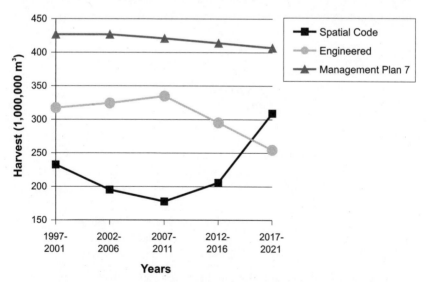

Figure 1. Apparent gap between aspatial harvest allocation and spatially available harvest within the Adam-Eve Landscape Unit, Tree Farm Licence 39, Vancouver Island, BC.

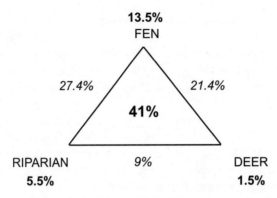

Figure 2. Reserve area contributions to the harvest gap (individual element impacts at vertexes, combination impacts on the sides, and total impact in the middle) within the Adam-Eve Landscape Unit, Tree Farm Licence 39, Vancouver Island, BC.

contribution of the reserve areas (Forest Ecosystem Network [FEN], deer winter range and spring forage areas, and riparian reserves) and then assessed the harvest rate constraints (adjacency, visual quality objective [VQO] alteration, and VQO green-up). This was done by either turning the constraint off, as with adjacency, or modifying the constraint to levels agreed by the Working Group as acceptable for analytical purposes, as in the case of VQOs. The VQO green-up requirement was assessed at 5 m and 3 m and the VQO alteration levels were assessed at 2.5/10/22 and 10/20/35 (percent allowable alterations for retention/partial retention/modification).

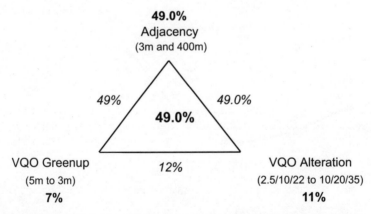

Figure 3. Rate constraint contributions to the harvest gap (individual element impacts at vertexes, combination impacts on the sides, and total impact in the middle) within the Adam-Eve Landscape Unit, Tree Farm Licence 39, Vancouver Island, BC.

Table 1

Socio-economic cost estimates of reserve components within the Adam-Eve Landscape Unit, Tree Farm Licence 39, Vancouver Island, BC.

	Km^3/year	Economic activity	Number of jobs	Government revenue
Forest ecosystem network	35.1	$12,285,000	64	$2,176,200
Riparian	14.3	$5,005,000	67	$886,600
Deer	3.8	$1,330,000	18	$235,600
All	106.9	$37,415,000	500	$6,627,800

We used MIDAS (Lockwood and Moore 1992), a spatially explicit harvest-scheduling tool developed by MacMillan Bloedel, to assess the spatial constraints and FSSIM, a wood supply analysis model used by the Province of British Columbia, to assess the aspatial constraints. MIDAS employs a simulated annealing algorithm to schedule pre-defined harvest blocks. The results of this analysis indicated that the aspatial to spatial gap ranged from 90,000 m^3 to 250,000 m^3 per year over the next twenty-five years (Figure 1). This is compared to a target aspatial Annual Allowable Cut (AAC) of 400,000 m^3 to 425,000 m^3 per year.

The major contributors to this gap were the Forest Ecosystem Network (FEN) and adjacency constraints (Figures 2 and 3). The Forest Ecosystem Network impact is self-explanatory as it consists of a large permanent reserve of productive forest mostly in old growth. The adjacency impact is less obvious, but it results in the dispersion of the harvest (400 m block separation) and a reduction in the rate-of-harvest (3 m green-up requirement). To assess the socio-economic costs of these spatial constraints, we applied the socio-economic multipliers established for the Tree Farm Licence (TFL) 44 Management Plan (MacMillan Bloedel 1997). These estimates are presented in Tables 1 and 2.

Table 2

Socio-economic cost estimates for the harvest rate constraints within the Adam-Eve Landscape Unit, Tree Farm Licence 39, Vancouver Island, BC.

	Km^3/year	Economic activity	Number of jobs	Government revenue
Adjacency greenup	135.1	$47,285,000	632	$8,376,200
Aesthetics				
(visually effective greenup)	20.4	$7,140,000	95	$1,264,800
Visual alteration	31.2	$10,920,000	146	$1,934,400
All	135.2	$47,320,000	633	$8,382,400

Spatial Silviculture Investment Plan

The intent of developing the Spatial Silviculture Investment Plan (SSIP) was to target silviculture investments on those areas where they would provide the greatest benefit in closing the gap between the assigned aspatial harvest level and the spatially available harvest. The gap and pinchpoint analyses indicated that the two largest spatial constraints in the Adam-Eve landscape unit were the Forest Ecosystem Network and adjacency. The Spatial Silviculture Investment Plan was then devised with three explicit components to meet these issues – two to address the Forest Ecosystem Network issue (Old Growth Recruitment and Riparian Restoration) and one to address adjacency (Rapid Early Establishment).

Old-Growth Recruitment

In addressing the issue of the impact of the Forest Ecosystem Network reserve area, the Working Group discussed the function the network was supposed to provide and agreed that two key elements were old-growth ecological representation and connectivity. We also agreed that a permanent network was not the only means of maintaining connectivity at the landscape level. To meet the Forest Ecosystem Network functions, both components should be explicitly addressed in the Spatial Silviculture Investment Plan. The Old-Growth Recruitment portion of the Spatial Silviculture Investment Plan was devised to address ecological representation and the Riparian Restoration component to address connectivity.

Divisional engineers and the district Forest Ecosystem specialist in the Working Group devised the current draft Forest Ecosystem Network for the Adam-Eve landscape unit. While it was done with the best local knowledge, it was not guided by an analysis of ecological representation as no ecosystem mapping was available. To address the representation requirement we had to provide an analysis.

The Working Group agreed that we should use a surrogate approach based on existing mapping in the adjacent landscape unit. To do this we used the classification tree method to develop a predictive model based on forest inventory attributes. Site association was fit to species, variant, aspect, elevation, and slope position. The 59,200 observations (5 ha grids split by stand and biogeoclimatic variant) were split randomly into two halves. The model was developed on one half and tested on the other.

Our assessment indicated a 40 percent misclassification error for site association. We were, however, within one site unit of the correct classification approximately 80 percent of the time. We felt this was accurate enough for strategic planning purposes and applied the model to the Adam-Eve landscape unit.

We then intersected the site series surrogate coverage with the forest inventory to determine where the existing reserves did not meet the old-growth targets in the Biodiversity Guidebook of the Forest Practices Code of British Columbia (BC Forest Service 1995). This created a short list of target site series that required old-growth recruitment. The next step was to search for recruitment candidates. Two solutions were possible: either capture existing old growth in these site series or, if there was no old growth, to treat mature stands with silvicultural treatments that would provide old-growth attributes sooner than letting the stands age.

We developed an Arcview application for old-growth recruitment candidacy. This application targeted individual site series and searched for existing reserved stands. It then indicated stands of the same site series within 600 m of those targets. The operator was instructed to select areas in old growth first and mature growth second to fill the recruitment requirement. If mature growth was selected, these stands were then assigned a specific silvicultural treatment type based on age and density. The objective of the treatment types was to create old-growth attributes rapidly.

In this landscape unit we were able to find all of the necessary old-growth areas for all deficient site series with old growth within 600 m of existing old-growth reserves. A total of approximately 1,700 ha was targeted for reserves.

Riparian Restoration

This component of the Spatial Silviculture Investment Plan was devised to address the connectivity requirement of the Forest Ecosystem Network. It is clear from the Biodiversity Guidebook of the Forest Practices Code of British Columbia that the backbone of a Forest Ecosystem Network is the riparian reserve network and that it can provide a significant contribution to connectivity for some organisms. One of the serious problems in meeting Forest Ecosystem Network requirements in landscapes with a long pre-Code harvest history (such as the Adam-Eve) is that much of the riparian reserve system may have been harvested and now consists of early seral stands. To improve these reserve areas and subsequently the connectivity function of the Forest Ecosystem Network, silvicultural treatments can be applied to enhance and speed up the acquisition of riparian attributes faster than could be achieved by simply aging the forest.

To assign treatment areas we used an approach similar to that used in old-growth recruitment strategies. We used the same Arcview application as for old-growth recruitment candidacy, but in this case we restricted the search to blocks with any portion in a riparian buffer. We sorted all stands by age class and density and assigned silvicultural treatments as follows.

Age	Treatment
15-25	variable density spacing with deciduous retention
40-55	thinning with snag recruitment and cavity creation
80-90	variable density thinning and snag retention
150-200	variable retention harvesting with snag retention

The riparian reserve portion of each block was selected as the target area for recruitment. Stand density and a 20 percent buffer were applied for purposes of estimating treatment costs. A total of approximately 950 ha was identified.

Rapid Early Establishment

The remaining spatial constraint issue to address in the Spatial Silviculture Investment Plan was adjacency. The Working Group discussed silvicultural strategies and decided that a Rapid Early Establishment strategy had to be devised that applied genetics, fertilization, and large planting stock to achieve green-up targets sooner than with natural seed-in or conventional planting programs. We assumed that this would incur an incremental cost of approximately $200 per hectare and provide a two-year benefit on achieving green-up.

We used MIDAS to schedule rapid early establishment treatments. MIDAS selected only those forest stands where applying the treatment would provide a positive net present value on the investment. The results indicated

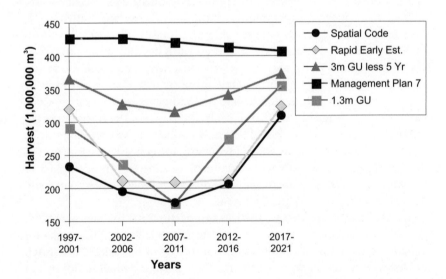

Figure 4. Sensitivity analysis of reducing adjacency GU (green-up) constraints with silvicultural interventions within the Adam-Eve Landscape Unit, Tree Farm Licence 39, Vancouver Island, BC.

that approximately 7,800 ha could be treated over the next 25 years. We also ran a variety of options to assess the sensitivity of addressing green-up. Finally, we examined reducing the green-up rule to 1.3 m and achieving green-up five years faster. The results are presented in Figure 4.

The analysis indicated that:

- fibre supply will dip in 15 years and the dip will be difficult to avoid
- reducing the green-up height to 1.3 m to close the gap would be more effective than investment in the Rapid Early Establishment Strategy
- a heroic silvicultural intervention that might achieve a five-year jump on green-up would reduce the gap only by about half.

This analysis did not address the 400 m proximity issue, which could have a major impact through the cost of dispersion.

Next Steps
This phase of the project provided a good analytical framework for the future landscape units in Block 2 and provided an excellent opportunity to learn. What we learned in this process was that:

- this type of work is not possible without powerful computer tools that deal with spatial data
- data preparation is key to success and can be expensive and time consuming
- the assumptions in the strategic analysis need to be tested in an adaptive management framework.

During the analytical phase of this project MacMillan Bloedel made a significant change in its forest management direction. Some of the impetus for this change results from this analysis and from a strong recognition that coastal forest operations in British Columbia will lose their social licence to harvest old-growth forests if they did not address clearcutting and old-growth conservation.

The change in forest management approach by MacMillan Bloedel will mean adopting variable retention silvicultural systems applied in a framework of stewardship zones. The change will effectively eliminate clearcutting as a harvest system employed by MacMillan Bloedel. This will be phased in over the next five years and will include:

- developing harvest criteria for safety, ecology, economics, and regeneration
- adopting forest certification
- implementing an international advisory panel.

The change in management direction means that some of the work conducted during the first phase of the pilot project will be revisited. The impacts and pinchpoints identified in the Adam-Eve landscape unit are the result of regulatory requirements under the Forest Practices Code of British Columbia directed at reducing the impact of clearcutting. Adjacency, greenup, block size, and Forest Ecosystem Networks are administrative constructs used to disperse the harvest, conserve old growth, and reduce the aggregation of early seral forest types. With the adoption of variable retention harvest systems and stewardship zoning, we have a new set of tools to meet landscape objectives and these may significantly reduce the need for administrative rules directed at clearcutting. If this is true it could have a major impact on closing the aspatial to spatial harvest gap identified in the Adam-Eve landscape unit.

References

BC Forest Service. 1995. Biodiversity guidebook: Forest Practices Code of British Columbia. BC Ministry of Forests, Victoria, BC.

Carey, A.B. 1989. Wildlife associated with old-growth forests in the PNW. Natural Areas Journal 9(3):151-162.

Carey, A.B., and R.O. Curtis. 1996. Conservation of biodiversity: A useful paradigm for forest ecosystem management. Wildlife Society Bulletin 24(4):610-620.

Daust, D.K., and F.L. Bunnell. 1992. Predicting biological diversity on forest land in British Columbia. Northwest Environmental Journal 6(2):191-192.

Dunsworth, B.G., and S.M. Northway. 1998. Spatial assessment of habitat supply and harvest values as a means of evaluating conservation strategies: A case study. Pp. 315-330 *in* P. Bachmnan, M. Kohl and R. Paivinen (eds.). Assessment of biodiversity for improved forest planning. European Forest Institute Proceedings No. 18, October 1996. Kluwer Academic Publishers, Monte Verde, Switzerland.

Ehrlich, P.R. 1988. The loss of diversity: Causes and consequences. Pp. 21-27 *in* E.O. Wilson and F.M. Peter (eds.). Biodiversity. National Academy Press, Washington, DC.

Ehrlich, P.R., and H.A. Mooney. 1983. Extinction, substitution and ecosystem services. BioScience 33(4):248-254.

Fenger, M.A., E.H. Miller, J.F. Johnson, and E.J.R. Williams (eds.). 1993. Our living legacy. Royal British Columbia Museum, Victoria, BC.

Green, R.N., and K. Klinka. 1994. A field guide to site identification and interpretation for the Vancouver Forest Region. British Columbia Ministry of Forests Research Branch. Crown Publications, Victoria, BC.

Haley, D. 1996. Paying the piper. Prepared for: Working with the B.C. Forest Practices Code. Insight Information Inc., Vancouver, BC.

Hansen, A.J., T.A. Spies, F.J. Swanson, and J.L. Ohmann. 1991. Conserving biological diversity in managed forests: Lessons. Bioscience 41(6):382-392.

Kohm, K., and J.F. Franklin (eds.). 1997. Creating a forestry for the 21st century: The science of ecosystem management. Island Press, Seattle, WA.

Lockwood, C., and T. Moore. 1992. Harvest scheduling with spatial constraints: A simulated annealing approach. Canadian Journal of Forest Research 23:468-478.

MacMillan Bloedel (Corporate Forestry Division). 1997. TFL 44 socio-economic analysis. *In* Management Plan #3, TFL 44, Appendix 3, Section 4. MacMillan Bloedel Ltd., Nanaimo, BC.

Meidinger, D., and J. Pojar (eds). 1991. Ecosystems of British Columbia. Special Report Series No. 6. British Columbia Ministry of Forests, Victoria, BC.

Nelson, J.D., and T. Shannon. 1994. Cost and timber supply assessment of the coastal biodiversity guidelines. Unpublished report. Faculty of Forestry, University of British Columbia, Vancouver, BC.

Province of British Columbia. 1996. Forest Practices of Code of British Columbia Act, Revised Statutes of British Columbia, Chapter 159.

United Nations. 1992. Convention on biological diversity.

van Kooten, G.C. 1994. Economics of biodiversity and preservation of forestlands in British Columbia. FRDA Publication No. 21, Victoria, BC.

Wilson, E.O. 1988. The current state of biological diversity. Pp. 3-18 *in* E.O. Wilson and F.M. Peter (eds.). Biodiversity. National Academy Press, Washington DC.

Wilson, E.O., and F.M. Peter (eds.). 1988. Biodiversity. National Academy Press, Washington DC.

A Process for Ecosystem Management at a Landscape Scale

Jonathan B. Haufler, Carolyn A. Mehl, and Gary J. Roloff

Abstract

We describe a process for ecosystem management that emphasizes a coarse filter approach for ecological communities but also has a fine filter component to check for viability of selected species. The coarse filter is designed to identify threshold levels of ecological communities to meet biological diversity and other ecological objectives. The fine filter or species viability assessment is applied as a check on the effectiveness of the coarse filter. The process involves the following steps: delineate the planning landscape, develop an ecosystem diversity matrix, describe a matrix of historical disturbance regimes, determine adequate ecological representation of ecological communities, identify social and economic objectives, quantify the existing landscape, determine desired future conditions, identify and implement required actions, and monitor in an adaptive management format. This requires collaborative efforts to determine compatible classification systems and data as well as to provide a sound framework for quantifying ecological objectives. Investigations of bird, small mammal, reptile, and amphibian populations have helped confirm the validity of the coarse filter. This process will provide a landscape plan that will meet the ecological objectives of ecosystem management while also providing a framework for more focused discussions on social and economic objectives.

Introduction

Ecosystem management is an approach to landscape planning and management that strives, through an integrated avenue, to meet ecological, social, and economic objectives. The ecological objectives of ecosystem management can be varied, but generally they include maintaining and enhancing biological diversity and ecosystem integrity (Grumbine 1994). Biological diversity includes the consideration of genetic, species, and community diversity maintained or enhanced at a landscape scale. Under this objective, ecosystem integrity is included at the community level. The ecological objectives require that ecosystem management operates at larger planning landscapes than typically considered under previous management approaches, and they require a sophisticated understanding and analysis of the landscape from an ecological perspective. We present an approach to ecosystem management based on a coarse filter (ecological community-based approach) that also incorporates the ability to consider the needs of individual species (fine filter approach), and we describe a process for implementing it. While most of our discussion is on meeting the ecological objectives of ecosystem management, we also show how social and economic objectives are included in the process. General descriptions of this process have previously been published (Haufler et al. 1996, 1999).

Strategies for Ecosystem Management

Ecosystem management, at present, is used to describe nearly any land management initiative that considers more than one factor in the planning or implementation process. Few initiatives actually attempt to integrate the ecological objectives of biological diversity and ecosystem integrity with social and economic objectives. Very few initiatives even attempt to fully address the ecological objectives, at least for more than one ecological community type or a few selected species. Various strategies for meeting the ecological objectives, particularly the objective of maintaining or enhancing biological diversity, have been proposed. Haufler (1999a, 1999b) described and contrasted a number of these proposed approaches. Here we describe an approach for ecosystem management at a landscape scale that, while using a coarse filter approach, also incorporates a fine filter (or species assessment) to address ecological objectives and integrate social and economic objectives.

Coarse Filter Approach

A coarse filter approach attempts to meet the ecological objectives of ecosystem management by classifying, describing, and quantifying an appropriate mix of ecological communities across a planning landscape to provide for ecosystem integrity and to support biological diversity. For this approach to work, the coarse filter must fully and adequately account for the distribution of biological diversity and ecological processes within the planning landscape. Effective coarse filters include past ecosystem dynamics, present conditions, and future dynamics. Effective coarse filters must also have sufficient detail to account for past ecosystem dynamics, yet if coarse filters become too complex, they become infeasible to operate. The process for ecosystem management that we describe uses a tool to help define the coarse filter called the ecosystem diversity matrix (Haufler 1994). Our approach focuses primarily on the coarse filter, but we also recognize the need to check the coarse filter with an assessment of the viability of selected species.

The Ecosystem Management Process

The ecosystem management process we have described (Haufler et al. 1999) identifies the following steps:

- delineate the extent and boundaries of the planning landscape
- classify the coarse filter by using the ecosystem diversity matrix (EDM)
- quantify existing conditions by using the ecosystem diversity matrix
- determine the historical range of variability within the ecosystem diversity matrix
- quantify adequate ecological representation of the ecosystem diversity matrix
- check for adequate ecological representation with a species assessment
- identify social and economic concerns
- integrate adequate ecological representation with economic and social concerns to determine desired future conditions within the ecosystem diversity matrix
- compare desired conditions with existing conditions, identify needed management actions, and implement, monitor, and evaluate outcomes, adjusting as necessary.

Delineating a planning landscape is a critical step to ecosystem management initiatives. For the planning landscape to address ecological objectives, it needs to be delineated by ecological boundaries. These boundaries should provide commonality for the coarse filter by fitting into a hierarchical ecological classification. The planning landscape needs to be large enough to address the objectives of ecosystem integrity and biological diversity with the recognition that the viability needs of a few species may not be completely met within one planning landscape. We have found that the National Hierarchy of Ecological Units (Ecomap 1993) provides a useful hierarchy for landscape delineation, and we feel that the section level or aggregates of sub-sections of this hierarchy function well as appropriate planning landscapes.

We have devoted considerable effort to describing ecosystem diversity matrices to serve as the coarse filters for ecosystem management planning. We envision several matrices functioning together to form the complete coarse filter. For example, we have identified four matrices that, when integrated, form the coarse filter for describing the Idaho Southern Batholith Landscape. These four are a forest ecosystem diversity matrix, a grassland/shrub ecosystem diversity matrix, a riparian/wetland ecosystem diversity matrix, and an aquatic ecosystem diversity matrix. We will provide further description and definition of the forest ecosystem diversity matrix in this paper, although all four of the matrices need to be incorporated to fully meet ecological objectives.

Vegetation growth stage	Habitat type class			
	Dry Douglas Fir	Dry Grand Fir	Cool Moist Grand Fir	Warm Dry Subalpine Fir
Seedling/Sapling				
Understory ⇨ Small tree *Fire Regime* Medium tree Large tree				
Small tree Medium ⇦ *Stand Replacing* tree *Fire Regime* Large tree Old Growth				
Total				

Figure 1. A simplified example of the forest ecosystem diversity matrix for the Idaho Southern Batholith planning landscape. Habitat type classes are groupings of habitat types (Steele et al. 1981). Vegetation growth stages are successional trajectories under two different fire regimes: frequent low-intensity understory fires and less frequent high-intensity stand-replacing fires.

The framework for the forest ecosystem diversity matrix for Idaho is described in Haufler et al. (1996, 1999). Figure 1 is a simplified version of the full matrix and shows habitat type classes in the columns and vegetation growth stages in the rows. This matrix identifies two different fire regimes as the major structuring element of the vegetation growth stages. Matrices developed for other landscapes where fire was not a dominant disturbance can use other structuring elements, for example, shade tolerance of the overstory species.

The forest ecosystem diversity matrix characterizes combinations of habitat type classes and vegetation growth stages as individual cells of the matrix (ecological units) for the planning landscape. The distribution of ecological units can be displayed by mapping habitat type classes and vegetation growth stages for forested areas. The overlay of these two maps produces polygons of ecological units of the forest ecosystem diversity matrix. Summing the areas within all polygons for each cell of the matrix produces and quantifies the existing conditions in the ecosystem diversity matrix (Figure 2).

As mentioned, the forest ecosystem diversity matrix also serves to frame the historical range of variability (Morgan et al. 1994) for ecological units.

Vegetation growth stage		Habitat type class			
		Dry Douglas Fir	Dry Grand Fir	Cool Moist Grand Fir	Warm Dry Subalpine Fir
Seedling/Sapling		26,495	6,882	11,695	72,322
Understory Fire Regime ⇨	Small tree	1,200	120	0	120
	Medium tree	800	87	0	44
	Large tree	1,200	288	0	0
Small tree		79,500	12,955	26,200	168,016
Medium tree	⇦ *Stand Replacing Fire Regime*	54,600	12,145	25,100	77,328
Large tree		55,400	5,668	8,910	39,129
Old Growth		2,000	46	4,005	5,400
Total		221,195	38,191	75,910	362,359

Figure 2. An example of the forest ecosystem diversity matrix for existing conditions. Values in each cell are hectares of that ecological unit occurring within the planning landscape. Total hectares for a column are a fixed feature of the landscape. Vegetation growth stages within a column will change in amounts with additional disturbance or with successional dynamics.

Information on historical disturbance regimes, modelling efforts, and/or use of expert analysis and opinion can be used to construct a historical range of variability ecosystem diversity matrix for the planning landscape. An example of such a matrix for forest ecosystems of the Idaho Southern Batholith landscape is shown in Figure 3.

The historical range of variability ecosystem diversity matrix identifies those forest ecological units occurring under historical disturbance regimes that would be required under our coarse filter approach if ecological objectives are to be met. For this coarse filter to function for ecosystem management, economic and social objectives must also be integrated. This means that the historical range of variability ecosystem diversity matrix must be taken a step further to identify the needed amounts of each ecological unit in the matrix. This "needed amount" has been termed adequate ecological representation (Haufler 1994), or a sufficient amount and distribution of all inherent ecological communities to support viable populations of all native species that were historically viable within the planning landscape. Haufler et al. (1999) describe how adequate ecological representation can be calculated as a set percentage (they identify 10 percent) of the maximum of the

Vegetation growth stage		Habitat type class			
		Dry Douglas Fir	Dry Grand Fir	Cool Moist Grand Fir	Warm Dry Subalpine Fir
Seedling/Sapling		0-4	0-6	5-16	0-11
Understory Fire Regime ⇨	Small tree	0-4	0-11	0-12	0-4
	Medium tree	3-22	1-167	2-26	2-8
	Large tree	59-99	66-99	19-59	3-8
Small tree		0	0	1-9	8-26
Medium tree	⇦ *Stand Replacing Fire Regime*	0	0	2-19	12-26
Large tree		0	0	8-26	5-15
Old Growth		0	0-1	1-5	2-6
Total		221,195	38,191	75,910	362,359

Figure 3. An example of a historical range of variability matrix for forest ecosystems in the Idaho Southern Batholith landscape. Values are the percentage of a particular vegetation growth stage within a habitat type class that might occur within the planning landscape over approximately 400 years prior to major anthropogenic alterations.

historical range of variability for each ecological unit in the ecosystem diversity matrix. This allows coarse filter thresholds to be quantified independently of any fine filter assessments, a critical step if the coarse filter approach is to avoid the problems caused by the complexity of fine filter approaches. An example of the calculation of adequate ecological representation is presented in Figure 4. Based on this example, the entire adequate ecological representation ecosystem diversity matrix can be portrayed (Figure 5). This matrix represents threshold levels for ecological objectives. The amounts of each ecological unit in the matrix must always be above this amount if ecological objectives are to be met.

The amounts and distributions of the ecological units provided through adequate ecological representation can be checked with a species assessment. The species assessment is designed to ensure viability. The description of this process is provided in Haufler et al. (1996, 1999) with more specifics on the habitat-based approach to viability assessment in Roloff and Haufler (1997).

Economic and social objectives need to be evaluated relative to the ecosystem diversity matrix. Many of these objectives will involve spatial analyses

Existing Conditions EDM (ha)	Historical Range of Variability EDM (%)	Historical Range of Variability EDM (ha)	AER EDM (10% of max HRV) (ha)
Dry Grand Fir HTC	Dry Grand Fir HTC	Dry Grand Fir HTC	Dry Grand Fir HTC
6,882	0-4	0-1,528	153
120	0-11	0-4,201	420
87	0-16	382-6,111	611
288	**66-99**	**25,206-37,809**	**3,781**
12,955	0	0	0
12,145	0	0	0
5,668	0	0	0
46	0	0-382	38
38,191	38,191	38,191	38,191

Figure 4. An example of the calculation of adequate ecological representation (AER) for one particular ecological unit (dry grand fir) within the forest ecosystem diversity matrix. The first column shows the dry grand fir habitat type class (HTC), the existing amounts of each vegetation growth stage within this class, and, in boldface type, the total area of the landscape comprised of that habitat type class. The second column lists the historical range of variability for each vegetation growth stage within the dry grand fir habitat type class. The third column represents the range in amount of the large tree with understory fire ecological unit that occurred under historical disturbance regimes. The fourth column represents 10 per cent of the maximum of the area in the bolded ecological unit that occurred historically; it represents a proposed threshold for adequate ecological representation.

of the landscape to determine areas suitable for access, recreation use, viewsheds, or other desired outputs. These objectives, coupled with the adequate ecological representation ecosystem diversity matrix, form the basis for the desired conditions ecosystem diversity matrix. This matrix, when compared to the existing conditions ecosystem diversity matrix, will show where additional amounts of any ecological unit are needed to meet ecological, economic, or social objectives. The final steps in the process are to identify the actions required to provide the needed amounts of each ecological unit over time, to monitor the outcomes, and to adjust where necessary.

Validation of the Coarse Filter Process

The coarse filter approach discussed here has been examined for its ability to account for the abundance of selected species throughout planning landscapes. In Idaho, an analysis of small mammal populations using selected

Vegetation growth stage		Habitat type class			
		Dry Douglas Fir	Dry Grand Fir	Cool Moist Grand Fir	Warm Dry Subalpine Fir
Seedling/Sapling		885	153	1,215	3,986
Understory Fire Regime ⇨	Small tree	885	420	911	1,449
	Medium tree	4,866	611	1,974	2,899
	Large tree	21,898	3,781	4,479	2,899
Small tree		0	0	683	9,421
Medium tree	⇦ *Stand Replacing Fire Regime*	0	0	1,442	9,421
Large tree		0	0	1,974	5,435
Old Growth		0	38	380	2,174
Total		221,195	38,191	75,910	362,359

Figure 5. An example of the adequate ecological representation ecosystem diversity matrix for the Idaho Southern Batholith planning landscape.

ecological units was conducted as one test (Knapp 1997). Breeding bird populations have been examined in a number of ecological units to determine the functioning of the forest ecosystem diversity matrix in describing bird distributions and abundances (Sallabanks et al. forthcoming).

Conclusions

This paper presents an overview of an ecosystem management process emphasizing forested ecosystems in central Idaho, implemented in three planning landscapes, to date, by Boise Cascade Corporation. The coarse filter and its application to meeting ecological objectives is being validated in all three landscapes. Additional analyses using the species assessments will occur as checks on the application of these methods. At present, the process and use of the ecosystem diversity matrix appear to offer an effective and feasible approach to ecosystem management at a planning landscape scale.

References

Ecomap. 1993. National hierarchical framework of ecological units. US Department of Agriculture (USDA) Forest Service, Washington, DC.

Grumbine, R.E. 1994. What is ecosystem management? Conservation Biology 8:27-38.

Haufler, J.B. 1994. An ecological framework for planning for forest health. Journal of Sustainable Forestry 2:307-316.

Haufler, J.B. 1999a. Strategies for conserving terrestrial biological diversity. Pp. 17-30 *in* R.K. Baydack, H. Campa III, and J. B. Haufler (eds.). Practical approaches to the conservation of biological diversity. Island Press, Washington, DC.

Haufler, J.B. 1999b. Contrasting approaches for the conservation of biological diversity. Pp. 219-232 *in* R.K. Baydack, H. Campa III, and J.B. Haufler (eds.). Practical approaches to the conservation of biological diversity. Island Press, Washington, DC.

Haufler, J.B., C.A. Mehl, and G.R. Roloff. 1996. Using a coarse filter approach with a species assessment for ecosystem management. Wildlife Society Bulletin 24:200-208.

Haufler, J.B., C.A. Mehl, and G.R. Roloff. 1999. Conserving biological diversity using a coarse filter approach with a species assessment. Pp. 107-125 *in* R.K. Baydack, H. Campa III, and J.B. Haufler (eds.). Practical approaches to the conservation of biological diversity. Island Press, Washington, DC.

Knapp, B.E. 1997. Small mammal, reptile, amphibian, and bat species associations with habitat type classes and successional stages in west central Idaho. M.S. thesis, Michigan State University, East Lansing, MI.

Morgan, P., G.H. Aplet, J.B. Haufler, H.C. Humphries, M.M. Moore, and W.D. Wilson. 1994. Historical range of variability: A useful tool for evaluating ecosystem change. Journal of Sustainable Forestry 2:87-111.

Roloff, G.J., and J.B. Haufler. 1997. Establishing population viability planning objectives based on habitat potentials. Wildlife Society Bulletin 25:895-904.

Sallabanks, R., J.B. Haufler, and C.A. Mehl. Using avifauna to demonstrate a forest planning tool for the Idaho Southern Batholith Landscape. Forthcoming.

Steele, R., R.D. Pfister, R.A. Ryker, and J.A. Kittams. 1981. Forest habitat types of central Idaho. General Technical Report INT-114, US Department of Agriculture (USDA) Forest Service, Intermountain Forest and Range Experiment Station, Ogden, UT.

Forest Ecosystem Management without Regulation

Daryll Hebert

Abstract

Alberta and British Columbia present two alternative scenarios for forest ecosystem management. The Alberta government allowed Alberta Pacific Forest Industries to undertake research into ecological management, develop new planning protocols, and implement a newly designed Detailed Forest Management Plan. The concept of forest ecosystem management evolved from the broad input of company personnel, university researchers, government personnel, and the general public. The concept is being implemented while it is still evolving. The company research program evolved into the Network of Centres of Excellence on Sustainable Forest Management, which has 25 universities and 75 of Canada's best researchers working on many aspects of future forest ecosystem management programs.

In British Columbia, guidebooks and codes were developed without company input. Similarly, regional plans and implementation strategies were developed unilaterally by government, allowing written input, which remained unused, on completed documents. Research, although extensive, was not necessarily directed to forest ecosystem management problems. Guidebook direction became qualitative estimates of the real world. Since guidebook development excluded the implementers and operational planners, there was little connection to reality and the myriad of consequences that followed.

Introduction

Although biologists, the public, and other groups have long argued for a broader range of values from our forests, the Brundtland Commission (World Commission on Environment and Development 1987) gave the argument its needed strength by providing this definition of sustainable development: "Sustainable development is development that meets the needs of the present without compromising the ability of future generations to meet their own needs."

Since then, however, definitions of sustainability and sustainable development have proliferated to the point where they are used in almost every forest management context. According to Sample (1991), "The protection and sustainable management of forest ecosystems seems destined to become the 'conventional wisdom' in forest resource management in the twenty-first century." The definitions promise everything to everybody, often legitimizing many unsustainable operations and often providing a false sense of security.

Understanding Process

It would seem that sustainability results from two factors:

- an understanding of natural systems, especially forest turnover rates and the frequency distribution of patch size, structural attributes, and spatial pattern
- the ability to implement science using operational practices that do not destroy or minimize altering these natural systems.

Maintaining basic ecological process can be seen as the establishment of baseline carrying capacity. Although we also view the attainment of sustainability as the integration of social, economic, cultural, and ecological objectives, including the first three (social, economic, and cultural) objectives actually lowers carrying capacity. Thus sustainability can be highly variable and probably won't be attained unless we have a goal for a future forest that initially represents natural ecological process and provides some method to monitor its achievement.

Sustainability lies in the methodology (keep and use all the parts) used to compare natural disturbance processes and harvesting. It must allow a change in existing sustained yield procedures to a process that contains an understanding of the overall ecological system and some means of measuring success. The bottom line is that we must agree upon essential elements.

Roads Travelled

In the quest for sustainability, three main routes appear to be under examination:

(1) sustained yield with some options
(2) sustained yield with extensive constraints
(3) science-based ecological management.

At present in British Columbia the development of extensive constraints in the form of legislation, regulation, and guidelines drives a system based on limited input from science and industry. In contrast, the ecologically based

systems developing in Alberta were initiated and driven by the forest industry (Alberta Pacific in particular) and were supported by extensive cooperative research.

At issue in both situations is when and how to implement a socially and
ecologically workable system. To accomplish an implementable procedure,
we need a description and set of objectives for a future forest and some
agreed-upon essential elements.

Examples

Partial Retention/Residual Material

Historically, clearcutting and block size have been two of the most contro-
versial issues in forest management. To date, they have been dealt with as
independent issues and almost solely as stand issues.

In fire disturbance systems, especially where stand-replacing fires play a
role, residual material has a relationship with patch size and is dependent
on severity of disturbance. Thus the science behind residual material is actually a landscape feature that involves patch size, cumulative patch residual material, pulses of snags from wild fire, salvage logging, the use of
prescribed fire, and physical disturbance, among other things. We must also
consider the utility of providing residual material for biological purposes
while recognizing changes in plant species composition, differences in successional pathways, and the temporal aspects of convergence of pathways.

Currently in British Columbia, block size is restricted to the small end of
the gradient, residual material is dealt with as a stand issue, adjacency is the
driving attribute, and severity of disturbance is ignored or reduced.

Ungulate Winter Range

Partial retention at a stand level without considering the range and integration of patch size and severity has also affected ability to understand the
development and production of ungulate winter range.

Historically, ungulate winter ranges in the southern Rocky Mountain
Trench in British Columbia were produced by large, severe fires with little
standing live material for cover (e.g., Premier Ridge, Wigwam Flats, Fontaine
Basin). Thus patch size, shape, severity, and residual material were all interrelated. Specific sites, south slopes in particular, produced high quality (capability) winter ranges through the alteration of their successional pathways
and species composition while producing extensive delays in regeneration.
Cool, wetter north slopes provided much of the post-fire residual material
for cover.

Current ungulate winter range guidelines in British Columbia severely
restrict patch size variation and have severe reductions on disturbance. Harvest openings are restricted in size, include extensive cover which is not in

short supply, and, most importantly, restrict severity of disturbance to the point where species composition and successional pathways are not changed. In effect, harvest openings do not represent palatable forage production, and ungulate production is severely curtailed. This produces a lose-lose situation for both ungulates and the forest industry.

By selecting specific south slopes for harvest and land manipulation, which includes patch size, shape, and severity variability, we increase returns per unit effort for elk *(Cervus elaphus)*. This means that areas for elk having lower return per unit effort can be freed for timber harvest. The scenario produces a win-win situation for both elk and timber.

Caribou Requirements

Mountain caribou *(Rangifer tarandus caribou)* in the southern Rocky Mountain Trench in British Columbia have received special attention because of their seasonal reliance on mature forests for arboreal lichen, their primary food source in winter. The current guidelines for caribou, as listed in the Kootenay-Boundary Implementation Strategy (Kootenay Inter-Agency Management Committee 1997), indicate that:

- All "parkland" caribou habitat should be reserved from harvesting or road building.
- In each landscape unit, retain 70% of productive forest above the 1994 operability line in age class 8 or greater.
- In each landscape unit in the ESSF (Englemann Spruce Subalpine fir zone; as defined in Braumandl and Curran 1992), retain 40% of the operable forest on slopes < 80% in age class 8 or greater. One quarter of this is to be age class 9. Also 1/4 of 40% is harvestable through alternative silvicultural systems to maintain caribou habitat.
- In each landscape unit in the ICH (Interior Cedar Hemlock zone; Braumandl and Curran 1992), retain 40% of the operable forest on slopes < 80% in age class 8 or greater. One quarter of this is to be age class 9.
- Retain 40% of the operable forest on slopes < 80% in age class 6 or greater in the MS (Montane Spruce; Braumandl and Curran 1992).

Assessment of a large watershed (the Goat River, 700 km^2) in the southern Purcell Mountains in British Columbia suggested that the caribou 1 and 2 mapped areas (core and supplementary habitat) did not correspond very well with five years of radio-telemetry locations. We determined that 75 to 80 percent of the radio-telemetry locations in the Goat River fell on the inoperable land base (on slopes of 20 to 30 percent) rather than on the operable land base, as restricted by the guidelines.

Applying the guidelines in watersheds with caribou essentially removed up to 98 percent of the operable land base, curtailing all harvesting opportunities. In addition, the guidelines were applied to the entire provincial mountain caribou land base, a distance of approximately 1,500 to 2,000 km, in circumstances where caribou movements and habitat requirements differed significantly. Without flexibility in the guidelines, in many locations the application significantly overemphasized the intent.

Discussion

The development of a forest ecosystem management program is often based on some definition of sustainability. Most definitions attempt to satisfy everyone. To make the program work, it must contain specific essential elements and it must be implementable.

A comparison of British Columbia and Alberta suggests that guidelines or constraints should not precede the science. As Bunnell (1998) indicated, "Next time try data," suggesting that a review of the current science is a minimum approach to new management. As well, developing constraints around each individual attribute overemphasizes the effect of the constraints because of their additive nature. The current guidelines de-emphasize the interrelationship of several attributes such as patch size, shape, site disturbance, and partial retention.

Initially, developing concepts and philosophy about future forests is key to building a systems approach to ecological management of our forests. Without the proper model, constraints at an attribute level hinder the entire process because they encourage the cessation of thinking and encourage merely following rules.

Forest ecosystem management programs are complex visions of a wide array of natural processes. It has become clearer during the past five years that, although we might be on the right track, formulating strict rules too early in the process can be extremely counterproductive. Our ability to employ adaptive management is restricted by its early developmental history.

Stewardship is an attitude. The fact that the Alberta industry initiated much of the thinking and research surrounding forest ecosystem management probably gives it a higher chance of succeeding. The support and freedom given to Alberta Pacific by the Alberta government during the developmental phase of ecological management induced a cooperative effort by government, industry, and academia throughout the network of Centres of Excellence on Sustainable Forest Management. In all situations, our lack of an adequate monitoring system is slowing the development of ecologically based planning models.

Management by constraint often suggests adversarial approaches to problems. It also suggests that we know answers to complex problems when we don't. In British Columbia, it allowed limited input by the forest industry, reduced the problem-solving atmosphere, and kept many people and much money on the wrong track for a significant period of time.

Although adaptive management is a key strategy in resource management, it is unclear how it operates, it is slow to operate having significant
lag time, and it often generates adversarial feelings. When constraints imposed by management opinion require extensive scientific information to
change them, it is similar to placing a rebuttal to a headline on the last page
of a newspaper.

References

Braumandl, T.F., and M.P. Curran. 1992. A field guide for site identification and interpretation for the Nelson Forest Region. British Columbia Ministry of Forests, Victoria, BC.

Bunnell, F.L. 1998. Next time try data: A plea for variety in forest practices. Interior Forest
Conference, Investing today in tomorrow's forests, Keynote speech, 24-25 March 1997,
Kamloops, BC.

Kootenay Inter-Agency Management Committee. 1997. Kootenay-Boundary Implementation Strategy. Nelson, BC.

Sample, V.A. 1991. Bridging resource use and sustainability: Evolving concepts of both
conservation and forest resource management. Presented at the Society of American Foresters National Convention, Pacific Rim Forestry: Bridging the world, 4 - 7 August 1991,
San Francisco, CA.

World Commission on Environment and Development (Brundtland Commission). 1987.
Our Common Future. Oxford University Press, New York, NY.

Certification and Sustainable Forest Management: A British Columbia Forest Industry Perspective

Dan Jepsen

Introduction

The British Columbia forest sector has experienced significant pressure over the past 10 years to pursue, on the forest lands it manages, better forest management practices, increased protected areas, and better protected fish and wildlife habitat. These pressures have brought revolutionary changes to our industry. Few will question the significant progress we have made. We must anticipate, however, that due to the dynamics of the complex ecosystems in which we operate and shifting public and customer expectations, change will continue to be a part of forest management in British Columbia.

In a mere 10 years we have moved from having less than six percent of the province preserved in parks to almost 12 percent. This is approximately four times the amount of land settled in British Columbia. The Forest Practices Code of British Columbia, revised in 1996 (Province of British Columbia 1996), is one of the most thorough and intensive legislative tools for guiding forest management in the world. The Code provided a much-needed boost in public confidence in British Columbia's forest practices but it significantly increased the management burden, added significant costs to operations, and decreased overall flexibility. The recent Asian financial crisis, coupled with softening markets for British Columbia's forest products in the global marketplace, has added further uncertainty to our industry. The development and implementation of the Code has been an important tool for demonstrating to the public and customers that we in British Columbia are serious about sustainable forest management.

Issues such as clearcutting, old growth harvesting, and the sustainability of the forest resource continue to dominate the public debate. As a result, the Canadian forest industry and more specifically the British Columbia forest industry have come under increasing pressure from the marketplace to independently confirm our commitment to environmentally responsible and sustainable forest management – in other words, third party independent certification.

Forest companies are rising to the challenge by taking the necessary steps to achieve certification of their forest management practices and their environmental management systems. By certifying these aspects of forest operations, companies will be able to prove, with third party verification, that their operations are managed according to principles of sustainable management that conform to internationally accepted standards.

Certification is becoming an important tool for marketing and selling forest products in the global marketplace. Certification can provide customers with independent assurance that their forest product purchases come from well-managed and sustainable forests.

Certification/Sustainable Forest Management

To clarify "certification" and "sustainable forest management," the following two definitions are provided:

Certification: "To attest as being true or as represented or as meeting a standard" (*Webster's Ninth New Collegiate Dictionary*).

Sustainable development: "Development that meets the needs of the present without compromising the ability of future generations to meet their own needs" (World Commission on Environment and Development 1987).

The three main certification systems available to British Columbia forest companies are:

(1) CSA International (formerly Canadian Standards Association) Sustainable Forest Management CAN/CSA Z809-96 SFM Registration
(2) Forest Stewardship Council (FSC) Certification
(3) International Organization for Standardization (ISO) 14001 Environmental Management System Certification (or Registration).

While all three voluntary systems may eventually lead to some form of certification, each initiative has different purposes. The CSA International and Forest Stewardship Council approaches provide certification of a company's sustainable forest management (SFM) while ISO 14001 is a certification of a company's environmental management system (EMS). All three processes have some complementary elements yet they also have different objectives and outcomes. Table 1 provides a summary of the three systems.

Table 1

Comparison of forest certification systems available in Canada.

CSA International	Forest Stewardship Council (FSC)	International Organization for Standardization (ISO14001)
Purpose is to provide credible and recognized process for certifying sustainable forestry in Canada	Purpose is to promote environmentally appropriate, socially beneficial, and economically viable forest management	Purpose is to demonstrate commitment to sustainable forest management through confirmation of environmental management system
Based on six criteria	Based on ten principles	Based on company-specific organizational structure
National standards with international link (ISO) using local values, goals, and indicators	International principles and criteria with standards developed at regional level	International standard. System-based, not compliance-based
Operational but no registrations at this time	Partly operational. Certifications exist but only one regional standard has been approved (Sweden)	Operational with numerous registrations throughout the world
ISO 14000 Series compatibility	No official link with ISO 14000 Series	CSA International compatible
Structured to conform with Canadian Council of Forest Ministers criteria	Does not conform with Canadian Council of Forest Ministers criteria	Designed to address environmental issues
International buyer groups have not generally endorsed	Endorsed by international buyer groups	Commonly accepted by international buyer groups
Registrars independently accredited	Certifiers accredited by Forest Stewardship Council	Registrars independently accredited

Before making a commitment to proceed with certification, forest companies should consider a number of factors that include the following:

- implementation and ongoing administrative and operational costs
- company institutional philosophy. A perception exists that the company loses control of the process (i.e., "the auditor dictates the goals and objectives" and measures performance)
- the commitment necessary to make changes to meet certification requirements
- does the certification system under consideration meet the requirements of customers and consumers?
- what is actually certified must be made clear
- public versus private forest land
- volume-based versus area-based tenure complications
- marketplace demand for certified products and motivation to pay extra for certified products
- product labelling
- chain of custody.

CSA International Sustainable Forest Management System

The CSA International Sustainable Forest Management system was approved in 1996 as Canada's national Sustainable Forest Management system standard. The system is based on the Canadian Council of Forest Ministers' (CCFM) six criteria. The Canadian Council of Forest Ministers' goal for development of Canada's forest is "to maintain and enhance the long term health of our forest ecosystems for the benefit of all living things, both nationally and globally, while providing environmental, economic, social and cultural opportunities for the benefit of present and future generations" (Government of Canada 1998).

The system permits the development of local values, goals, and indicators that are established by the forest company that is applying for certification for a defined forest area. Local indicators must provide measurable features to show criteria are being met. The indicators are then fine-tuned through mandatory extensive public consultation.

The main components of the CSA International Sustainable Forest Management system include:

- commitment to appropriate policy
- public participation in the development and implementation of criteria and indicators and system components such as planning and review
- commitment to continual improvement.

The goal of the CSA International Sustainable Forest Management system is to conform with the criteria of the Canadian Council of Forest Ministers by developing measurable indicators that reflect sustainable forest management practices.

The CSA International system was designed to be compatible with ISO 14001 and as a result many synergies exist between the two standards. ISO 14001, however, is a generic environmental management system while the CSA International system relates specifically to forestry. In many cases the CSA International requirements go beyond what is required for ISO 14001.

Western Forest Products Limited retained KPMG Management Consulting in 1996 to conduct a CSA International Sustainable Forest Management Gap Analysis. The Gap Analysis of Western Forest Product Limited's operations confirmed areas that required attention and is providing an excellent reference document in addressing the gaps and bringing our environmental management system up to standard. At this time Western Forest Products Limited is working on the ISO 14001 certification as a building block to more active pursuit of the CSA International standard.

Forest Stewardship Council (FSC)

Founded in 1993, the Forest Stewardship Council is an international, non-profit organization with headquarters in Oaxaca, Mexico. The Forest Stewardship Council operates a voluntary accreditation program for organizations that provide forestry certification; it does not undertake the certification itself. The Forest Stewardship Council certification system is based on 10 principles that reflect the goal of promoting environmentally appropriate, socially beneficial, and economically viable management of forests. Regional standards are formulated based on the ten principles. In the absence of a regional standard as in British Columbia, however, the certifier's own audit protocol with input from stakeholders forms the basis of the standard.

To date approximately 10 million hectares of commercial forest land worldwide have been certified using the Forest Stewardship Council system. In Canada, the largest land base to receive certification to date is J.D. Irving Limited's Black Brook District in northwestern New Brunswick.

Western Forest Products Limited began assessing readiness for certification in early 1997 in response to increasing customer interest in products from sustainably managed forests.

Western Forest Products Limited retained SGS Forestry Limited of the United Kingdom, one of only six accredited certifiers worldwide, to undertake the pre-assessment under their Forest Stewardship Council accredited Qualifor program. The pre-assessment was completed in the spring of 1998 and compared Western Forest Products Limited's operations against the

10 Forest Stewardship Council Principles and Criteria. The SGS Forestry Limited pre-assessment provided confidence that Western Forest Products Limited had reason to proceed with Forest Stewardship Council certification. Western Forest Products Limited publicly announced its intent to proceed with Forest Stewardship Council certification in early June 1998. The decision to proceed was well received by customers who are interested in independent verification of Western Forest Products Limited's forest stewardship and management practices.

Not surprisingly, the public commitment to pursue Forest Stewardship Council certification has met with some criticism. There are those who argue that the firm should not pursue Forest Stewardship Council certification until regional standards are in place. Western Forest Products Limited is in a unique position to help advance the process of standards development in British Columbia. Western Forest Products Limited has continued to work with SGS Forestry Limited over the past months. SGS Forestry Limited has completed a draft Forest Stewardship Council checklist for Western Forest Products Limited's assessment, through intensive consultation with a group of respected British Columbia academics and a First Nations Hereditary Chief. The draft checklist has been distributed to over 100 international, national, regional, and local stakeholders for input and comment. The total number of people or groups who have received packages of either the checklist or Western Forest Products Limited's statement of forest management is now well over 200. Input on the first draft of the checklist is being carefully reviewed by SGS Forestry Limited to produce a revised SGS Forestry Limited British Columbia Forest Stewardship Council checklist. This will be reviewed by interested stakeholders, the British Columbia Forest Stewardship Council Steering Committee, and representatives of Forest Stewardship Council Canada and the international secretariat. We anticipate that SGS Forestry Limited's British Columbia Forest Stewardship Council checklist, which would be used in Western Forest Products Limited's main assessment, will be ready in 1999.

International Organization for Standardization (ISO 14001)
The ISO 14001 standard allows forest companies to demonstrate their commitment to sustainable forest management and achieve certification (registration). This standard examines environmental management by assessing the entire organizational system or framework to determine how environmental issues fit the overall management of a company. The standard enables a company to review the significant environmental aspects of its operations and then apply a set of procedures and controls to manage these aspects. This is a risk-based strategy and one that allows companies to meet their legal requirements (namely the Forest Practices Code of British Columbia) while demonstrating due diligence and risk management. ISO 14001

is a standard that allows forest companies to manage risk. It also allows companies to set their own targets for continual improvement with respect to environmental aspects. This can include extra regulatory commitments.

The ISO 14001 standard enables forest companies to achieve continual improvement by incorporating all operational levels including planning, implementation and organization, checking and corrective action, and management review. Although a whole series of 14000 documents exist, ISO 14001 is the only standard against which an organization's environmental management system may be audited for registration. To aid forestry organizations in developing an environmental management system to cover the scope of forestry activities, ISO 14061 has been created as an informative reference document. It is intended for use in conjunction with ISO 14001 by providing a link between the environmental management system and the range of forestry applications that a forestry organization can consider when implementing its system.

The main components of ISO 14001 are to:

- establish an Environmental Policy
- establish goals and objectives that reflect the Environmental Policy
- establish a formalized Environmental Management System (EMS)
- measure performance against Environmental Policy and the Environmental Management System.

Adopted as a final standard in late 1996, ISO 14001 accreditation processes are now underway for registrars, auditors, and trainers. At this time new registrars are becoming accredited and there are indications that many forestry companies will pursue ISO 14001.

ISO 14001 has not yet received the same recognition as the Forest Stewardship Council and the CSA International Sustainable Forest Management system, but some opportunity seems to be available for integration with the CSA International Sustainable Forest Management standards. It is still unknown what the response of the market will be to this particular standard and whether environmental groups will support it.

Unresolved Issues and Concerns with Certification Systems in Canada

Forest Stewardship Council (FSC) Unresolved Issues and Concerns

- political process
- implications on clearcutting practices unclear
- principle #9; awaiting international vote of new draft revision of "High Conservation Value Forests"

- no regional standards in place
- demanding with regard to environmental and social requirements; weak on economic viability; need balanced decision-making and representation at regional levels
- current British Columbia treaty process and related uncertainty.

CSA International Sustainable Forest Management
Unresolved Issues and Concerns

- concern that this system relies heavily on local public input to identify indicators and establish performance objectives for a wide range of social, economic, and ecological parameters; leaves sustainable forest management interpretation to local multi-stakeholder-driven process
- not well received by environmental non-governmental organizations. How will this affect credibility?
- standards are extremely technical; lack of prescriptive language in indicators
- not generally embraced by customers/consumers in the international marketplace due to lack of knowledge
- does not provide label for products.

International Organization for Standardization (ISO) 14000
Unresolved Issues and Concerns

- not well received by environmental groups
- for maximum success must apply to woodland operations, not just to offices
- does not provide label for products
- not generally embraced by customers/consumers in the international marketplace at this time.

Conclusion
Despite considerable efforts by forest companies and the British Columbia government to change forest management practices and actively demonstrate good forest land stewardship, the global market is requesting a more formalized process for independently demonstrating our commitment to sustainable forest management. Certification by an internationally recognized, independent auditor can provide the objective support necessary for the British Columbia forest industry to demonstrate that its practices meet or exceed international standards. Certification is a means of demonstrating sustainable forest management according to standards that the public and British Columbia forest products customers will accept. The overall goal

of certification is to develop measurable indicators that allow a more objective assessment of our sustainability efforts through third party verification that is easily interpreted by customers and the public alike.

Western Forest Products Limited is working to achieve certification of its managed forests under all three systems. At Western Forest Products Limited, responsible forest stewardship is essential to survival and success as a company. Western Forest Products Limited is committed to sustainable forest management on the public and private land in its care.

References

Government of Canada. 1998. Canada Forest Accord. National Forest Congress, 29 April to 1 May, Ottawa, ON. On-line: <http://nfc.forest.ca>.

Province of British Columbia. 1996. Forest Practices Code of British Columbia Act, Revised Statues of British Columbia, Chapter 159.

World Commission on Environment and Development. 1987. Our common future. Oxford University Press, New York, NY.

British Columbia Forestry and the Forest Stewardship Council's Certification Framework

Marcelo Levy

Abstract

The Forest Stewardship Council (FSC) is an international, independent, non-profit, non-governmental organization. Its mission is to promote environmentally appropriate, socially beneficial, and economically viable management of the world's forests. It does so through a voluntary accreditation program for certification of forest management. The Forest Stewardship Council's main functions are to accredit certification bodies worldwide and support national and/or regional initiatives to develop forest management standards within the framework of the Forest Stewardship Council Principles and Criteria of Forest Management.

The National Initiatives Program decentralizes the work of the Forest Stewardship Council and fosters local involvement in the development of standards. Through the Accreditation Program the Forest Stewardship Council evaluates, accredits, and monitors forest product certifiers to ensure that market place claims are consistent and reliable. These programs make the Forest Stewardship Council certification program flexible and responsive to local ecological, social, and economic circumstances while at the same time providing a label recognized worldwide that assures consumers that wood products come from a well-managed forest.

Certification aims to link the environmentally/socially conscious consumer to producers willing to incorporate consumer concerns into their management practices. The two main goals of certification are to improve forest management and to provide market access for certified products.

In this paper, I provide a background to forestry certification, the Forest Stewardship Council, and British Columbia's Forest Stewardship Council initiatives to develop regional standards. I discuss the implications of certification to forestry in British Columbia.

Background

Certification of forest products and management practices is still a relatively new process that emerged in the late 1980s after calls for boycotting the use of certain wood products. This certification program aims at linking the environmentally and/or socially conscious consumer to producers willing to incorporate those concerns into their management practices. Implicit in a certification program are the following assumptions: consumer purchasing decisions can be influenced by differentiating products based on environmental and social concerns; producers' practices can be influenced by market-driven incentives arising from such concerns; increased price and/or market share will provide producers a sufficient margin to adopt improved management practices (Cabarle 1994).

The Forest Stewardship Council (FSC), an international, independent, non-profit, non-governmental organization, has as its mission promoting environmentally appropriate, socially beneficial, and economically viable management of the world's forests. It does so through a voluntary accreditation program for certification of forest management (Forest Stewardship Council 1994a). The Forest Stewardship Council was founded in 1993 by a diverse group of representatives from environmental institutions, the timber trade, the forestry profession, indigenous peoples' organizations, community forestry groups, and forest product certification organizations from 25 countries. This diverse group developed the Forest Stewardship Council's Principles and Criteria that apply to tropical, temperate, and boreal forests. The Principles and Criteria cover broad issues such as land tenure, the reduction of environmental impacts, optimal use of forest products, and written management plans.

The Forest Stewardship Council's main functions are to accredit and evaluate certification bodies worldwide and to support the development of national and regional forest management standards within the framework of the Forest Stewardship Council's Principles and Criteria. Through its accreditation program, the Forest Stewardship Council evaluates, accredits, and monitors forest product certifiers to ensure that marketplace claims carry a consistent and reliable set of values as defined by clear principles and criteria. The National Initiatives Program decentralizes the work of the Forest Stewardship Council and fosters local involvement in the development of national or regional standards. These two programs make the Forest Stewardship Council's certification framework flexible and responsive to local ecological, social, and economic circumstances while at the same time providing a label, recognized worldwide, that assures consumers that wood products come from a well-managed forest. The inclusive and transparent manner in which the Forest Stewardship Council operates offers a promising and practical alternative to help define what good forest management is and to communicate this message effectively to increasingly environmentally and socially concerned consumers worldwide.

The Forest Stewardship Council Certification Framework

The Forest Stewardship Council has been set up to oversee a credible process of independent certification of well-managed forests. The Forest Stewardship Council fulfills this role through the Accreditation Program, the National Initiatives Program, and through a series of policies to address specific issues such as certification of composite products and group certification. Forest certification within the Forest Stewardship Council framework involves assessing forest management operations against ecological, social, and economic standards. The Forest Stewardship Council includes chain of custody certification by which the products from any forest certified under this process would then be eligible to carry a label in the shop or yard where they are sold, ensuring that they have come from a certified forest.

The Accreditation Program

Accreditation refers to the process of evaluating, endorsing, and monitoring organizations that independently conduct forest management assessments and chain of custody audits. The Forest Stewardship Council is itself an accreditation body, i.e., a "certifier of certifiers." The role of accreditation is to provide credible assurance that certifiers are competent and independent in providing specified certification services. An accreditation body achieves this by examining and evaluating the certifier's organizational structure, practices, and resources against operational norms. Once a certifier is accredited, the accreditation body continues to monitor and, if necessary, regulate the activities of certification bodies within their area of competence. The Forest Stewardship Council bases its evaluation of a certifier's organizational competency on adherence to the Forest Stewardship Council Guidelines for Certifiers and the Forest Stewardship Council's Principles and Criteria of Forest Stewardship (Forest Stewardship Council 1997b). The main objectives of accreditation are "to provide consistency among certifiers and standards, to ensure credibility of certification programs to the public, and to verify the integrity of a certifier's claims" (Ervin et al. 1996).

With the exception of Sweden and the United Kingdom, whose national standards have been approved by the Forest Stewardship Council, certifiers are currently using the Principles and Criteria to conduct their assessments. Once regional or national standards are approved by the Forest Stewardship Council, certifiers are required to use those standards as the basis for their assessments. Additionally, the Forest Stewardship Council requires certifiers to consult with local stakeholders and with local Forest Stewardship

Council initiatives that are developing standards for a given area (Forest Stewardship Council 1994b). These procedures ensure that "while the certification and accreditation decisions are based on global principles, they are grounded in local circumstances" (Ervin et al. 1996).

The National Initiatives Program

The main objective of the National Initiatives Program is to "decentralize the work of the Forest Stewardship Council and to encourage local participation" (Forest Stewardship Council 1997a). The Forest Stewardship Council is a fairly decentralized organization that encourages and promotes the establishment of Forest Stewardship Council national or regional initiatives that agree with the Forest Stewardship Council's mission. The main function of national or regional initiatives is to develop standards within the global framework of the Forest Stewardship Council's Principles and Criteria. The objective of this structure is to "have consistency in standards setting so that standards developed in different countries can be evaluated in the global context" (Forest Stewardship Council 1997a). For certification to be a credible and rigorous process, forest management audits should be based on locally applicable standards. The existence of locally defined forest management standards contributes to a fair, transparent, and systematic certification process (Forest Stewardship Council 1998a).

The national or regional initiatives must comply with a series of protocols and guidelines. The Forest Stewardship Council then establishes formal agreements with regional groups and endorses them as Forest Stewardship Council Working Groups. The Forest Stewardship Council Working Groups must have a composition similar in balance to the Forest Stewardship Council Board of Directors (i.e., equal representation of at least environmental, social, and economic interests; in Canada, Aboriginal representation was added). The Forest Stewardship Council Working Groups must remain transparent and participatory, allowing access to the process from any interested party. They must also have clear grievance procedures established for both the working group as well as for any forest management standards or documents that are developed (Forest Stewardship Council 1995).

The Standards Development Process

The Working Group develops national or regional standards according to the procedures outlined in the "Forest Stewardship Council Process Guidelines for Developing Regional Certification Standards." The document states that "to ensure the consistency and integrity of standards in different regions around the world, the FSC formally endorses those standards which clearly meet all FSC requirements, including the process leading to their development" (Forest Stewardship Council 1998a).

The requirements for Forest Stewardship Council endorsement of standards include:

- Compatibility with the Forest Stewardship Council Principles and Criteria. At a minimum, regional standards must show how each principle and criterion is met.
- Compatibility with local circumstances. The regional Forest Stewardship Council standards are expected to provide clarity, definition, and consensus. They will be much more detailed than the Principles and Criteria. They must be fully compatible, but they cannot be less strict. They may be more strict if regional laws or a regional consensus justifies it.
- Consultative process. Participants are entitled to shared ownership of the process, fair decision-making procedures, transparency and accountability, adequate participation and representation, and clear grievance procedures.
- Harmonization. Regional standards will provide compatibility with similar and/or neighbouring regions.

The Canadian Context
The Forest Stewardship Council Canadian Initiative was founded in January 1996 and was endorsed by the Forest Stewardship Council in May 1998. A Board of Directors was elected by Forest Stewardship Council members based in Canada. Formal agreements exist between Forest Stewardship Council Canada and the Forest Stewardship Council. While one of the key functions of the working group is to develop standards, it was decided that because Canada is a large, diverse country, standards would be developed at regional (i.e., sub-national) levels. In this context, the main tasks of Forest Stewardship Council-Canada have been to support regional efforts of standards development, to raise funds, to increase the Forest Stewardship Council profile in Canada, to increase membership, and to further develop the organization. Forest Stewardship Council-Canada must provide national endorsement to the regional standards before they are submitted to the head office of the Forest Stewardship Council in Mexico. Forest Stewardship Council-Canada requires that regional groups follow Forest Stewardship Council procedures in their standards development and have a structure similar to Forest Stewardship Council-Canada. At this stage, three active Forest Stewardship Council regional groups are engaged in standards development processes: Maritimes, Great Lakes-St. Lawrence (Ontario), and British Columbia.

The Maritimes region has been working for over two years. At the outset, through a wide public consultation process, a Regional Committee was elected with representatives from nine categories (including the required but not limited to Aboriginal, economic, environmental, and social interests). A Standards Writing Committee was appointed that added technical advisors from two government departments. Regional approval of the standards rests with the elected Maritimes Regional Committee. Their third draft of the standards was reviewed by the Forest Stewardship Council-Canada Board, which required further work on harmonization with neighbouring regions. It was expected that the standards would be ready to be submitted to the Forest Stewardship Council International Board at their January 1999 meeting.

The Great Lakes-St. Lawrence Regional Group (Ontario) has been working for almost two years, has undergone two rounds of public consultation, and is getting ready to field test their standards. Their structure is similar to the Maritimes Region as they have a Steering Committee that represents various parties in the region and a technical committee that is developing the standards. Field testing occurred in the Spring of 1999 so that standards would be ready for submission to the Forest Stewardship Council in Mexico at their June 1999 International Board Meeting.

The British Columbia Regional Initiative was founded in the summer of 1996 and an interim steering committee composed of Forest Stewardship Council members or applicants who are resident in British Columbia was established at the April 1997 meeting. By spring 1998, the steering committee began establishing protocols and procedures for the standards process in British Columbia while at the same time encouraging membership in the Forest Stewardship Council and making appointments to fill the vacant Steering Committee seats. Following the model of the Maritimes Region, a Standards Drafting Committee was established at the September 1997 meeting. This committee worked to prepare a first draft of the standards by late fall 1998, to be released for public comment. The standards will cover both coastal and interior British Columbia.

The British Columbia Regional Steering Committee makes a recommendation regarding a British Columbia standard to Forest Stewardship Council-Canada, the body officially recognized by the Forest Stewardship Council internationally. Forest Stewardship Council-Canada then forwards its recommendation to the Board of Forest Stewardship Council international for approval.

Forest Management Certification and Regional Initiatives

The Forest Stewardship Council has a process for providing certification in the absence of endorsed national or regional standards. Certifiers can use generic standards (approved by the Forest Stewardship Council) that they must, through a consultative process, tailor to the region where the assessment will take place. Certifiers are required to consult with endorsed Working Groups (in Canada, Forest Stewardship Council-Canada), with regional initiatives, with Forest Stewardship Council members in the region, and with a broad range of stakeholders. In this way, the generic standards are adapted to local conditions. These standards are then used to audit forest operations in the region. Once regional or national standards are formally endorsed by the Forest Stewardship Council at the international level, all certifiers are required to use those standards as the basis for the assessments. Companies certified with the generic standards will have to comply with the new regional standards within one year.

Other Forest Stewardship Council Policies

To facilitate the certification of small landholdings, the Forest Stewardship Council has developed a policy for group certification. Under the Group Certification Program, woodlot owners can have their woodlots managed jointly by a group entity (an association, a cooperative, or by appointing an individual responsible for the group). The group entity holds any group certificate that is issued. The group entity is responsible to the certification body for ensuring that the requirements of the Forest Stewardship Council Principles and Criteria for Forest Stewardship are met in the forest lands covered by the certificate. Although small land holdings were concerned about the cost of certification, it is important to note that about a third of the total number of certificates worldwide were issued to operations smaller than 1,000 hectares.

Another important policy is the one guiding the use of the Forest Stewardship Council logo on products. For solid wood products, the requirement is that one hundred percent of the product is certified by a Forest Stewardship Council accredited certifier as coming from a certified forest. For other products, at least seventy percent by volume (assembled products such as furniture and musical instruments) or seventy percent by weight (pulp and paper products) of the wood and/or virgin fibre contained in the product is certified by a Forest Stewardship Council accredited certifier as coming from a certified forest. When the products include recycled or non-wood materials, the policy states that "pulp, paper and other assembled products which contain up to 75% of recycled and non wood materials may carry the Forest Stewardship Council Logo as long as the remainder meets the 70 percent certified content" (Forest Stewardship Council 1997b).

Conclusion

The Forest Stewardship Council's approach provides a system that is accountable to the various stakeholders in a region or country by involving them in an equitable and transparent standards development process while at the same time providing a framework for consistent worldwide certification.

A complex process such as forest certification can be successful only if those who have a stake in forest management are involved. The Forest Stewardship Council provides a framework that includes internationally agreed

Principles and Criteria and local involvement at the standards development stage.

The accreditation program ensures that certifiers are competent and that they will evaluate forest management operations consistently on a global scale. To take into account local conditions, certifiers are required to use regional or national standards endorsed by the Forest Stewardship Council as they become available. In the absence of local standards, certifiers are required to consult with stakeholders in a region to incorporate local ecological, social, cultural, and economic circumstances. In this way, certifiers can assess forest management operations in light of local values within a global framework.

The Forest Stewardship Council system provides consumers with the opportunity to express their values through their purchasing habits. The key to the success of forest certification is credibility. "Ultimately, consumers will determine the credibility of an accreditation system according to how well the system represents their particular values" (Ervin et al. 1996). The Forest Stewardship Council process is transparent and inclusive. It involves local communities in the development of standards, issues an internationally recognized label to well-managed forests, and provides consumers with options to support good forest management through their purchasing habits. The Forest Stewardship Council's approach effectively links local ecological, social, and cultural environments to the global responsibility for the health of forests worldwide.

References

Cabarle, B. 1994. Forest management certification and the Forest Stewardship Council. Pp. 102-112 *in* Natural Resources Canada, Timber supply in Canada: Challenges and choices.

Ervin, J., C. Elliott, B. Cabarle, and T. Synnott. 1996. Accreditation process. Pp. 42-53 *in* V. Viana, J. Ervin, R. Donovan; C. Elliott, and H. Gholz (eds.). Certification of forest products: Issues and perspectives.

Forest Stewardship Council. 1994a. Statutes. Forest Stewardship Council - Canada, Toronto, ON. On-Line: <http://www.fscoax.org/principal.htm>.

Forest Stewardship Council. 1994b. FSC Guidelines for certification bodies. *In* FSC Statutes Appendix A. Forest Stewardship Council - Canada, Toronto, ON. On-Line: <http://www.fscoax.org/principal.htm>.

Forest Stewardship Council. 1995. Protocols for endorsing national initiatives. Forest Stewardship Council - Canada, Toronto, ON. On-Line: <http://www.fscoax.org/principal.htm>.

Forest Stewardship Council. 1996 (revised). Principles and criteria for Forest Stewardship. Forest Stewardship Council - Canada, Toronto, ON. On-Line: <http://www.fscoax.org/principal.htm>.

Forest Stewardship Council. 1997a. Status of national and regional certification initiatives. Forest Stewardship Council - Canada, Toronto, ON. On-Line: <http://www.fscoax.org/principal.htm>.

Forest Stewardship Council. 1997b. Accreditation manual (Draft). Forest Stewardship Council - Canada, Toronto, ON. On-Line: <http://www.fscoax.org/principal.htm>.

Forest Stewardship Council. 1997c. Board decision on percentage based claims. Forest Stewardship Council - Canada, Toronto, ON. On-Line: <http://www.fscoax.org/principal.htm>.

Forest Stewardship Council. 1998a (revised). Process guidelines for developing national or regional certification standards. Forest Stewardship Council - Canada, Toronto, ON. On-Line: <http://www.fscoax.org/principal.htm>.

Forest Stewardship Council. 1998b. Group certification: FSC guidelines for certification bodies. Forest Stewardship Council - Canada, Toronto, Ontario. On-Line: <http://www.fscoax.org/principal.htm>.

The Effect of Patch Size on Timber Supply and Landscape Structure

John Nelson and Ralph Wells

Abstract

The ATLAS/SIMFOR simulation models are used to forecast timber supply and landscape pattern response to six patch management options on a 31,000-ha landscape unit in the Golden Timber Supply Area in southeast British Columbia, Canada. The six options are as follows: (1) no opening size limits, (2) 40-ha openings with 20-year adjacency, (3) 100-ha openings with strict adjacency, (4) 100-ha openings with flexible adjacency, (5) 100-ha openings with strict adjacency and long rotations, and (6) 100-ha openings with flexible adjacency and long rotations. We conclude that when attempting to force a target patch size distribution onto a very different landscape pattern, we should expect decreased harvest levels, especially in the short term. If we work with the existing landscape when introducing new patterns that contain a variety of patch sizes, and if we are prepared to compromise on the target structure, our conclusion is that we can achieve increased yields in the short term by abandoning the adjacency rule. Regardless of the patch size strategy, rotation ages of approximately 100 years do not lead to large, old patches. We observed that the forest developed a normal age-class structure on the harvested land (in zero to 100 years), while isolated reserves and inoperable lands grew very old – with a huge age gap between the two (i.e., a generation gap). This is of concern because it forecloses options to recruit old growth should this be necessary for conservation or in response to natural disturbances. If we want large, old patches on a continuous basis, located throughout the forest, then longer rotations and an appropriate patch size distribution are necessary. Most importantly, we need to think of patch size, adjacency, and rotation ages as objectives rather than "hard" constraints and be prepared to deviate from these objectives from time to time.

Introduction

Over the past 30 years, the typical landscape cutting pattern in British Columbia has evolved through three stages: (1) progressive clearcutting, (2) the three-pass cut/leave system, and (3) small openings combined with adjacency and green-up. The move towards smaller openings was first driven by biological and environmental concerns and later by aesthetic and other social values. In moving from one extreme to another, however, we have created new environmental and social problems. These include the availability of short-term timber supply, managing access to dispersed harvest areas, and forest fragmentation.

In 1995, following several years of negotiation, the government of British Columbia released regional land use plans prepared by the Commission on Resources and the Environment (CORE 1995a,b). These plans specify land use objectives for approximately eight million ha of the Nelson Forest Region. Part of the implementation strategy is to explore forest management alternatives, including partial cutting and patch size distributions. This paper examines patch size alternatives in Landscape Unit 26 of the Golden Timber Supply Area located in southeast British Columbia, Canada. This is a challenging landscape for work of this nature due to historical cutting patterns. It was chosen as a case study where patch management alternatives could be simulated with the ATLAS/SIMFOR planning models.

The objective of the study is to investigate how alternative patch management strategies affect timber supply and landscape structure. We are interested in quantifying the relative differences among patch management alternatives rather than establishing absolute harvest levels for the study site. To explore patch management, six simulations are made: (1) no limits on opening size, (2) 40-ha openings with strict adjacency, (3) 100-ha openings with strict adjacency, (4) 100-ha openings with strict adjacency and long rotations, (5) 100-ha openings with flexible adjacency, and (6) 100-ha openings with flexible adjacency and long rotations. We first describe Landscape Unit 26 and the six patch management simulations. Second, we present results and discuss the patch size simulations on timber supply. Next we discuss the effects of patch management options on landscape structure and finally we present conclusions and recommendations for further analysis.

Methodology

Description of Landscape Unit 26 and the Data

Landscape Unit 26 has a disturbance history of fire and harvesting that includes both large and small clearcuts (Figure 1). The unit totals 31,729 ha, of which 16,166 ha (51 percent) is operable and 15,563 ha (49 percent) is inoperable or reserved from harvest. Approximately 85% of the inoperable area is contained in large polygons above the timber line (Figure 1). In the ATLAS/SIMFOR project, 2,271 harvest units averaging eight ha were manually designed for the area. The objective of the harvest unit design phase was to create small, feasible harvest units that, if necessary, could be combined into larger openings. Many of these small units were grouped to form larger openings (40 and 100 ha) for the computer simulations.

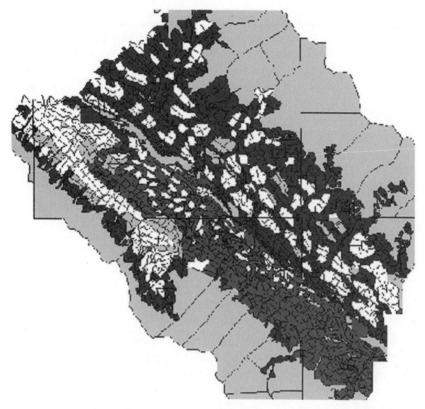

Figure 1. Map of the study area showing polygon boundaries: Landscape Unit 26, Golden Forest District, southeast British Columbia. Lighter shades represent young stands; darker shades represent older stands.

Description of the Patch Size Simulations

All simulations are for 240 years, using 10-year planning periods. The objective is to maximize short-term harvest levels, subject to a controlled decline in harvest levels to the long-term, steady-state yield. In most simulations, the priority is to harvest oldest stands first. We do not model natural disturbance in the simulations, so stands not harvested continue to age over time. It is not our intent to accurately predict future forests in Landscape Unit 26. Rather, we wish to evaluate the implications of various policy options if they are carried out over an extended time. We do, however, discuss the implications of natural disturbance for the various policies, especially in relation to recruitment of old stands.

In total, we generated six simulations to investigate how patch size affects timber supply and landscape structure (Table 1).

Runs 1 and 2 establish baselines to which the patch management options can be compared. Run 1 (*unconstrained*) has no opening or adjacency

Table 1

Summary of constraints and parameters used in six ATLAS/SIMFOR simulations within Landscape Unit 26, Golden Forest District, southeast British Columbia.

Run	Descriptive code	Block size[1] (ha)	20-year adjacency/ green-up	Minimum harvest age in years (existing/regen)
1	*unconstrained*	n/a	No	100/100
2	*40 ha_blocks*	40/10	Yes	100/100
3	*hard_constraints*	100/10	Yes	100/100
4	*soft_constraints*	100/10	No	100/100
5	*hard_constraints_long_rotation*	100/10	Yes	120/180
6	*soft_constraints_long_rotation*	100/10	No	120/180

1 Wildlife biologists specified 10-ha openings to be used in the valley bottom in runs 2-6. The 40- and 100-ha openings are located on the upslopes of the study site.

constraints; it is free to harvest polygons subject to only harvest flow constraints. Run 2 (*40_ha blocks*) represents the status quo in British Columbia, 40-ha blocks with 20-year adjacency. The 40-ha openings are aggregations of the 10-ha openings and are confined to the upslope portions of the site. In the valley bottom, we specified 10-ha openings for all runs. This represents about 45 percent of the productive area of Landscape Unit 26. Runs 3 through 6 use both 10-ha and 100-ha openings. The 100-ha openings are on the upslope portion of the site and are aggregations of the 10-ha polygons.

Run 3 (*hard_constraints*) strictly enforces adjacency and the requirement that all polygons within the 100-ha openings are harvested simultaneously. This run models the consequences of forcing a new pattern on the landscape without compromise. To avoid the strict adjacency rule and the requirement that a block be completely harvested in one time period, run 4 (*soft_constaints*) has the 100-ha blocks sorted into five zones based on non-adjacency. This leads to a scheduling problem similar to the four-colour theorem method described by Nelson and Errico (1993). The harvest priority is then zone 1, zone 2, zone 3, and so on as shown in the example in Figure 2. Within these zones, the harvest priority was oldest first. The result is a cutting pattern very similar to the adjacency rule although minor violations can and do occur. Within the 100-ha openings, only those polygons above the minimum harvest age are cut. In the case of 100-ha blocks that contain both young and old stands, this results in openings that are smaller than the 100-ha target. Run 4 models the consequences of gradually implementing a new pattern on the landscape (i.e., compromises on opening size and adjacency).

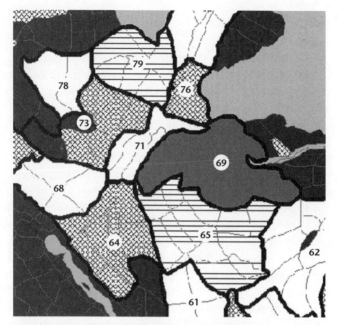

Figure 2. An example of 100-ha blocks sorted by a non-adjacency rule that establishes harvest priorities. In this example from Landscape Unit 26, Golden Forest District, southeast British Columbia, double hatched lines represent first priority blocks, horizontal hatches represent second priority blocks, no hatches represents third priority blocks, and so on.

Runs 5 and 6 are similar to runs 3 and 4 except that, in an effort to continuously recruit stands into the older age classes, we introduced long rotation ages. In the absence of long rotations we observed that the only old stands remaining at the end of the planning horizon were those contained within reserves, and the age class structure of the remaining stands closely approximates a normal forest. These old forests are at risk to natural disturbance and, if lost, could take 50 to 100 years to replace unless the recruitment issue is explicitly addressed. In run 6 (*soft_constraints_long rotations*), we manually assigned about 30 percent of the 100-ha blocks to stands that are either eligible or already on long rotations. The purpose was to provide a continuous flow of stands through the 100 to 180 year age class. We needed about 30 percent of the 100-ha blocks on long rotations to achieve this target. This results in cutting about 1.67 percent of this area every decade (30 percent of the area divided by 18 age classes; i.e., a uniform age class distribution for the long-rotation stands).

Results and Discussion

Timber Supply

Harvest schedules for the six simulations are shown in Figure 3, and the timber volumes (expressed as percentages relative to run 2 [*40_ha blocks*]) are summarized in Figure 4. Run 1 (*unconstrained*) produces the highest volumes, especially in the short and medium term. Run 3 (*hard_constraints*) produced a significant increase in the short term (13 percent) because the 100-ha blocks have freed numerous polygons that were previously constrained by adjacency. This short-term gain comes at the expense of the medium term (-15 percent) when diverse ages within the remaining 100-ha blocks are limiting timber supply. Run 4 (*soft_constraints*) generated a large increase in the short term (25 percent) and a modest increase in the medium term (4 percent) since it was able to harvest all polygons above the minimum rotation age within each 100-ha block. As Figure 3 shows, the harvest schedules for runs 1 through 4 begin to converge near the end of the planning horizon when the forest has reached its target structure and the large patches are cycling with little or no structural limits. All harvest schedules have low points near decades eight to nine that are caused by our objective to maximize short-term volume combined with the shortage of mature timber during the transition from existing stands to regenerated stands. In the presence of opening size constraints and longer rotation ages, these low points become more pronounced.

Run 5 (*hard_constraints_long_rotations*) increased existing and regeneration rotation ages from 100 years to 120 years for existing and 180 years for regeneration. The rotation age for existing stands was not increased to 180

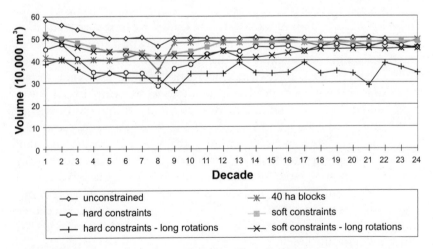

Figure 3. Harvest schedules for six ATLAS/SIMFOR simulations within Landscape Unit 26, Golden Forest District, southeast British Columbia.

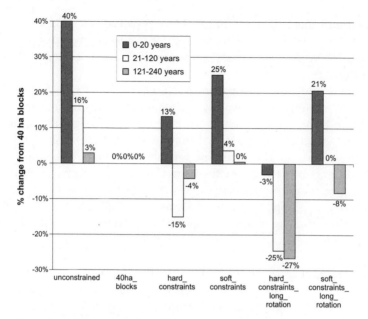

Figure 4. Timber volume produced relative to run 2 (*40ha_blocks*) for the short, medium, and long term using ATLAS/SIMFOR simulation modeling in Landscape Unit 26, Golden Forest District, southeast British Columbia.

years because this proved too restrictive. It was felt that achieving the structural target within the second rotation was a reasonable compromise. Therefore short-term volumes are relatively close to the base case (-3 percent), but the long-term and very long-term volumes are much lower than the base case (-25 percent for long term and -27 percent for very long term). These long-term reductions were expected and correlate well with the reduction in mean-annual increment associated with the incremental change in rotation age.

The soft constraints in runs 4 and 6 produced a variety of opening sizes and patches. The harvest priority disperses the harvest similar to the adjacency rule. Adjacency, however, is not strictly enforced nor is the requirement that the larger, 100-ha blocks be harvested in their entirety. This results in higher timber harvests because mature polygons that are arranged in an approximately adjacent pattern can be harvested immediately. By not forcing the 100-ha block pattern, we have considerable flexibility to harvest within the existing landscape structure. The resulting pattern is a compromise between the current landscape structure and the 100-ha blocks.

For run 6, it was difficult to establish and maintain a uniform age class structure in the 100- to 180-year group. The beginning inventory and the long rotations lead to an age class gap in this group around 120 years that persists until about 180 years. In the absence of an adequate age class structure in the initial inventory, there is no feasible solution to this problem. Beyond 180 years, the age class problem resolves itself, leading to the desired structure.

Attempting to force a new pattern on the landscape, such as large patches over small openings, is equivalent to adding constraints to the harvest. In Landscape Unit 26, the impacts of the large opening constraints are greater than the current adjacency rule. Wallin et al. (1994) reported similar trends in their study on altering landscape structure. The more rigidly these constraints are applied, the greater the impacts on timber supply. In our study, only 55 percent of the productive area was subjected to the large blocks. If large blocks were applied to the remaining 45 percent, and we assume that this area has a similar initial landscape pattern, the impacts on timber supply would be greater than those reported here.

Landscape Structure

In the absence of harvesting constraints, the year 1 landscape pattern of seral patches is largely preserved in run 1 (*unconstrained*; Figure 5a). The pattern is distributed over a younger range of seral stages because of the 100-year-rotation length applied. In run 2 (*40_ha blocks*; Figure 5b) block size limitations result in a highly homogenous pattern, with most patches ranging from 10 to 40 ha. In runs 3 through 6, the landscape pattern of seral stages reflects the 10-ha to 100-ha block pattern that was the objective for these scenarios (Figure 5c-5f).

Because the same reserve strategy was imposed in all runs, the area and distribution of old stands (> 250 years) was nearly identical for all runs (Figure 5) and represents about 11 percent by area in all scenarios. This reserve amount falls within the range of current guidelines in British Columbia (BC Forest Service 1995). Much of the reserve area occurs in high elevation inoperable stands (Figure 5), which is also reflective of current guidelines (BC Forest Service 1995). Thus the reserve strategy used in our scenarios is a plausible one in the current regulatory environment in British Columbia. In reality, the amount of old reserve would be less since some stands would eventually be lost through natural disturbance.

When we evaluated seral stage distributions we found they did not vary substantially for most seral stages. For stands < 121 years, seral stage distributions were similar among runs. This reflects the normalizing of age classes expected in an area harvested under an even timber flow strategy. As noted above, because old stands (> 250 years) are based on static reserves, they are also similar for all runs.

Unlike other seral stages, area in late seral stands (121 to 250 years) varied substantially among runs (Figure 6). After 20 periods, run 1 (*unconstrained*) had almost no area in late seral stands. This is not a surprise because the

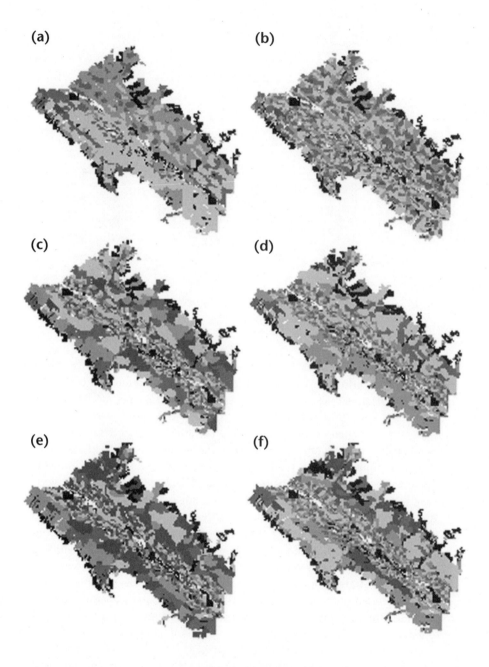

Figure 5. Seral patterns at period 20 of ATLAS/SIMFOR simulations for (a) *unconstrained*, (b) *40ha_blocks*, (c) *hard_constraints*, (d) *soft_constraints*, (e) *hard_constraints_long_rotation*, and (f) *soft_constraints_long_rotation* within Landscape Unit 26, Golden Forest District, southeast British Columbia. Lighter shades represent young stands; darker shades represent older stands.

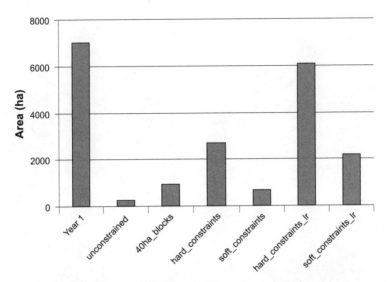

Figure 6. Area in late seral stands (121-250 years) resulting from six ATLAS/SIMFOR simulations within Landscape Unit 26, Golden Forest District, southeast British Columbia.

rotation age applied was 100 years, so over time we expect ages to range between zero and 100 years. Runs 2 and 4 (*40_ha blocks* and *soft contraints*) also had relatively less area in late seral stands, again because stands are harvested on a 100-year rotation. Runs 3 and 5 (*hard_constraints* and *hard_constraints_long rotation*) had the greatest area in late seral stands. In run 3, this occurs despite the 100-year rotation age because hard adjacency rules in conjunction with the block size pattern constrain harvest, allowing some stands to recruit into the late seral age class. In run 5 (*hard_constraints_ long_rotation*), a combination of block size, adjacency, and longer rotation age combine to result in substantially more area in late seral stands than in all other scenarios. Run 6 (*soft_constraints_long rotation*) yielded an amount of late seral age class similar to run 4 (*hard_constraints*). The primary difference is that in run 6, late seral patches were achieved through an explicit recruitment strategy (long rotation), while in run 4 late seral patches were an unintended outcome. As seen in the previous section, a substantial decline in timber supply occurred in the short- to medium-term in run 4 (*hard_constraints*) relative to run 6 (*soft_constraints_long_rotation*) (Figure 3).

During our analyses we observed that in Landscape Unit 26, periods 3 and 10 were significant periods for recruitment of large patches into the late seral age class (Figure 7). The patch that recruits in period 3 results from a large fire-origin patch; the patch that recruits in period 10 is from a large clear-cut harvest. These periods represent substantial opportunity to influence both landscape pattern and seral stage distributions for this landscape

Period 3   **Period 10**

Period 8

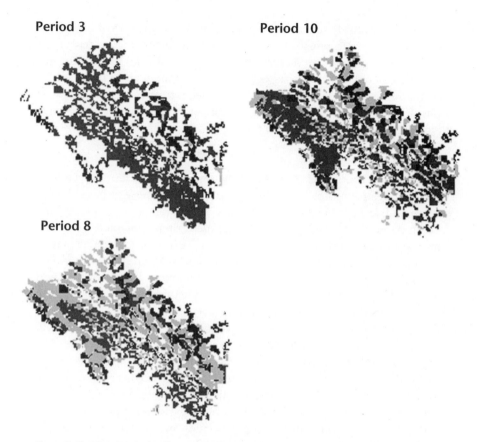

Figure 7. Seral patterns for three periods resulting from ATLAS/SIMFOR simulations using 40-ha block and 20-year adjacency constraints (Run 2: *40ha_blocks*) within Landscape Unit 26, Golden Forest District, southeast British Columbia. Lighter shades represent younger stands; darker shades represent older stands.

unit. Conversely, by period 8, a short-term decline in both timber supply (Figure 3) and late seral patches (Figure 7) was observed in all scenarios as harvesting of the large patch that recruited in period 3 was completed before the recruitment of the large patch in period 10.

Three primary lessons emerge from evaluation of landscape pattern in the six scenarios. First, a 40-block size limitation with strict adjacency will, in the absence of other disturbances, lead to a highly homogeneous block size distribution. This pattern, once established, will take substantial time to change should a homogeneous block pattern be undesirable. Moreover, as our analyses show, changing landscape pattern can have significant consequences for timber supply. Second, without a strategy to maintain late seral stands, a "generation gap" will develop between rotation-age stands and old reserves. This generation gap is cause for concern because static reserves are unlikely to remain intact over extended periods, due to natural disturbances. To replace these stands lost by disturbance, a century or more may be required for recruitment where few late seral stands exist in a landscape. Finally, temporal aspects of patch dynamics are as important as spatial ones: in Landscape Unit 26 patch recruitment into late seral age class in periods 3 and 10 represents critical points for management of large old patches in this landscape unit.

Conclusions

In Landscape Unit 26, we found that the impacts imposed by strict adherence to large patch size constraints are greater than those imposed by the current adjacency constraints. This is due to the fragmented nature of the existing forest, which is far from the imposed patch structure. The fundamental conclusion that follows is that forcing a new pattern (large blocks) onto an existing landscape pattern (small blocks) is equivalent to adding constraints, and harvest levels will decline. The more the existing pattern deviates from the target pattern, the greater the impacts on timber supply, especially in the short term. If we are fortunate enough to have surplus timber, the harvest, to offset short-term deficits in the developed areas, will quickly disperse to undeveloped areas. We have seen this effect before, first when the three-pass cut/leave system was introduced on top of progressive clearcutting and again when small openings and the adjacency rule were introduced on top of the three-pass system. The key to maintaining or increasing short-term yields is to work with the existing landscape structure and to be flexible when introducing new landscape patterns. For example, by relaxing the patch size constraints through our soft constraints, we were able to increase harvest levels beyond the base case and generate a dispersed harvest with a variety of patch sizes.

If manually designed blocks are used, then an opening/patch size distribution should be used rather than the more uniform opening size (approximately 100 ha) that was used in this study. Most importantly, patch rules must be thought of as patch objectives. This allows us to temporarily deviate from the target structure to avoid scheduling catastrophes that typically arise when "hard" rules are applied. Patch management options must be analyzed spatially, and for long time periods. What we do today may well foreclose tomorrow's options. Short-term plans (i.e., five-year plans) that identify immediate gains from the design of large patches are incomplete. The long term must also be addressed to clearly identify the consequences.

Harvest schedules designed to create a distribution of patch sizes where rotations are limited to 100 years or less result in patches that have a very limited opportunity to recruit to older age classes (i.e., the generation gap). This undermines the point of creating a patch size distribution in the first place. Landscapes harvested intensively with normal rotations will eventually develop a gap in age class between zero to 100 years (harvested areas) and 200+ years (old-growth reserves). This should be of concern to managers because it limits options for recruiting old growth if recruiting old growth became necessary for conservation and/or coping with natural disturbances. Long rotations were effective for maintaining late seral stands in this analysis. We suggest that partial cuts (appropriately designed) in late seral stands could mitigate impacts of long rotations on timber supply while maintaining or enhancing development of old growth attributes. Generalized rules designed to maintain seral stage and patch size objectives are unlikely to achieve desirable landscape patterns on their own. Rather, an explicit requirement to allow some patches to "escape" harvest may be required to allow recruitment of late seral stands (and the associated recruitment opportunities for old growth).

In Landscape Unit 26, period 8 represents a substantial loss of late mature and contiguous old stands for almost all scenarios and as such should be an area of concern for managers. Conversely periods 3 and 10 represent opportunities for old-growth-related strategies as two large blocks in the landscape unit recruit to the late mature age class. An important conclusion can be drawn from this: the legacy of the age class and patch structure (regardless of origin) found in a given landscape unit will have a significant influence, for a substantial time, on the patch dynamics resulting from any harvest scenario. As such, general conclusions independent of specific landscapes about harvesting alternatives on wood supply and patch dynamics should be made with caution. Ultimately, managers should develop harvest patterns and timber volume flows appropriate to the landscape at hand.

Acknowledgments
Forest Renewal British Columbia, the British Columbia Forest Service, the Natural Sciences and Engineering Research Council of Canada, and the Sustainable Forest Management Research Network provided funding for this project. We gratefully acknowledge the technical and analytical support provided by graduate student Mark Boyland and the conceptual contributions towards soft adjacency constraints by Reg Davis of Interior Reforestation Ltd. This is research contribution R-41 of the Centre for Applied Conservation Biology.

References
BC Forest Service. 1995. Biodiversity guidebook: Forest Practices Code of British Columbia. BC Ministry of Forests, Victoria, BC.

CORE. 1995a. West Kootenay-Boundary land use plan. Commission on Resources and the Environment, BC Land Use Coordination Office, Victoria, BC.

CORE. 1995b. East Kootenay land use plan. Commission on Resources and the Environment, BC Land Use Coordination Office, Victoria, BC.

Nelson, J.D., and Errico, D. 1993. Multiple-pass harvesting and spatial constraints: An old technique applied to a new problem. Forest Science 39(1):1-15.

Wallin, D.O., Swanson, F.J., and B. Marks. 1994. Landscape pattern response to changes in landscape pattern generation: Land-use legacies in forestry. Ecological Applications 4(3):569-580.

An Ecosystem Management Investment Model: Applications to Planning Issues in British Columbia

J.C. Niziolomski, Guoliang Liu, and Chad Wardman

Abstract

Numerous government agencies and industrial companies are embracing the theory of ecosystem management as the desired approach to sustainably managed forested landscapes. To develop and analyse ecosystem management scenarios a process is required that can (1) hierarchically connect strategic-based objectives and operational activities, (2) integrate spatial and temporal scales, (3) address data scales, (4) include modeling techniques for basic constraint-based simulation as well as optimization analysis, (5) be able to design the landscape towards a desired state, and (6) be flexible enough to accept adaptive changes in the landbase and in management practices.

This paper describes the ongoing development of a hierarchical landscape management approach, FORUM (Forest Optimization and Resource Utilization Management), and its focus on ecosystem management as a comparison of investments in the forested landscape. We will compare planning scenarios with varying resource priorities and address their social, ecological, and economic implications. We will also discuss the potential applications of FORUM to various planning issues currently facing forest planners and managers in British Columbia.

Introduction

Many types of models have been popular for use in forest management and can exist in many forms with varying applications such as word models, physical models, picture models, flowcharts, graphs, mathematical models, and computer models (Kimmins 1987). Models can be developed for the purposes of definition, prediction, or prescription. In the forest industry, many of these model types are a part of everyday forest management activities. Each model is developed for a certain need and has inherent benefits over other models. In forest landscape planning, the complexity of landscape level analysis and the difficulty in managing numerous resource requirements demand that prescriptive models (models that assist in strategy development) be used to assist forest managers and planners in developing the steps essential for reaching landscape-level objectives.

Current legislation, regulation, and land-use planning processes in British Columbia have provided a need for spatially and temporally linked analytical and modeling techniques. The forest industry and regulatory agencies must manage the forest landbase for timber and non-timber objectives (BC Forest Service 1996). With a variety of management objectives for unique geographic areas within a landbase, it becomes very difficult for planners and managers to schedule activities that consider the ecological, economic, and social requirements both now and into the future.

Traditionally, resource analysis and timber supply modeling procedures have focused on a "bottom-up," constraint-based descriptive simulation approach. These techniques provide numerous opportunities for sensitivity analysis and assessment of proposed management strategies on various timber and non-timber resources. While these tools have been effective at testing different scenarios, they have not provided an effective method to develop tactics to reach the "desired future condition." A desired future condition is that land or resource condition that is expected to result if the landscape goals and objectives are fully achieved.

Determining the desired future condition is not a simple task. To quantify this condition, planners and policy makers must consider the region's ecosystem, ecology, economics, social interests, and industrial needs. This process has brought about concerns over sustainability and, more recently, the concept of sustainable ecosystem management. The concept of ecosystem management is an ecological approach to natural resource management that assures the sustainability of productive and healthy ecosystems by blending social, economic, and biophysical values. Ecosystem Design is one method currently being developed that is intended to direct future operational planning through a commitment to sustainable forest management and the ability to address complex and conflicting resource values (Slocan Forest Products 1998). Ecosystem Design is a sub-landscape-level planning approach that is intended to assist in identifying the desired future condition as well as link landscape-level plans and stand-level operational plans.

Analysis tools must now progress to the principles of forest ecosystem management and consider non-timber values such as biodiversity, watershed values, and visually sensitive areas in a spatial manner while accounting for tactical timber harvest scheduling. These issues have directed the development of Forest Optimization and Resource Utilization Management (FORUM). FORUM recognizes the spatial and temporal variability that forest managers must consider. Perspectives and values, as diverse from place to place as the people that inhabit these regions, are apt to change as new knowledge emerges, regulations change, or resource emphasis shifts. This ethno-geographical diversity facilitates the development of a universal strategy that can evaluate various management scenarios based on the array of regulations, special management areas, economic requirements, and social perceptions that are present in any geographical region. We have termed this strategy the forest ecosystem investment approach. To understand the point at which the forest ecosystem investment approach begins, a discussion of our current FORUM approach and its basic principles is necessary.

Forest Optimization and Resource Utilization Management (FORUM) Planning

FORUM is a comprehensive approach aimed at defining forest planning solutions. It incorporates a broad range of ecosystem management principles and techniques to assist managers in developing "optimal" management strategies.

FORUM uses a hierarchical or "top down" approach to achieve landscape objectives. At the strategic planning level, overall landscape goals and objectives are established that provide guidelines for the timing and spatial configuration of timber harvesting at the tactical (i.e., 20-year plan) and operational (i.e., development plan) levels. With all planning phases linked both spatially and temporally, plans can be easily adapted at any level as regulations change, the landbase is altered, objectives change, or better information is acquired. Because ecosystems are dynamic, adaptability is an important requirement for any analytical tool used in forest ecosystem management.

Traditional management approaches have attempted to indirectly create a "desirable" ecosystem using linear decision-making processes using constraints that are imposed by a host of regulations.

> In this linear approach, *forest* feedback control does not exist, since there is no stated *forest* objective as a basis for evaluating forest response. Probably the harvest level will be achieved and the regulations will be followed. The forest that emerges, however, will not be measurable against a target. Management has not focussed on achieving a forest condition – a specific age class structure, for example – as a basis for sustaining timber and non-timber flows. In such a case, *forest* management does not exist. (Wardoyo and Jordan 1996)

Though linear, non-objective-based modeling approaches can be used in FORUM to model forest management scenarios, reflect current or planned management strategies, or provide a contrast between differing management philosophies, FORUM also uses different management approaches. At the centre of FORUM is a target-oriented approach, where a course of action is developed that will lead to the desired future conditions through the application of simulated annealing algorithms.

Simulated annealing uses an element of randomness to globally search the solution space through multiple iterations. The objective function evaluates the relative "success" of each random solution. Simulated annealing always accepts solutions that improve the objective function but can accept inferior solutions in the short term to avoid convergence on local optima. As the number of iterations increases, the process becomes more stringent and only solutions that improve the objective function are accepted until an optimal solution is achieved (Lockwood and Moore 1993; Liu 1995).

FORUM interfaces directly with GIS data, overlaying numerous resource layers to create resultant polygons with unique attribute lists. This reduces the database size restrictions and processing times, which inhibit project progression. As the model runs through numerous iterations, these resultant polygons are amalgamated into various sized harvest units. This gives great flexibility in harvest scheduling as adjacent resultant polygons can be added or subtracted from the harvest opening through successive iterations. The patterns of amalgamation are defined by the principles of landscape dynamics and geographically referenced forest structure attributes.

The forest landscape results from the relationship between pattern and process (Crow and Gustafson 1997). The temporal variability and spatial scale of disturbance in the forest landscape creates the vegetation pattern we see on the landscape. It is therefore possible to represent any landscape level non-timber resource using two indicators, patch size and distributions (Liu 1999). For example, setting a maximum target level for early age classes (e.g., 20 percent < 35 years) and keeping openings small (< 40 ha) within watersheds may achieve specific water quality objectives. Also, wildlife habitat cover objectives can be met by ensuring that more than 70 percent of the wildlife corridor is greater than 100 years in age.

Since it is the distribution of patch size within different age classes that creates biological and structural diversity at the landscape level (BC Forest Service 1995, 1998), it is essential that landscape models integrate patch and age class management strategies. Not only must the analytical tool be able to define an age group linked to an area, it must also be able to track the development of the patch as an evolution of these mosaics over time for the entire landscape. Figure 1 shows how targets can be set by specifying the desired patch size distribution within each age class.

Most non-timber resource objectives can be incorporated using patch and age class objectives. Each resource layer contributes to the overall evaluation function, and the harvest schedule with the least deviation from the desired future condition for all resource objectives (the near-optimal solution) is then chosen. Tolerance levels can also be set to allow for the natural

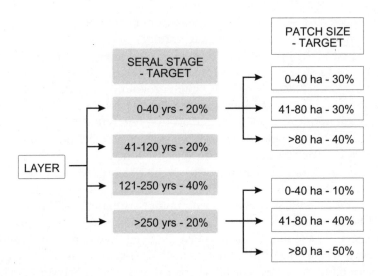

Figure 1. An example of patch size and age-class (seral stage) integration within a forest management context.

variability that occurs in a dynamic ecosystem. The target-oriented approach allows the user to set long-term non-timber objectives that we attempt to achieve over time (Figure 2). There also must be a way to link the resources together to ensure that the option that best meets all of the resource objectives is chosen. This is done by applying weights, or priorities, to resource objectives.

Establishing Priorities by Using Weightings

Differing social, economic, and ecological issues and priorities govern how discrete landscape units are managed. This leads to an approach that can address the relative importance of a multitude of resource objectives and is the key to trade-offs between the harvest level and the length of time required to achieve patch or age class distribution targets. A typical weighting scheme that balances timber and non-timber resources may look similar to the following equation:

$$\text{Objective Function} = 1.7 \text{ (block size distribution)} + 1.5 \text{ (patch)} + 1.2 \text{ (age class)} + 1.0 \text{ (volume)} + 0.3 \text{ (even flow)}$$

where block size distribution, volume, and even flow represent timber values; patch and age class represent ecological values; social values generally result from a combination of the two (e.g., employment, recreation access).

Within patch and age class objectives are the multitude of non-timber resources that include but are not limited to:

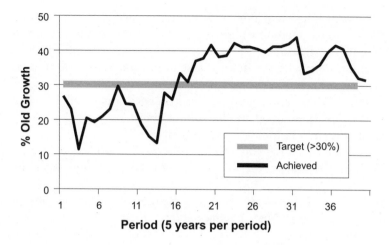

Figure 2. Modifying the landscape from an initial condition to a desired state. In this example, the desired future condition requires that a minimum of 30 per cent of the landscape be in an old-growth state. The achieved value represents an investment in the landscape as well as trade-offs with other resources. The values result from feedback mechanisms that evaluated and modified previous solutions and incorporate strategy to ensure the long-term maintenance of the old growth target.

- biodiversity objectives
- wildlife habitat
- visual quality
- water resources management (equivalent clearcut areas).

Some objectives are relatively difficult for the model to achieve (e.g., block size distribution) or are relatively more important than other resource values and receive higher weightings. Weightings are easily changed in FORUM so that differences in management priorities can be accounted for in the analysis. The simulated annealing function is then used to calculate penalties based on the defined weighting scheme, with higher weighting resources contributing higher penalties for deviations from the targets.

Forest Ecosystem Investment Approach

In response to the need to develop an approach that can evaluate different feasible planning scenarios based on an established criteria, we have begun to develop a methodology within the FORUM planning framework by which forest ecosystem objectives or investments can be quantified. We have called this the Forest Ecosystem Investment Model (FEIM).

In developing the Forest Ecosystem Investment Model we determined that several features must be incorporated into the approach:

- *Adaptability:* The evaluation criteria must be flexible enough that its application is not bound by time or space (i.e., the approach can be used at anytime for anywhere).
- *Versatility:* The approach must be able to incorporate a wide range of resource criteria.
- *Relativity:* Within any landscape, some resource criteria will be relatively more important than others. The approach must have the ability to apply higher priorities to these resource objectives.
- *Reportability:* The user interface must be easily understandable, highly visual, and accurately reflect the underlying reporting functions from the model.
- *Interactive ability:* Changing priority weightings and starting subsequent runs must be at the user interface and be directly linked to the model evaluation criteria.

Given the above requirements, Figure 3 outlines the flows and linkages of the proposed Forest Ecosystem Investment Model.

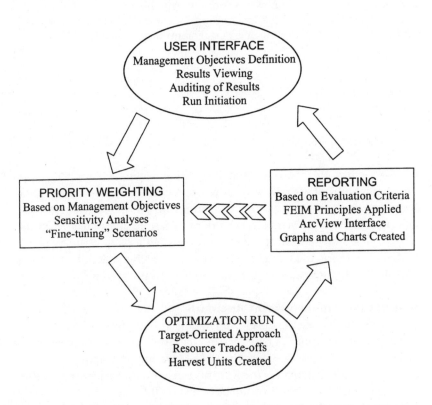

Figure 3. Flow chart illustrating linkages within the Forest Ecosystem Investment Model.

Figure 4. Defining sustainability using interactive resource spheres. The three spheres represent values, goals, and needs from three different perspectives. Only where these spheres overlap can a sustainable solution appropriate for all spheres be found.

Using Indicators for Evaluation

Sustainable forest management can be defined as "forest management regimes that maintain the productive and renewal capacities, as well as the genetic, species and ecological diversity of forest ecosystems" (Dunster and Dunster 1996). If forest management is thought of as three intersecting spheres – economic, ecological, and social – then a sustainable solution would be located within the area where these three spheres of influence intersect (Figure 4).

Within "sustainability" we can find numerous, if not infinite, possible solutions, but all solutions that are sustainable are not necessarily desirable. For example, high harvest levels in Clayoquot Sound, British Columbia, are "sustainable"; social and ecological pressures dominate in this region, however, and economic considerations are minimized due to the high tourism and old-growth values in this region (Scientific Panel for Sustainable Forest Practices in Clayoquot Sound 1995). Conversely, in intensively managed ecosystems where the community depends on the forest industry for its survival, economic considerations play a much larger role in guiding forest management decisions, and ecological and eco-social considerations are more limited.

The desired future condition must be a sustainable condition. But what sustainable condition of the infinite options available best represents the desired future condition? We have begun developing a process and evaluation criteria to answer this question.

Table 1

A list of potential social, ecological, and economic indicators for evaluating landscape planning scenarios.

Value spheres	Indicators
Economic	Timber volume, species, stem size distribution, job creation, unemployment, distance to mill, cutting systems, stumpage rates, road construction length, growth rates, silviculture investment return.
Social	Employment, visual quality, recreation (fishing, hunting, hiking, camping, rafting, access to remote areas), birdwatching, water quality, perception of appropriate forest stewardship.
Ecological	Biodiversity (composition and structure, patch size and seral distributions, abundance of exotic and specialist species, area by ecotype [Bonar 1995]), habitat and corridor management, watersheds, riparian, wildlife trees, landscape metrics, old-growth.

Currently, the FORUM model can assist in the preparation of a "sustainable" solution (defined by the manager) and can place varying priorities on different resource values (e.g., economic vs. ecological), but it cannot quickly evaluate the resource trade-offs among different solutions. The Forest Ecosystem Investment Model will allow forest ecosystem managers to customize resource weighting priorities based on local resource management objectives, work with visual and interactive decision support tools that report resource trade-offs associated with a particular management scenario, and modify scenarios in subsequent analyses.

One of the greatest difficulties with creating a Forest Ecosystem Investment Model is defining the criteria by which different landscapes are evaluated. For example, how are penalties applied when the equivalent clearcut area limit is exceeded in a watershed? Does the penalty increase linearly or exponentially, and at what rate? What combination of factors influences overall recreation satisfaction? What set of indicators is used for economic evaluation? How does one address conflicting values in the same value sphere such as the employment rate and anti-logging values?

If time, resources, and existing knowledge allow, we can develop complex ecological, economic, and social evaluation criteria to answer these questions. Table 1 lists some criteria for evaluating social, ecological, and economic satisfaction.

Study Area

Landscape Unit 16 in Tree Farm Licence 3 (Slocan Forest Products) was used as a sample dataset to test the concepts of the Forest Ecosystem Investment Model (FEIM).

Tree Farm Licence 3 (TFL 3) is located in the Nelson Forest Region (Arrow Forest District) near the village of Slocan in southeast British Columbia. The TFL comprises an area of 79,800 hectares and is located predominantly within the Interior Cedar Hemlock (ICH) and Engelmann Spruce-Subalpine Fir (ESSF) biogeoclimatic zones (Braumandl and Curran 1992). These zones receive substantial precipitation and contain some of the most productive growing sites in the British Columbia interior. Landscape Unit 16 (LU 16) is the only landscape unit of the three present in the TFL that is entirely within the TFL boundaries. Due to varied topography and climate, LU 16 consists of a wide variety of tree species including Englemann spruce *(Picea engelmannii)*, subalpine fir *(Abies lasiocarpa)*, Douglas-fir *(Pseudotsuga menziesii)*, western hemlock *(Tgusa heterophylla)*, western larch *(Larix occidentalis)*, lodgepole pine *(Pinus contorta)*, western redcedar *(Thuja plicata)*, and deciduous tree species.

Evaluation Criteria

FORUM currently tracks the performance of each resource layer throughout the planning horizon. The changes in the performance of each resource resulting from changes in resource weighting reveal the trade-offs among differing scenarios. The key is to link to this information and display it in a meaningful way for managers and planners to understand. The ability to have quick forest ecosystem decision support tools will enable planners and managers to make informed decisions when scheduling current and future forest management activities.

To convey the concepts, we have greatly simplified the complexity of landscape interactions by using timber volume as an indicator for economic investment, domestic watersheds for social values (denoting water quality), and old-growth percentage in natural disturbance type 1 (NDT 1; as defined in BC Forest Service 1995) as an indicator for ecological values. Domestic

Table 2

Setting priorities (weighting) of parameters used in the Forest Optimization and Resource Utilization Management model for landscape unit 16 in southeast British Columbia.

	Weight		
Parameter	Scenario 1 (Ecological)	Scenario 2 (Compromise)	Scenario 3 (Economic)
Total volume flow	1	1	1
Patch size distribution	16	1.6	0.16
Age class distribution	15	1.5	0.15
Even volume flow	0.2	0.2	0.2
Cutblock size	2	2	2

watersheds and old growth were chosen because these are currently inconsistent with the desired state. The priorities applied for each scenario are shown in Table 2.

Cutblock size distributions and volume flow tolerance were held constant as these are important at the operational level. The low flow tolerance gave the model the flexibility to cut higher proportions in some years, allowing the algorithm to find the best strategy without significant flow constraints. With an increase or decrease in one resource level we expect a resultant reaction from other resource values. For example, higher rates of cut should result in lower amounts of old growth.

To compare the various scenarios, we selected simple evaluation criteria: variation in timber harvest and years to reach target conditions. As presented, timber can provide a link to economic value while the time to reach the target condition represents the measure of social and ecological value. The domestic watershed indicator is a sum of seven different watersheds. The total time for these watersheds to reach the desired condition is used as the evaluation criteria.

Results and Discussion
After running the FORUM target-based model, we summarized the performance of all three differently weighted scenarios. FORUM output formats

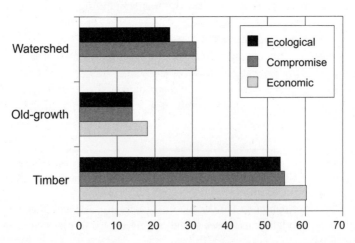

Figure 5. Results of three differently weighted management options from the Forest Optimization and Resource Utilization Management model in Landscape Unit 26 in southeast British Columbia. Watershed values represent the proportion violation for eight separate sub-basins calculated at each five-year period and spanning the first 50 years. Old-growth is the number of years to achieve the desired state. Timber values are the average annual volume flow in thousands of cubic metres over the entire planning horizon.

allowed these results to be summarized graphically to show the variation in levels between the differently weighted scenarios (Figure 5).

Timber flow increased by 13 percent between the economic (60,300 ha) and compromise (54,500 ha) options and by 2.1 percent between the compromise and ecological (53,400) options. Overall, watershed values took 155 years to achieve the economic and compromise options versus 120 years for the ecological scenario. Old-growth performance was similar for all management scenarios, though the ecological and compromise options achieved the desired state 20 years faster (70 years total) than the economic alternative (90 years total). Although these results represent a rather simplistic analysis, the intent is to show the methodology behind the Forest Ecosystem Investment Model.

Overall, resource trade-offs were minimal between the ecological and economic results. This is because of the tendency of FORUM to always manage towards the desired future condition even though the parameter weightings differ. Also, FORUM timber levels are an output of the model and are determined by the productive capability of the landscape. Therefore, as we increase the priority on timber flow, the maximum level cannot exceed what the landbase can support. Even low weightings for non-timber resources provide direction for landscape design and set minimum and maximum levels for targets. Although the economic option generated 13 percent more timber volume, it did not greatly affect the time required to achieve non-timber resource objectives especially considering that the resource indicators we chose were the values that initially deviated most from the desired state.

Applications to Forest Ecosystem Management

In response to the need for an approach capable of integrating a wide range of quantitative and qualitative resource values to determine the investment trade-offs associated with multiple planning scenarios, we have begun to develop the forest ecosystem investment approach. The concepts described in this paper, though still in the developmental stages, have the potential to be applied to a wide range of forestry and ecological management practices. These include but are not limited to the following examples:

- *Landscape unit planning:* Landscape unit plans are derived through landscape level modeling exercises based largely on local, regional, and provincial regulation, expert opinion, and local perception. Often there is no way to evaluate whether the modelled result is a desirable solution and meets long-term planning objectives. The Forest Ecosystem Investment Model combined with the flexibility and power of FORUM planning solutions can provide not only multiple options but also weigh the relative effectiveness of each to achieve the short- and long-term management objectives.

- *Scenario planning and sensitivity analysis:* Scenario planning, which incorporates sensitivity analysis, is a "gaming" approach that determines the sensitivity that changes in landscape objectives (e.g., increasing green-up age by 10 percent) have on timber flow levels. The Forest Ecosystem Investment Model can greatly enhance this process by determining not only the sensitivity of timber flow levels to changes but also the impact of any timber or non-timber resource to changes in any other resource value.
- *Operational investment analysis:* Operational forest activities are often poorly directed and the multiple alternatives never considered. By linking operational activities to a range of indicators such as species selection, stem size distributions, market trends, and silvicultural systems, more informed decisions on cutting permit development and cutblock locations could be quickly evaluated and profits could be increased.
- *Ecological strategy development:* Multiple strategies could be analysed and the one that best reflects the management intent can be chosen. For example, if an old-growth connectivity network with a maximum of forest interior, minimal edge, and fractality is desired, then various connectivity configurations could be tested and the response to these indicators evaluated.
- *Fire hazard investment strategy:* Currently, there is no definitive methodology for maximizing forest fire fighting strategy to maximize effectiveness and minimize cost. Often fires are fought, at great expense, based on where they encroach upon settled areas while thousands of hectares of forest land are sacrificed. A fire hazard investment model based on the Forest Ecosystem Investment Model concepts could weigh social, economic, and ecological factors and develop strategies based on desired indicators.

Making informed management decisions requires that numerous alternatives be considered and the relative merits of each weighed. To do this efficiently, managers require the proper tools. FORUM allows managers and planners to use either traditional or innovative tools and concepts to help derive an optimal solution for their forest estate. The Forest Ecosystem Investment Model will provide the means to evaluate and adapt these initial solutions to transform them into a truly optimized forest planning solution.

References

BC Forest Service. 1995. Biodiversity Guidebook: Forest Practices Code of British Columbia. British Columbia Ministry of Forests, Victoria, BC.

BC Forest Service. 1996. Higher level plans: Policy and procedures. Forest Practices Code of British Columbia. British Columbia Ministry of Forests and Ministry of Environment, Victoria, BC.

BC Forest Service. 1998. Age class stages across forested landscapes: Relationships to biodiversity. Extension Note, Part 7 of 7. British Columbia Ministry of Forests and Ministry of Environment, Victoria, BC.

Bonar, R. 1995. Biodiversity indicators for forest managers – an example. Discussion Paper.

Braumandl, T.F., and M.P. Curran. 1992. A field guide for site identification and interpretation for the Nelson Forest Region. British Columbia Ministry of Forests, Victoria, BC.

Crow, T.R., and E.J. Gustafson. 1997. Ecosystem management: Managing natural resources in time and space. *In* K.A. Kohm and J.F. Franklin (eds.). Creating a forestry for the 21st century: The science of ecosystem management, Island Press. Washington, DC.

Dunster, J., and K. Dunster. 1996. Dictionary of natural resource management. UBC Press, Vancouver, BC.

Kimmins, J.P. 1987. Forest ecology. Macmillan Publishing, New York, NY.

Liu, G. 1995. Evolution programs, simulated annealing and hill climbing applied to harvest scheduling problems. M.Sc. thesis, University of British Columbia, Vancouver, BC.

Liu, G. 1999. A target oriented approach for forest harvest scheduling and landscape modification. Ph.D. thesis (in progress), University of British Columbia, Vancouver, BC.

Lockwood, C., and T. Moore. 1993. Harvest scheduling with adjacency constraints: A simulated annealing approach. Canadian Journal of Forest Research 23:468-478.

Scientific Panel for Sustainable Forest Practices in Clayoquot Sound. 1995. Report 5, Sustainable ecosystem management in Clayoquot Sound: Planning and practices. British Columbia Ministry of Environment, Lands and Parks, Victoria, BC.

Slocan Forest Products. 1998. Bonanza Face ecosystem design. Slocan Forest Products, Slocan, BC.

Wardoyo, W., and G.A. Jordan. 1996. Measuring and assessing management of forested landscapes. The Forestry Chronicle 72(6):639-645.

Applying the Natural Variability Concept: Towards Desired Future Conditions

Russ Parsons, Penny Morgan, and Peter Landres

Abstract

Natural resource managers increasingly use the natural variability concept in landscape assessment and planning. Also called the historical range of variability, range of natural variation, and reference variability, the concept was developed by applied scientists and managers seeking a legally and politically defensible reference for setting management goals that sustain ecosystems and conserve biodiversity. This coarse-filter approach argues that restoring or managing ecosystems within the bounds of past structure, composition, and disturbance regimes is most likely to sustain the viability of diverse species. In the Interior Columbia River Basin Ecosystem Management Project, a broad regional assessment, departures from historical (pre-European) vegetation composition were used as one of several criteria for identifying areas of low ecological integrity.

Natural variability is defined as "the ecological conditions, and the spatial and temporal variation in these conditions, that are relatively unaffected by people, within a period of time and geographical area appropriate to an expressed goal" (Landres et al. 1999). The natural variability concept can be useful for understanding dynamic ecosystems, for evaluating changes in ecosystems over time, for placing priorities on management actions, and for assessing risk and hazard. A comparison of past natural variability, present conditions, and desired future conditions will provide general direction for ecosystem management. When desired future conditions lie outside of past natural variability, people must accept both the costs of sustaining unprecedented conditions and the increased risk of unpredictable and often unacceptable changes in ecosystems. The magnitude of the departure may determine the degree of difficulty and uncertainty that might be associated with attempting to maintain the system outside of its normal limits.

The natural variability concept can be useful in ecosystem management, but it is not a panacea to solve all problems. When information about past conditions is insufficient, when present conditions are substantially different from past conditions, or when management focus is on isolated reserves, it may be inappropriate to use the natural variability concept as a driving force in ecosystem management. If risk and hazard associated with past disturbances are socially or politically unacceptable, it may be inappropriate to manage for the full range of natural variability. In many cases, a narrower range of conditions, or management variability, must be defined to guide management direction to decrease the probability of extreme events that defined the range of past natural variability.

Introduction

In recent years there has been increasing interest in developing an understanding of past ecological conditions for use as a reference in contemporary management. A number of terms such as "natural variability," "range of natural variation," "natural range of variability," "historical range of variability," and "reference variability" have been used for the same concept. As some potential exists for misinterpretation of the terms "historical" and "range," we suggest that the phrase "natural variability" is the most clear, and we will refer to the concept in this way throughout this paper. The purpose of this paper is to present a brief review of the natural variability concept, discuss the utility and appropriate use of the concept, and consider its practical application in ecosystem management with specific attention to the relationship between natural variability and desired future conditions.

Brief Review of the Concept

Definition

The natural variability concept is widely used in natural resource management (e.g., Poff et al. 1997; Skinner and Chang 1996; Brown et al. 1994; Gauthier et al. 1996; Ripple 1994) but must be carefully defined and applied to be successful for specific management scenarios. In this paper we define natural variability as: "The ecological conditions, and the spatial and temporal variation in these conditions, that are relatively unaffected by people, within a period of time and geographical area appropriate to an expressed goal" (Landres et al. 1999).

While the concept can also be used to describe variability in ecosystem processes (e.g., fire frequency and severity, flooding, insect infestations), it is perhaps easier to visualize in terms of ecosystem structure. For example, consider natural variability in the geographical extent of a particular type of vegetation such as old, single-story ponderosa pine (Figure 1). Within the spatial bounds of a particular ecosystem, old, single-story ponderosa pine forest may occupy a portion of the total area over time, depending on dynamic factors such as fire frequency and severity, insect infestations, and climatic variability. To adequately represent past conditions, one must consider the range, mean, and standard deviation, as well as other appropriate

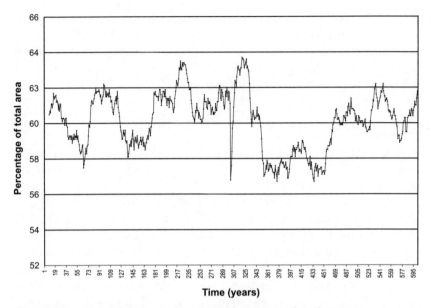

Figure 1. Natural variability over time in the proportion of total ecosystem area occupied by "old growth" ponderosa pine forest (P. Morgan, R.A. Parsons, J. Hauffler, and B. Holt, unpublished data 1998, Natural variability estimates for the Idaho Southern Batholith).

measures, of the percentage of total area occupied rather than the point estimate of area occupied at any given time.

Importance in Ecosystem Management

In today's prevailing climate of litigation, particularly over threatened and endangered species issues, managers need a legally defensible strategy for proactively addressing habitat and population viability. The natural variability concept is a key part of this strategy. This is one reason it is emphasized by United States agencies implementing ecosystem management (Swanson et al. 1993; Morgan et al. 1994; Kaufman et al. 1994; Manley et al.1995; Landres et al. 1999). Developing an understanding of reference conditions is a required critical step in agency protocols for "Ecosystem Management at the Watershed Level" (US Department of Agriculture and US Department of the Interior 1995).

Although the natural variability concept has most commonly been applied when management goals are to restore or maintain historical conditions, it can be useful in almost any management scenario, even when restoration or conservation are not principal objectives. Natural variability is an expression of the natural tendency of a system. Managing natural systems within the constraints of their natural bounds rather than outside of them will require fewer inputs, because to do so takes advantage of the natural tendency of the system (Allen and Hoeskstra 1992). Thus, even in situations in which the desired condition lies outside the historical bounds, it is useful to know the historical bounds to approximate the departure between desired and historical conditions. The magnitude of the departure will approximate the difficulty that may be encountered in attempting to maintain the system outside of its natural tendency.

The natural variability concept can thus be used for different purposes and towards different management objectives. For this reason, it may provide a common ground for discussion of management issues across ownership boundaries, a key concept in ecosystem management.

Principal Premises Behind the Natural Variability Concept

The natural variability concept is firmly grounded in ecological theory and is based on a number of key premises (Landres et al. 1999):

- Natural variability is a useful reference for evaluating the influence of anthropogenic change in ecological systems including lakes (Smol 1992), commodity production lands (Morgan et al. 1994), and protected areas such as wilderness (Haila 1995).
- Contemporary human influence has diminished the viability of many species adapted to past or historical conditions and processes (Swanson et al. 1994).
- Approximating the variety of historical conditions provides a coarse-filter management strategy that is likely to sustain viable populations of many species about which we know too little (Hunter et al. 1989).
- Working within the bounds of the system as approximated by natural variability is easier, requires fewer external subsidies, and is more cost-effective than trying to sustain unprecedented conditions (Allen and Hoeskstra 1992).
- Analysis over longer time frames and at different spatial scales helps us to understand what constrains and drives ecosystem change (Allen and Hoekstra 1992).
- Disturbances have a strong and lasting influence on species, communities, and ecosystems (White 1979; Sousa 1985) and have been called a "key structuring process" at midscales (i.e., the scale of forest stands) (Holling 1992).
- Variability is important. Reducing spatial variability typically results in declining biological diversity (Petraitis et al. 1989), increased vulnerability to insects, pathogens, or other disturbances (Lehmkuhl et al. 1994), and decreased resiliency to subsequent disturbances (White and Harrod 1997).

Quantifying Natural Variability

Natural variability estimates will be meaningful only if they are developed for well-defined regions with explicit spatial and temporal bounds. The selection of these bounds is extremely important to the overall usefulness of the concept and should be done with great care. Temporal bounds should be defined in such a way as to minimize the influence of long-term climate change (e.g., glacial versus interglacial periods) for the geographic region in question while capturing the natural variability of the conditions being studied. Temporal bounds should be appropriate to management objectives and should consider the degree of past human influence, quality and length of record, presence of exotic species, and known fluctuations in climate. Spatial extent should be determined based on similar factors. As much as possible, natural variability should be assessed for areas over which climatic, edaphic, topographic, and biogeographic factors are relatively consistent (Morgan et al. 1994). In the Interior Columbia Basin Ecosystem Management Project, for example, natural variability was considered within the temporal bounds of the last 2,000 years (Hann et al. 1997), a period in which conditions are believed to have been relatively consistent, both climatically and in terms of Native American influence (Schoonmaker and Foster 1991). Large areas such as regions should be divided into smaller subunits with reasonably consistent biophysical conditions such as subregions (ECOMAP 1993) or potential vegetation types (Hann et al. 1997). Natural variability should then be estimated separately within each subunit.

Once temporal and spatial bounds are defined, it will be possible to quantify natural variability for the area in question. The accuracy and precision of estimates depend on the quantity of data involved, and the degree of confidence data from points and small areas can be extrapolated to larger areas. For example, an understanding of fire history of an area might be built from fire scar and tree ring data at particular locations. As it is rarely possible to develop natural variability estimates based on comprehensive data, some degree of extrapolation or inference will always be involved. For this reason, it is of utmost importance that all assumptions used to construct natural variability estimates be explicitly and carefully stated. Failure to do so may substantially limit the effectiveness and defensibility of the estimates and will obstruct future endeavors to refine them.

Natural variability can be described in a number of ways. For most applications, it is useful to know common statistical estimates of the variables of interest, such as range, mean, median, standard deviation, coefficient of variation, percentiles, and skewness. In analyzing patch dynamics it is useful to describe certain parameters in spatial terms, such as spatial arrangement, sizes and shapes, continuity, and contagion. Disturbance patterns such as fire and flood regimes will need to be described in terms of size and severity, frequency, and spatial distribution.

Range is often used to describe natural variability and to evaluate whether current conditions are beyond the bounds of natural conditions. Different interpretations of the term necessitate that it be clearly explained to avoid confusion. Some researchers may interpret range as a general description of conditions. In statistical terms, however, range refers to the difference between the maximum and minimum values for a given attribute of interest. Extreme events will tend to define the bounds of ecological conditions, but it may be difficult to define spatial and temporal limits for such events. For this reason, we suggest that it is inappropriate to use the range of natural variability *exclusively* as a guide for management decision-making. Range of natural variability will be more useful when considered in conjunction with other parameters such as means, standard deviations, spike values, and trend.

Utility of the Concept

The term *natural variability* has been used in a variety of ecosystems and for a number of purposes, but generally speaking it has been used in two ways: to improve our understanding of ecosystem dynamics and to evaluate changes between the past and the present.

By examining past conditions, we gain an understanding of the processes, driving forces, and overall behavior that characterized a given ecosystem in the past. The natural variability concept has been applied in improving our understanding of old-growth forest ecosystems in the west (Ripple 1994; Camp et al. 1997), mid-west (Mladenoff and Pastor 1993), southwest (Covington and Moore 1997), and northwest (Lesica 1996; Lertzman et al. 1997). It has also been used to better understand processes and dynamics of boreal forest (Gauthier et al. 1996) and of mid-western forest (Baker 1992a). An increased understanding of past conditions can offer insight into contemporary ecosystem conditions.

The natural variability concept has been useful as a framework for evaluating change between the past and the present, particularly in terms of impacts of altered disturbance regimes. In fire ecology, for example, it has been used to evaluate changes in the structure and composition of forest ecosystems resulting from altered fire regimes (Skinner and Chang 1996). In hydrology and limnology, impacts of altered flow regimes have been assessed for the Everglades (Harwell 1997) and for the Colorado River (Poff et al. 1997). It has been used to assess and monitor ecosystem conditions such as eastern deciduous forests (Keddy and Drummond 1996). Such assessments can enable researchers to analyze the type and degree of change and can help identify causes and potential consequences of change. In many cases results of such analyses can be used to direct management. For example, departure from historical conditions has been used to set priorities for management actions (Caprio et al. 1997; Covington and Moore 1994b).

An additional way in which the concept may be useful is in communicating the dynamics of complex ecosystems, a crucial step in developing a common understanding of the behavior of forest ecosystems. An improved understanding of the dynamic nature of ecosystems may help in identifying conflicts between overall ecosystem management goals and specific policies or practices. For example, an understanding that a given ecosystem has a historically higher frequency of fire disturbance may suggest a need for changes in fire management practices (such as increased prescribed burning, understory removal, or limited use of a "let-burn" policy) to manage accumulated fuels.

Appropriate Use of the Concept

The natural variability concept is often viewed as a key concept in ecosystem management. There are conditions, however, in which it may be inappropriate; in such cases, alternative approaches may be more effective.

Circumstances in which the natural variability concept may be inappropriate include the following (Landres et al. 1999).

(1) *When ecosystem functions and components have been substantially altered from historical conditions.* In the sagebrush grasslands of the interior Northwest, a widespread influx of cheatgrass and other exotic annual grasses has greatly altered fire disturbance patterns (D'Antonio and Vitousek 1992). One effect of this change in species composition has been a marked increase in fire frequency to the point that sagebrush, which was maintained because of less frequent fire, is unable to establish itself before the next fire. In the face of increased fire frequency as well as increased competition in growing space, the tendency in the new system is towards elimination of sagebrush and total domination by exotics (D'Antonio and Vitousek 1992), particularly in xeric areas (Tausch et al. 1993). In this case, it is likely that in some areas the system has crossed an essential threshold and can no longer function as before. Crossing such a threshold may result in a new steady state, or relative equilibrium, defined and maintained by its new components and processes (May 1977; Sprugel 1991; Landres 1992; Tausch et al. 1993). When a system has crossed a threshold into a new steady state, it will not naturally return to its former state. Thus it may be inappropriate to apply the natural variability concept until intensive restoration work is done (Wallin et al. 1996).

(2) *When information regarding historical conditions is too limited.* In many cases it is too time-consuming or expensive to get new data. Information on historical spatial patterns is especially difficult to find. If data on historical conditions is insufficient to determine the relationships

between historical components and their functions, the natural variability concept will be of little value. While it may not always be possible to establish quantitative estimates, however, qualitative descriptions can still be valuable.

(3) *When management interest is focused on isolated reserves rather than an entire system.* The natural variability concept is most useful when applied to an entire ecosystem or landscape. In cases in which management interest is restricted to a small fraction of the former landscape, such as an isolated nature reserve in a matrix of agricultural land, it may be unreasonable to assume that all ecosystem processes are intact, even if the protected area in question is apparently unchanged. In extreme cases, conditions within the reserve will be largely influenced by conditions outside the reserve (Saunders et al. 1991). The influence of the surrounding matrix on the reserve will depend on its spatial extent and arrangement and on the relative differences between them (Saunders et al. 1991). Although reserves have traditionally been designed to provide habitat for particular species, they have rarely been designed to include the range of past disturbances, as well as other ecosystem processes (Baker 1992b). Therefore caution should be exercised when applying the natural variability concept to small areas rather than to the whole ecosystem upon which the concept is based.

(4) *When risk and hazard associated with historical ecosystem behaviour is socially or politically unacceptable.* When disturbance events known to have occurred under historical conditions are socially or politically unacceptable (e.g., large floods or catastrophic fires), it may be inappropriate to use the natural variability concept to guide management direction. Other approaches to landscape management will be needed. The natural variability concept, however, may continue to be useful as a component in a management strategy, particularly in assessments of risk and probability of extreme events.

Practical Application: Natural Variability and Desired Future Conditions

As noted above, the concept of natural variability improves our understanding of ecosystem dynamics and provides a framework for comparison between past and present conditions. This comparison can be extended to include desired future conditions, an expression of target conditions preferred by decision makers.

Visualizing Natural Variability and Desired Future Conditions

Past natural variability, present conditions, and desired future conditions can perhaps be best visualized as a plate, a ball, and a bowl (Figure 2). The range of past conditions, or natural variability, can be considered as a flat surface (the plate) upon which current conditions (the ball) move around owing to the influence of dynamic factors and processes. The plate represents our best estimate of the natural variability of the system in the past; areas outside the plate could be defined as conditions that do not characterize the system in question. Thus, if current conditions (represented by the ball) are forced out of the bounds of the system, rapid and unpredictable change may result. This is analogous to the ball falling off the plate and bouncing in an unknown direction. In some situations the edge of the plate may have very real meaning in that, once conditions leave the known territory of past variability, they may cross a threshold over which return cannot be easily accomplished without intensive management action. Such a phenomenon is theorized to occur when there is a substantial change in species composition and ecosystem function because of invasion by aggressive exotics (Tausch et al. 1993). For example, in forest stands in which blister rust (*Cronartium ribicola*) has eliminated most of the white pine (*Pinus monticola*), Douglas-fir (*Pseudotsuga menziezii*) and grand fir (*Abies grandis*) have increased in abundance. This shift in species composition causes changes in ecosystem function because of higher susceptibility to root disease and differences in efficiency of water use (Byler et al. 1994).

In many cases, past conditions may have exhibited substantial variability. This variability could be represented with a very large plate or platter. Extreme events that defined the range of past variability such as rare, landscape-scale catastrophic fires or floods may be socially or politically unacceptable today. While such events may for all purposes be outside management control, target conditions can still be framed in such a way as to attempt to

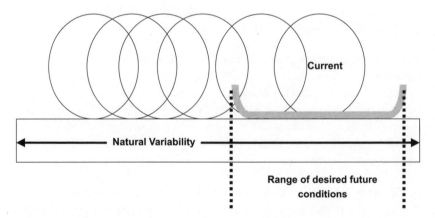

Figure 2. Visualizing natural variability, present conditions, and desired future conditions.

minimize them. Manley et al. (1995) recommend using a subset of the full range of natural variability, or "management variability," as a more practical method for applying natural variability concepts. Management variability will contain the range of permissible desired future conditions. It can be visualized as a bowl in which current conditions continue to move around, but with less total variability than in the past. Setting management targets as a subset of natural variability may allow managers to minimize extreme events while still maintaining ecosystem function.

Using Natural Variability to Guide Management Direction

A comparison of past conditions (characterized by natural variability), present conditions, and desired future conditions can clarify overall management direction (Figure 3). When desired future conditions lie within the bounds of past conditions, as is the case in most wilderness areas and national parks, two general management directions exist. When present conditions are within the bounds of natural variability for the system, management

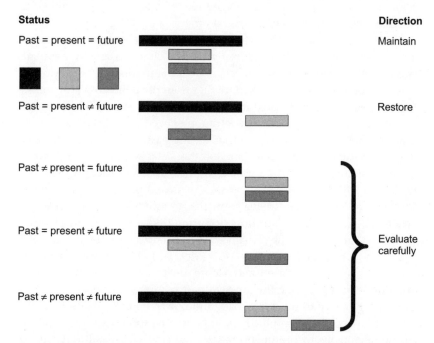

Figure 3. Comparison of past natural variability, current conditions, and desired future conditions and possible combinations of the three. The width of the bars represents the range of conditions for a given variable. Management direction can be clarified by a comparison of the three conditions. For those desired future conditions that lie outside the bounds of the system, the relative departure from past conditions may represent the degree to which unknown and potentially undesirable effects can be expected (from Landres et al. 1999).

direction should be to maintain the system in its current state. When present conditions are outside the natural variability, management direction should be to restore past conditions.

Although maintenance or restoration of past conditions are common management objectives, they are not the only possibilities. In many cases, desired future conditions will lie outside the bounds of natural variability, often for important social or economic reasons. For example, most modern commercial agriculture falls into the realm in which desired future conditions equal present conditions but are different from past conditions (Figure 3). We are able to maintain conditions outside the range of past conditions only through external inputs such as fertilizers, pesticides, and mechanical manipulation (e.g., tillage practices). It is possible as well that desired future conditions may be wholly different from either past or present conditions, such as the conversion of arid lands into reservoirs or golf courses. In these cases, while it is unlikely that components and processes of the original system will be maintained, the natural variability concept can be useful in clarifying the types and magnitudes of external inputs required to maintain the system outside its typical range.

Consider as an example the interior ponderosa pine forest of the western United States. In the past this forest type was characterized by frequent, generally low-intensity surface fires (Covington and Moore 1994a; Morgan 1994). Excluding low-intensity surface fires from the ponderosa pine forests of the inland West has resulted in unnaturally dense stands. These stands, having more ladder fuels and dense canopy cover, are substantially more vulnerable to stand-replacing fire than in the past. Such changes have so altered ecosystem structure and function that catastrophic fire now threatens the sustainability of the forest ecosystem in some places (Covington and Moore 1994a, 1994b). Comparisons of spatial patterns of fires in the southwestern United States have shown that current average fire sizes are three to six times larger than those occurring prior to settlement by Euro-Americans (Barrows 1978; Sweetnam and Betancourt 1990). Habitat for wildlife species such as the flammulated owl (Morgan 1994), northern goshawk, and other species (Covington and Moore 1994a) may be at risk. Current conditions thus lie outside the range of past natural variability.

Returning briefly to our plate, ball, and bowl, the ball is outside of the plate. The issues of catastrophic fires and potential species loss in this forest type are analogous to the uncontrolled bouncing of the ball outside the normal bounds of the system. If management objectives are to restore this forest to its past conditions, fires should be a part of management practices to support and maintain ecosystem processes and species dependent on fire disturbances.

If management objectives are to maintain the system in its current state, it will be necessary to accept the costs (external subsidies) and consequences (potential impacts) associated with maintaining the system outside of its normal range. Costs will include those for higher fire suppression; consequences will include potential long-term damage to streams, soils, aesthetics, commodity production, and higher risk to property and human safety due to increasingly likely crown fires (Covington and Moore 1994b). Thus, unless strong justification can be made for desired future conditions outside of the system bounds, management direction should attempt to restore past system conditions. This could include active management practices such as thinning from below, prescribed fire, and periodic tree harvest (Keane et al. 1990). The shape and bounds of the bowl, or range of desired future conditions, will be defined by the collective effect of the different management actions.

It is important to note that natural variability does not provide a step by step set of instructions for managers. It can be used to guide general management objectives, or overall policy, but not to define what a desired condition might be nor the detail of how to achieve such a condition (Millar 1996). Identifying desired future conditions and the means for achieving them should be based on operational, economic, and social realities of the management region in question. Haufler et al. (1996) recommend balancing human needs with those of other species by choosing desired future conditions that depart from natural variability only where needs of other species are met. Spatial allocation of such needs, such as setting aside some areas for particular purposes, offers one way to resolve conflicts.

Setting Priorities on Management Actions and Evaluating Risk

In addition to its potential use in identifying overall management direction appropriate to specified management objectives, the natural variability concept can be used in other ways. When natural variability estimates are made for a geographic region, the types, severities, and frequencies of disturbances are described. This information can be manipulated to systematically place priorities on management actions and evaluate risk and hazards caused by fire disturbance. At Sequoia and Kings Canyon National Park in California, for example, current conditions were compared to historic fire regimes to generate a map of departures, quantified as multiples of the natural fire return interval. This information was then used in conjunction with locations of important resources and areas of high visitor frequency to set priorities for management actions such as prescribed burns (Caprio et al. 1997). Information on the spatial distribution of fire origins was used to generate a probability-based risk assessment to classify geographic areas by likelihood of fire starts (fire risk), which could then be evaluated in terms of potential difficulty in fire suppression (fire hazard) (Caprio et al. 1997).

Conclusions

The natural variability concept has developed from several different disciplines toward a common need for coherent management direction across ownership boundaries and for different management objectives. Because of its interdisciplinary origins and flexible application, it is increasingly useful in ecosystem management. In addition to generating increased understanding and better communication of the dynamics of natural systems, it is useful in providing general direction to management and in assessing risk and hazard.

A certain danger exists with this tool, however. Because of the wide flexibility and generality of the concept, the potential exists for misapplication. It is highly recommended that users critically examine the situation to which the natural variability concept is to be applied. In some cases, the natural variability concept will be of limited use, particularly when present conditions are substantially different from those in the past or when information on past conditions is insufficient. If managers decide to use the natural variability concept, we recommend that the geographic and temporal extent used to define natural variability be thoroughly researched and carefully assigned and that assumptions used in building the estimates be explicitly stated so that their merits and impacts can be evaluated.

Acknowledgments
We appreciate financial support from the University of Idaho and the Boise Cascade Corporation.

References
Allen, T.F.H., and T.W. Hoekstra. 1992. Toward a unified ecology. Columbia University Press, New York, NY.

Baker, W.L. 1992a. Effects of settlement and fire suppression on landscape structure. Ecology 73:1879-1887.

Baker, W.L. 1992b. The landscape ecology of large disturbances in the design and management of nature reserves. Landscape Ecology 7(3):181-194.

Barrows, J. S. 1978. Lightning fires in southwestern forests. US Department of Agriculture (USDA) Forest Service, Northern Forest Fire Laboratory, unpublished report, Missoula, MT.

Brown, J.K., S.F. Arno, S.W. Barrett, and J.P. Menakis. 1994. Comparing the prescribed natural fire program with presettlement fires in the Selway-Bitterroot Wilderness. International Journal of Wildland Fire 4:157-168.

Byler, J.W., R.G. Krebill, S.K. Hagle, and S.J. Kegley. 1994. Health of the cedar-hemlock-western white pine forests of Idaho. Pp. 107-118 *in* D.M. Baumgartner, J.E. Lotam, and J.R. Tonn (eds.). Interior Cedar-Hemlock-White Pine forests: Ecology and management. Proceedings of a symposium, 2-4 March 1993. Washington State University Press, Pullman, WA.

Camp, A., C. Oliver, P. Hessburg, and R. Everett. 1997. Predicting late-successional fire refugia pre-dating European settlement in the Wenatchee Mountains. Forest Ecology and Management 95:63-77.

Caprio, A.C., C.M. Conover, M. Keifer, and P. Lineback. 1997. Fire management and GIS: A framework for identifying and prioritizing fire planning needs. In proceedings of 1997 ESRI Conference, San Diego, CA. Online. <esri.com/library/proc 97>.

Covington, W.W., and M.M. Moore. 1994a. Changes in southwestern ponderosa pine forests since Euro-American settlement. Journal of Forestry 92(1):39-47.

Covington, W.W., and M.M. Moore. 1994b. Postsettlement changes in natural fire regimes: Implications for restoration of old-growth ponderosa pine forests. Journal of Sustainable Forestry 2:153-181.

Covington, W.W., and M.M. Moore. 1997. Restoring ecosystem health in ponderosa pine forests of the Southwest. Journal of Forestry 94(4):23-29.

D'Antonio, C.M., and P.M. Vitousek. 1992. Biological invasions by exotic grasses, the grass/fire cycle and global change. Annual Review of Ecology and Systematics 23:63-87.

ECOMAP. 1993. National hierarchical framework of ecological units. United States Department of Agriculture (USDA) Forest Service, Washington, DC. Computer application, Version 1.10 published by Save the Planet Software in 1992.

Gauthier, S., A. Leduc, and Y. Bergeron. 1996. Forest dynamics modelling under natural fire cycles: A tool to define natural mosaic diversity for forest management. Environmental Monitoring and Assessment 39:417-434.

Haila, Y. 1995. Natural dynamics as a model for management: Is the analogue practicable? Pp. 9-26 *in* A.L. Sippola, P. Alaraudanjoki, B. Forbes, and V. Hallikainen (eds.). Northern wilderness areas: Ecology, sustainability, values. Arctic Centre Publications 7, Arctic Center, University of Lapland, Ravaniemi, Finland.

Haufler, J.B., C.A. Mehl, and G.J. Roloff. 1996. Using a coarse-filter approach with species assessment for ecosystem management. Wildlife Society Bulletin 24(2):200-208.

Hann, W.J., J.L. Jones, M.G. Karl, P.F. Hessburg, R.E. Keane, D.G. Long, J.P. Menakis, C.H. McNicoll, S.G. Leonard, R.A. Gravenmier, and B.G. Smith. 1997. Landscape dynamics of the Basin. Pp. 363-1055 *in* T.M. Quigley and S.J. Arbelbide (tech. eds.). An assessment of ecosystem components in the Interior Columbia Basin and portions of the Klamath and Great Basins, Volume II. General Technical Report PNW-GTR-405, US Department of Agriculture (USDA) Forest Service, Pacific Northwest Research Station, Portland, OR.

Harwell, M.A. 1997. Ecosystem management of south Florida. BioScience 47:499-512.

Holling, C.S. 1992. Cross-scale morphology, geometry, and dynamics of ecosystems. Ecological Monographs 62:447-502.

Hunter, M.L., G.L. Jacobson, and T. Webb. 1989. Paleoecology and the coarse-filter approach to maintaining biological diversity. Conservation Biology 2:375-385.

Kaufman, M.R., R.T. Graham, D.A. Boyce, W.H. Moir, L. Perry, R.T. Reynolds, R.L. Bassett, P. Mehlhop, C.B. Edminster, W.M. Block, and P.S. Corn. 1994. An ecological basis for ecosystem management. General Technical Report RM-GTR-246, US department of Agriculture (USDA) Forest Service, Rocky Mountain Research Station, Fort Collins, CO.

Keane, R.E., S.F. Arno, and J.K. Brown. 1990. Simulating cumulative fire effects in ponderosa pine/Douglas-fir forests. Ecology 71:189.

Keddy, P.A., and C.G. Drummond. 1996. Ecological properties for the evaluation, management, and restoration of temperate deciduous forest ecosystems. Ecological Applications 6:748-762.

Landres, P.B. 1992. Temporal scale perspectives in managing biological diversity. Transactions of the North American Wildlife and Natural Resources Conference 57:292-307.

Landres, P.B., P.S. White, G. Aplet, and A. Zimmermann. 1998. Naturalness and natural variability: Definitions, concepts, and strategies for wilderness management. Pp. 41-50 *in* D.L. Kulhavy and M.H. Legg (eds.). Wilderness and natural areas in eastern North America: Research, management, and planning. Center for Applied Studies, Stephan F. Austin State University, Nacogdoches, TX.

Landres, P.B., P. Morgan, and F.J. Swanson. 1999. Overview of the use of natural variability concepts in managing ecological systems. Ecological Applications 9(4): (November issue).

Lehmkuhl, J.F., P.F. Hessburg, R.L. Everett, M.H. Huff, and R.D. Ottmar. 1994. Historical and current forest landscapes of eastern Oregon and Washington. Part I: Vegetation pattern and insect and disease hazards. General Technical Report PNW-GTR-328, US Department of Agriculture (USDA) Forest Service, Pacific Northwest Research Station, Portland, OR.

Lertzman, K., T. Spies, and F. Swanson. 1997. From ecosystem dynamics to ecosystem management. Pp. 361-382 *in* P.K. Schoonmaker, B. von Hagen, and E.C. Wolf (eds.). The rain forests of home: Profile of a North American bioregion. Island Press, Washington, DC.

Lesica, P. 1996. Using fire history models to estimate proportions of old growth forest in northwestern Montana, USA. Biological Conservation 77:33-39.

Manley, P.N., G.E. Brogan, C. Cook, M.E. Flores, D.G. Fullmer, S. Husari, T.M. Jimerson, L.M. Lux, M.E. McCain, J.A. Rose, G. Schmitt, J.C. Schuyler, and M.J. Skinner. 1995. Sustaining ecosystems: A conceptual framework. General Technical Report R5-EM-TP-001, US Department of Agriculture (USDA) Forest Service, Pacific Southwest Region and Station, Berkeley, CA.

May, R.M. 1977. Thresholds and break points in ecosystems with a multiplicity of stable points. Nature 269:471-477.

Millar, C.I. 1996. The Mammoth-June Ecosystem Management Project, Inyo National Forest. Pp. 1273-1304 *in* Status of the Sierra Nevada, Volume II: Assessments and scientific basis for management options. Sierra Nevada Ecosystem Project, Final Report to Congress. Wildland Resources Center Report No. 37, Centers for Water and Wildland Resources, University of California, Davis, CA.

Mladenoff, D.J., and J. Pastor. 1993. Sustainable forest ecosystems in the northern hardwood and conifer region: Concepts and management. Pp. 145-180 *in* G.H. Aplet, J.T. Olsen, N. Johnson, and V.A. Sample (eds.). Defining sustainable forestry. Island Press, Washington, DC.

Morgan, P. 1994. Dynamics of ponderosa and Jeffrey pine forests. Chapter 5 *in* G.D. Hayward and J. Verner (eds.). Flammulated, boreal, and great gray owls in the United States: A technical conservation assessment. General Technical Report RM-253, US Department of Agriculture (USDA) Forest Service, Rocky Mountain Research Station, Fort Collins, CO.

Morgan, P., G.H. Aplet, J.B. Haufler, H.C. Humphries, M.M. Moore, and W.D. Wilson. 1994. Historical range of variability: A useful tool for evaluating ecosystem change. Journal of Sustainable Forestry 2:87-111.

Morgan, P., R.A. Parsons, J. Hauffler, and B. Holt. 1998. Natural variability estimates for the Idaho Southern Batholith. Unpublished data.

Petraitis, P.S., R.E. Latham, and R.A. Niesenbaum. 1989. The maintenance of species diversity by disturbance. Quarterly Review of Biology 64:393-418.

Poff, N.L., J.D. Allan, M.B. Bain, J.R. Karr, K.L. Prestegaard, B.D. Richter, R.E. Sparks, and J.C. Stromberg. 1997. The natural flow regime. BioScience 47:769-784.

Ripple, W.J. 1994. Historic spatial patterns of old forests in western Oregon. Journal of Forestry 92:45-49.

Saunders, D.A., R.J. Hobbs, and C.R. Margules. 1991. Biological consequences of ecosystem fragmentation: A review. Conservation Biology 5:18-32.

Schoonmaker, P.K., and D.R. Foster. 1991. Some implications of paleoecology for contemporary ecology. The Botanical Review 57:204-245.

Skinner, C.N., and C. Chang. 1996. Fire regimes, past and present. Pp. 1041-1070 *in* Status of the Sierra Nevada, Volume II: Assessments and scientific basis for management options. Sierra Nevada Ecosystem Project, Final Report to Congress. Wildland Resources Center Report No. 37, Centers for Water and Wildland Resources, University of California, Davis, CA.

Smol, J.P. 1992. Paleolimnology: An important tool for effective ecosystem management. Journal of Aquatic Ecosystem Health 1:49-58.

Sousa, W.P. 1985. The role of disturbance in natural communities. Annual Review of Ecology and Systematics 15:353-391.

Sprugel, D.G. 1991. Disturbance, equilibrium, and environmental variability: What is "natural" vegetation in a changing environment? Biological Conservation 58:1-18.

Swanson, F.J., J.A. Jones, D.O. Wallin, and J.H. Cissel. 1993. Natural variability – Implications for ecosystem management. Pp. 89-103 *in* M.E. Jensen and P.S. Bourgeron (eds.). Ecosystem management: Principles and applications, Volume II. Eastside forest ecosystem health assessment. US Department of Agriculture (USDA) Forest Service, Pacific Northwest Research Station, Forestry Sciences Laboratory, Wenatchee, WA.

Swanson, F.J., J.A. Jones, D.O. Wallin, and J.H. Cissel. 1994. Natural variability – Implications for ecosystem management. Pp. 80-94 *in* M.E. Jensen and P.S. Bourgeron (eds.). Ecosystem management: Principles and applications, Volume II. Eastside forest ecosystem health assessment. General Technical Report PNW-GTR-318, US Department of Agriculture (USDA) Forest Service, Pacific Northwest Research Station, Portland, OR.

Sweetnam, T.W., and J.L. Betancourt. 1990. Fire-southern oscillation relations in the Southwestern United States. Science 249:1017-1020

Tausch, R.J., P.E. Wigand, and J.W. Burkhardt. 1993. Viewpoint: Plant community thresholds, multiple steady states, and multiple successional pathways: Legacy of the Quaternary? Journal of Range Management 46(5):439-447.

US Department of Agriculture and US Department of the Interior. 1995. Ecosystem analysis at the watershed scale: Federal guide for watershed analysis. Version 2.2. US Government Printing Office 1995-689-120/21215, Region No. 10. Regional Ecosystem Office, Portland, OR.

Wallin, D.O., F.J. Swanson, B. Marks, J.H. Cissel, and J. Kertis. 1996. Comparison of managed and pre-settlement landscape dynamics in forests of the Pacific Northwest, USA. Forest Ecology and Management 85:291-309.

White, P.S. 1979. Pattern, process, and natural disturbance in vegetation. Botanical Review 45:229-297.

White, P.S., and J. Harrod. 1997. Disturbance and diversity in a landscape context. Pp. 128-159 *in* J.A. Bissonette (ed.). Wildlife and landscape ecology: Effects of pattern and scale. Springer-Verlag, New York, NY.

Application of Natural Disturbance Processes to a Landscape Plan: The Dry Warm Interior Cedar-Hemlock Subzone (ICHdw) near Kootenay Lake, British Columbia

Harry Quesnel and Heather Pinnell

Abstract

Literature review and detailed stand sampling were used to revise and update information on natural disturbances, stand types, and structural objectives for a landscape plan in the West Arm Demonstration Forest of southeast British Columbia. The driest sites of the dry warm Interior Cedar-Hemlock subzone (ICHdw) have not had stand-replacement fires for at least 244 to 319 years. Stand-maintaining fires occurred until a century ago and were followed by establishment of ponderosa pine (*Pinus ponderosa* Dougl. ex Laws.), Douglas-fir (*Pseudotsuga menziesii* [Mirb.] Franco), and lodgepole pine (*Pinus contorta* Dougl. ex. Loud.). The fire-free period has yielded higher density stands of mid to late seral species. A third site in the ICHdw had almost complete stand-replacement fire just over a century ago. All of these stands now have greater tree species diversity, lodgepole pine mortality resulting from an interaction between root disease and mountain pine beetle (*Dendroctonus ponderosae* Hopkins), and Douglas-fir affected by root disease and Douglas-fir beetles (*Dendroctonus pseudotsugae* Hopkins). Root disease is usually present at low levels in these stands. The natural disturbance patterns and stand structural objectives used in a Total Resource Design should be modified. Underburning or other disturbances that mimic low intensity fires are desirable for maintaining ponderosa pine and western larch (*Larix occidentalis* Nutt.) in the ICHdw. Managers in this subzone can best address forest health problems by managing for more open stands of ponderosa pine, western larch, and Douglas-fir.

Introduction

The main hypothesis of the Biodiversity Guidebook of the Forest Practices Code of British Columbia (BC Forest Service 1995a) is that biodiversity can be maintained by mimicking natural disturbance patterns. This hypothesis is untested and undemonstrated in British Columbia. Common disturbance statistics such as fire frequency provide average values which may not reflect actual stands nor address all socially desired objectives. An area with an average fire return interval of 150 years could have stands ranging in age from 50 to 250 years. Few stands may be 150 years old. We find many ecological and social reasons for not managing this area as if the entire area had the characteristics of 150-year-old stands. A further consideration is that the disturbance frequency and resulting patterns are controlled by changing climate regimes. Shugart (1998) recently reviewed the evidence for climate change at a variety of temporal scales and the resulting effect on vegetation patterns, concluding that the disturbance patterns of a landscape will be continuously changing.

The value of studying and understanding natural disturbance processes and the resulting patterns is to ensure ecologically sound management rather than to simply copy past disturbances. Thus silviculture prescriptions that retain western larch (*Larix occidentalis* Nutt.) and ponderosa pine (*Pinus ponderosa* Dougl. ex Laws.) as seed sources must also ensure that appropriate microsite conditions exist that allow these species to become established. Appropriate conditions may be created by a variety of silvicultural techniques, not just through large wildfires or slashburns.

Natural disturbance process and pattern information is also valuable for landscape-level planning. For example, decades of fire suppression can result in continuous and homogeneous stands where a mosaic of age classes and structures had existed. These homogeneous stands are at greater risk than diverse stands from more destructive fire, insect, and disease disturbance. Successful application and implementation of a landscape plan for an area requires an understanding of the following: local natural disturbance processes and patterns; the landscape and stand characteristics produced by these processes; and how forest management can integrate natural disturbance processes with silviculture, wildlife, and other objectives. Prescriptions developed within the context of a landscape ecology framework must be compatible with public values including water quality, visuals, opening size, and regeneration method. The key is to find the overlap between ecology and human values.

Landscape ecology is a subdiscipline of ecology that has principles appropriate for landscape planning. These principles are described in Forman and Godron (1986). Landscape ecology and planning have been combined and applied in the Mt. Hood National Forest of Oregon (Diaz and Apostal 1992). The method is described as forest landscape analysis and design. Landscape analysis and design, modified to fit the planning requirements in British Columbia, is called Total Resource Design (Duff 1994a).

The first application of Total Resource Design in British Columbia was the landscape-level plan for the West Arm Demonstration Forest (WADF) near Nelson, British Columbia (Duff 1994a). Details of the application of this design process in the West Arm Demonstration Forest are described in Duff (1994a,b) and Diaz and Bell (1997). One of the authors of this report

(Quesnel) was a participant in a workshop held during January, 1994, in Nelson, British Columbia. Several natural resource inventories and a list of local social values were integrated into a landscape-level plan. The results of an ecological analysis were used to develop ecological pattern objectives for different vegetation types in the West Arm Demonstration Forest. The vegetation types, based on biogeoclimatic ecosystem classification (Braumandl and Curran 1992), were further divided into stand types using hypothesized disturbance regimes and social values. For example, the dry warm Interior Cedar-Hemlock subzone (ICHdw) was divided into dry and moist stand types. Participants described each stand type, suggested appropriate forest management activities, and listed stand structural objectives.

The Total Resource Design plan, based on local expertise and existing resource inventories, lacked detailed literature review and field sampling. The goal of the present study on natural disturbances is to provide feedback to the current West Arm Demonstration Forest landscape management plan. Feedback is a part of adaptive management (Everett et al. 1994; Haney and Power 1996). Stand structural objectives for forest health are also required. We examined the natural disturbance literature for studies applicable to the biogeoclimatic variants for the West Arm Demonstration Forest. The main disturbance agents studied were insects, diseases, fire, wind, and geomorphologic processes. We noted stand structures and landscape features for each type of disturbance. The present study also included field sampling. This report focuses on the driest sites in the ICHdw.

Objectives

The objectives of this study are to:

(1) identify recent natural disturbance processes in the West Arm Demonstration Forest
(2) integrate forest health into stand structural objectives
(3) incorporate natural disturbance information from this project into the landscape plan developed in the Total Resource Design Workshop.

Methods

Study Area, Ecological Unit, and Plot Locations
The West Arm Demonstration Forest is situated near Kootenay Lake in southeast British Columbia. It encompasses 13,500 hectares and extends from lake level to the subalpine. The climax species on mesic sites in the ICHdw include western hemlock (*Tsuga heterophylla* [Raf.] Sarg.) and western redcedar (*Thuja plicata* Donn ex D. Don) (Braumandl and Curran 1992). The driest sites in this subzone are often dominated by early seral ponderosa pine and Douglas-fir (*Pseudotsuga menziesii* [Mirb.] Franco). Relative to the rest of British

Columbia, the ICHdw is characterized by hot, moist summers and mild winters with light snowfall. Mean annual precipitation in this subzone is 740 mm (range = 566 to 941 mm), mean summer precipitation is 243 mm (range = 189 to 300 mm), and mean annual snowfall is 239 mm (range = 122 to 495 mm).

We sampled three stands in the driest site series of the ICHdw. Prior to field sampling we reviewed forest cover and biogeoclimatic ecosystem maps for the main drainages in the West Arm Demonstration Forest to determine the range of common age classes in the ICHdw. We identified potential sampling sites at least 50 hectares in size having relatively good access, and we limited site selection to areas of operable forest. Prior to detailed stand sampling, we traversed stands to verify uniformity and approximate stand age.

Plot Layout and Measurements
We used transect sampling to select four or five plots per site. Plots were at least 50 m from roads, recent harvested areas, and significant changes in timber type. We used a random number between 50 and 150 to determine distance in metres between plots along a transect. While travelling between plots, we made observations on biogeoclimatic ecosystem classification, forest type, apparent stand age, and significant forest health symptoms.

At each plot we collected the following site information: slope, aspect, elevation, site series (Braumandl and Curran 1992), signs of fire or other disturbances, and observations on the ecology of individual tree species. We counted, by species, all trees < 1.3 m tall in a 2.52 m fixed radius plot. We established a 7.98 or 11.28 m fixed radius plot having the same plot centre as the 2.52 m radius plot for measuring trees > 1.3 m tall. For each tree in the larger radius plot we collected the following data: diameter at breast height (dbh), species, crown class, wildlife tree class (Backhouse 1993; Stewart and Todd 1995), and wildlife use. We recorded observations on forest health and damage. We measured heights for 6 to 12 live trees in each plot using clinometers and measured heights for each tree species and for different crown classes. Eight to 12 trees were aged in each plot. We collected ages from the oldest cohort or largest trees, from post-disturbance cohorts such as the largest lodgepole pines (*Pinus contorta* Dougl. ex. Loud.) in a stand with veteran Douglas-fir or ponderosa pine, or from a range of tree sizes, by species, where stand initiating or maintaining disturbances were not obvious. We sampled ages using increment borers at 0.3 m height or at 1.3 m heights where basal rot occurred. We cut with handsaws some suppressed or smaller diameter trees and used cross-sections to age these trees at germination point and 1.3 m height. We used ages from two heights on the same tree to determine correction factors for trees cored at dbh. Some trees could not be aged because they had rotten centres. We also examined cross sections for forest health symptoms, recovery or response to forest health agents, periods of release or suppression, and indications of post-harvest response potential. We aged some trees outside the plot if they were rare, large, old, or unique species not included in fixed area plots on a site. To provide better age estimates on sites with rotten-centred trees or where harvested areas were near a site, we collected sections or cores from nearby larger trees or stumps. We also collected sections or cores from recently cut stumps, fire-scarred trees, or larger trees located within the sample plots.

We collected forest health information from recently dead or suspect trees within or near a plot. This included observations of root collars, boles, and crown condition. Because root disease is expected to occur throughout the ICH zone and infection intensity cannot be reliably determined from above-ground symptoms, we restricted observations to estimating the role of root disease (i.e., degree of pathogenicity) in the stand.

Prior to leaving each site, we made a summary of stand history and development, current silvicultural practices on similar sites, strengths or weaknesses of current practices, and possible alternative silvicultural systems.

In the laboratory, personnel permanently glued increment cores on coreboards and sanded cross-sections with a belt sander using a sequence of coarse, medium, and fine sandpaper. We used a binocular dissecting microscope to count total age plus current 10- and 20-year increments. For cores or sections with rotten centres, we recorded age count as "minimum" age to indicate that total tree age was unknown but must be equal to or greater than the value recorded. We also noted periods of suppression and release.

Data Analysis and Report
We pooled and assessed plot data for each site. To help interpret stand development on each site we plotted tree ages, by species, in 20-year age classes. The plots included number of trees and were not on a per hectare basis as all trees in a plot were not sampled for age. The interpretation of stand development was also improved by plotting 10 cm diameter classes for the tree species on each site, with live and dead trees plotted separately. Because we measured all trees in a plot, these values were in trees/ha. We summarized data by tree species and site, for dbh, height, basal area, and volume. We made graphs and tables for tree species, tree mortality, wildlife tree class, crown class, and forest health agents for each site, plotting mean values by species for current 10- and 20-year increments.

We reviewed the current landscape plan for the West Arm Demonstration Forest. Using the results from literature review and stand sampling, we suggested modifications to the natural disturbance processes and patterns used to develop the Total Resource Design and revisions to the existing structural objectives. We also added structural objectives for forest health for each stand type.

Results

Stand Development and Forest Health on Dry Sites in the ICHdw of the West Arm Demonstration Forest

Kokanee Creek, Upper Slope Site

This site is characterized by open growing ponderosa pine veterans, abundant codominant Douglas-fir, and very little herb, shrub, and moss cover. The site has a 02 site series (Braumandl and Curran 1992). Shallow soils with thin forest floors predominate, scattered exposed bedrock is present, and this site is one of the driest within this biogeoclimatic variant. Moisture and nutrients appear limited on this site. The oldest trees on this site were

Table 1

Fire history for the Dry Warm Interior Cedar-Hemlock (ICHdw) subzone, Kokanee Creek, West Arm Demonstration Forest, southeast British Columbia.

| Century | Kokanee Creek, mid-slope (ICHdw/01a) | | | Kokanee Creek, upper slope (ICHdw/02) | | |
	Fire year	Years before present	Interval since last fire (yr)	Fire year	Years before present	Interval since last fire (yr)
2000						
			Fire-Free Period			
1900						
	1890	107	8			
	1882*	115	12	1884	113	18
	1870	127	4			
	1866	131	17	1866	131	43
	1849	148	6			
	1843	154	12			
	1831	166	19			
				1823	174	11
	1812	185		1812	185	16
1800						
				1796	201	34
				1762	235	
Number of fires	8			6		
Number of intervals	7			5		
Mean fire return interval	11.1			24.4		
SE (yrs)	5.55			13.5		
Range (yrs)	4-19			11-43		
Oldest tree (yrs)	244			319		

* First mining activity on Kootenay Lake, 1882-1883.

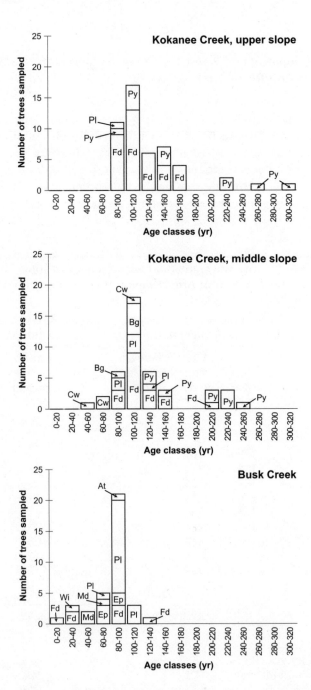

Figure 1. Age structure for three sites in the Dry Warm Interior Cedar-Hemlock (ICHdw) subzone. At = trembling aspen; Bg = grand fir; Cw = western redcedar; Ep = paper birch; Fd = Douglas-fir; Md = Douglas maple; Pl = lodgepole pine; Py = ponderosa pine; Wi = willow.

outside the fixed measurement plots and were 320- and 266-year-old ponderosa pines. Thus a stand-replacement fire has not occurred on this old-growth site for at least 320 years. Fire scar analysis of three trees indicates six stand-maintaining fire events occurred on this site during the 1700-1800s and have not occurred since 1884 (Table 1). The average fire return interval was 24 years (SE = 13.5 years). Fire events of 235, 185, 174, 131, and 113 years before present, corresponding to calendar years 1762, 1812, 1823, 1866, and 1884, were followed by recruitment of ponderosa pine and/or Douglas-fir (Figure 1). Lodgepole pine recruitment at approximately 100 years before present indicates an undocumented fire, as this species is fire-dependent for regeneration (Lotan and Critchfield 1990). The 100- to 120-year-ago time period also corresponds to significant recruitment of Douglas-fir and ponderosa pine (Figure 1).

Douglas-fir with lesser ponderosa pine dominates the current stand (Figure 2). Also present are lodgepole pine, grand fir (*Abies grandis* [Dougl. ex E. Don] Lindl.), and Douglas maple (*Acer glabrum* Torr. var. *douglasii* [Hook.] Dippel). The only species regenerating on this site is Douglas-fir, although grand fir is present in the smallest diameter classes (Figure 2; Table 2). Lodgepole and ponderosa pines are conspicuously missing from the smallest diameter classes of trees (Figure 2) and in the regeneration cohorts (Table 2). Thus a stand-establishment event over 320 years ago has been followed by a series of stand-maintaining fires, most fires are correlated with ponderosa pine or Douglas-fir recruitment, and the last fire event included some lodgepole pine regeneration. Fire has been missing from the site for at least 100 years. Maintaining the pine species on the site will eventually require a prescribed or natural fire, or another type of disturbance. Alternatively, grand fir has developed during the recent fire-free decades and active fire suppression would be required to maintain this species on the site.

Table 2

Tree regeneration on three sites in the Dry Warm Interior Cedar-Hemlock (ICHdw) subzone, West Arm Demonstration Forest, southeast British Columbia.

	Site		
	Kokanee Creek, mid-slope	Kokanee Creek, upper slope	Busk Creek
Species	(germinants/ha)*		
Douglas-fir	750	500	300
Grand fir	625	0	0
Trembling aspen	0	0	100
Western redcedar	375	0	0

* Living trees < 1.3 m tall.

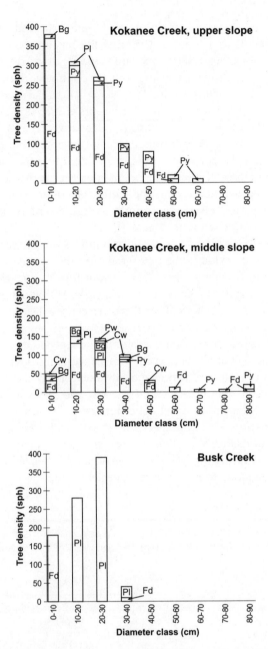

Figure 2. Diameter structure for live trees on three sites in the Dry Warm Interior Cedar-Hemlock (ICHdw) subzone. Douglas maple is present (< 20 sph) in the < 20 cm diameter classes at both Kokanee Creek sites. In the < 20 cm diameter classes, Busk Creek has Douglas maple (510 sph), paper birch (270 sph), willow (190 sph), and trembling aspen (50 sph). Bg = grand fir; Cw = western redcedar; Fd = Douglas-fir; Pl = lodgepole pine; Pw = western white pine; Py = ponderosa pine.

We observed few forest health agents on this site. Most mortality and damage is occurring in smaller (< 20 cm dbh) Douglas-fir (Figure 3). Fifty-two percent of these smaller trees are either dead, bent, thin crowned, or broken topped. Damage has been caused by competition and snow press, and most of these trees will not survive. In comparison, little and less threatening damage is occurring in larger diameter classes and consists mainly of small basal scars. We noted a large, dead Douglas-fir tree but it was fire-scarred and likely died during the last fire. No evidence of root disease or bark beetles is apparent in the larger diameter classes. Crowns of the largest Douglas-fir are full and healthy and exposed root systems do not have obvious signs of root disease. Ponderosa pine snags between plots are too decayed to determine cause of death.

Stand density appears to have increased during the last century because of fire exclusion. The dominant tree species has shifted from ponderosa pine to Douglas-fir as indicated by increased frequency of Douglas-fir in the small diameter classes (Figure 2). It appears that competition and snow damage are currently the main stand thinning agents. Stand structures on this site are developing like those in the northwestern United States where increasing frequency of Douglas-fir has resulted in greater incidence of defoliators, Douglas-fir mistletoe (*Arceuthobium douglasii* Engelmann), Douglas-fir beetles (*Dendroctonus pseudotsugae* Hopkins), and root disease (Hessburg et al. 1994). At present, defoliators are not common in the West Arm Demonstration Forest. A few have been trapped, but numbers indicate endemic levels only (Stewart 1994). Defoliation has not been observed. It is likely the West Arm Demonstration Forest does not at present have the appropriate climatic conditions for these defoliators to become epidemic (L. Unger, pers. comm. 1997). Douglas-fir mistletoe has not yet been observed in the West Arm Demonstration Forest and only occurs nearby in drier ecosystems along the east shore of Kootenay Lake between the towns of Creston and Gray Creek.

Although we observed neither bark beetles nor root diseases on this site, there is great concern in the West Arm Demonstration Forest about increasing risk of their presence. Douglas-fir bark beetle populations have been increasing throughout the West Arm Demonstration Forest in recent years. When comparing stand characteristics of this site to hazard parameters within the Bark Beetle Management Guidebook (BC Ministry of Forests and Ministry of Environment, Lands, and Parks 1995a), one notes several factors contributing to a high to very high hazard for Douglas-fir beetle outbreaks. These include: tree age (120 years), warm aspect, steep slopes (60 percent), moderate density (990 sph), high proportion of Douglas-fir (73 percent of total basal area), and declining vigour of the Douglas-fir. The last 10-year increment is 6.8 mm as compared to 10.2 mm during the previous decade (Figure 4). The diameter size classes for Douglas-fir are shifting

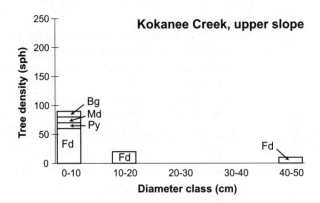

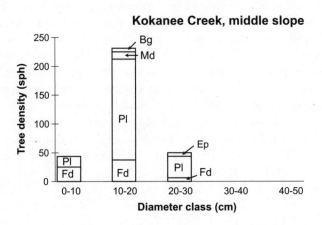

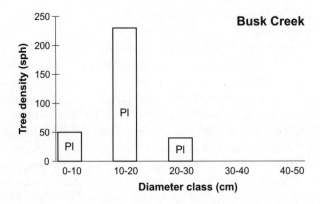

Figure 3. Diameter structure for dead trees on three sites in the Dry Warm Interior Cedar-Hemlock (ICHdw) subzone. Bg = grand fir; Ep = paper birch; Fd = Douglas-fir; Md = Douglas maple; Pl = lodgepole pine; Py = ponderosa pine.

toward high and very high hazard sizes (> 36 cm dbh) (Figure 2). Risk has also increased in recent years as beetle attack has occurred throughout the West Arm Demonstration Forest. Other factors that may be decreasing Douglas-fir vigour and increasing stand hazard include root disease, water stress, and breakage (Furniss et al. 1979). Introducing fire at this point could also increase stand hazard as scorched Douglas-fir are very attractive to beetles (Furniss 1965; Rudinsky 1966; Simon et al. 1994). Because beetles are also attracted to logging slash, stumps, and decked logs (Atkins and McMullen 1960; Lejeune et al. 1961), managers must exercise caution if partial cutting is planned for this site or if any harvesting is planned nearby. A final consideration regarding future bark beetle hazard is the effect of climate on dispersal of the insects. When temperatures are cool (< 20° C) and conditions are wet during flight times, as is often the case in the West Arm Demonstration Forest, beetles disperse nearby rather than great distances and will attack standing trees if downed material is not available (Atkins and McMullen 1960). This results in local build-ups and possible epidemics. In high hazard stands, windthrow can further increase risk of attack, and managers must exercise caution when planning partial cutting in areas prone to blowdown.

At present, root disease does not appear to be causing mortality at the Kokanee Creek upper slope site and the dry site conditions may reduce risk. Rhizomorphs stop growing if soil moisture falls below a critical level (Cruickshank et al. 1997). Root disease was identified, however, in the moister adjacent stand, and it may become active at this site if Douglas-fir and grand-fir densities increase.

Kokanee Creek, Mid-Slope Site

This site occurs on the middle to lower slopes directly below the previous site. It is characterized by deeper soils, denser stands, and more mesic vegetation species than the upper slope site. This mesic site is part of the moist 01a site series within the ICHdw subzone (Braumandl and Curran 1992). The oldest trees on this site were four ponderosa pines sampled for fire scars from outside the fixed plots. These trees were 218 to 244 years old and have many features associated with old-growth forests. A stand-replacement event, probably a fire, has not occurred on this site for at least 244 years. Fire scar analysis of these old ponderosa pines found eight separate fire events (Table 1). These events occurred, on average, every 11.1 years during the 1800s. The fire frequency is shorter on this site than on the very dry site directly above. The lower intensity fires from the middle slopes do not appear to have extended upslope. A fire event has not been documented on the Kokanee Creek mid-slope site since 1890 (Table 1). Most fire events were followed by Douglas-fir and ponderosa pine establishment with lodgepole pine established after the last three fires (Figure 1). Grand fir (116 years)

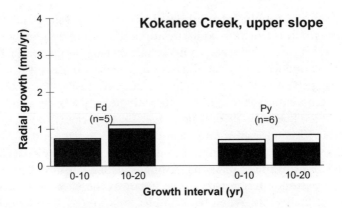

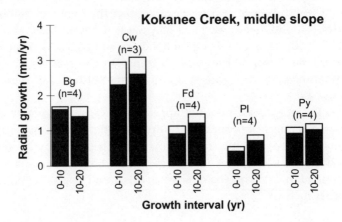

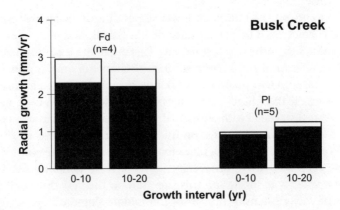

Figure 4. Radial growth for selected conifers on three sites in the Dry Warm Interior Cedar-Hemlock (ICHdw) subzone. Mean is indicated by solid bars, standard error by open bars. Bg = grand fir; Cw = western redcedar; Fd = Douglas-fir; Pl = lodgepole pine; Py = ponderosa pine.

germinated prior to the last two fires. Survival of this relatively more fire sensitive species (Agee 1993) suggests that the last two fires were spotty and uneven on the sites and that moister microsites were missed. Lack of fire has allowed western redcedar to establish during the last century as indicated by age and size class distribution and current regeneration (Figures 1 and 2; Table 2).

The present stand has veteran ponderosa pine and Douglas-fir. The Douglas-fir dominates all layers in the canopy and stagnant, dying lodgepole pine occurs in the subdominant canopy layer (Figures 2 and 3). Grand fir, western redcedar, paper birch (*Betula papyrifera* Marsh.), and Douglas maple occur in the intermediate layer (Figure 2). Regeneration includes Douglas-fir, grand fir, and western redcedar (Table 2). Age distribution by species is similar to diameter distributions (Figures 1 and 2). The lack of ponderosa pine in the understorey and regeneration cohorts emphasizes the need for fire or another type of disturbance to maintain this species on the site. Alternatively, lack of fire has been beneficial for establishment of grand fir and western redcedar.

The most obvious forest health problem on this site is lodgepole pine mortality. Sixty-nine percent (125 sph) of mature lodgepole pine are dead (Figures 2 and 3). The cause of mortality is suspected to be an interaction between *Armillaria* root disease (*Armillaria ostoyae* [Romagn.] Herink) and mountain pine beetle (*Dendroctonus ponderosae* Hopkins). Less than one percent of the lodgepole pine also have *Atropellis* canker (*Atropellis piniphila* [Weir] Lohman and Cash). *Armillaria* was also confirmed in a dead Douglas-fir (20.5 cm dbh) that was then attacked by Douglas-fir beetles. Two adjacent Douglas-fir have thinning crowns indicating potential root disease. Root disease is more active here than at the upslope site. This may be because the mid-slope site at Kokanee Creek is moister with deeper soils and greater tree densities. The highest incidence of root disease has been found in more productive habitat types in studies in the Lolo National Forest (Byler et al. 1987) and in the Idaho Panhandle National Forest (Hagle 1985). Byler et al. (1987) found the highest incidence of root disease in Douglas-fir, grand fir, western redcedar, and western hemlock stands that are comparable to Interior Cedar Hemlock stands in the West Arm Demonstration Forest. Lodgepole pine vigour, as observed in incremental growth, is rapidly declining (Figure 4). This tree species has passed its maximum growth potential on this middle slope and root disease is creating gaps for understorey release. Drought conditions in 1994 may also have increased susceptibility to root disease. Western redcedar and grand fir have the greatest incremental growth (Figure 4). Lodgepole pine will continue to decline and the more shade-tolerant species will release and persist.

The more frequent occurrence of fires during the previous century would have maintained lower stand densities, and mortality caused by root disease may not have been obvious. The current developmental patterns in this stand may follow those in the northwest United States where increased tree densities combined with a shift in tree species are resulting in increased pathogenicity of root disease (Hessburg et al. 1994). At this time, pathogenicity on the mid-slope site appears low as nearly all mortality is occurring in low-vigour lodgepole pine. Without disturbance, persistent and highly pathogenic root disease centres may develop.

Douglas-fir beetle hazard may be lower on this site than at the upslope site, due to proportionately fewer Douglas-fir. Hazard remains high, however, because of overall high stand density (838 sph), warm aspect, moderate slope, and low growth rate of large-diameter trees (Figure 4). During field sampling, we found beetles in trap trees between plots. As root disease is often associated with Douglas-fir beetle attack (James and Goheen 1981), these interactions may result in increased mortality over time. Because very few living lodgepole pine remain, hazard from mountain pine beetle is very low.

Busk Creek

This xeric site has shallow soils with bedrock outcrops, very thin forest floors, and a western aspect. The vegetation, soil, and other site characteristics typify a 02 site series in the ICHdw variant (Braumandl and Curran 1992). Moisture stress may exist on this site. Extensive occurrence of charcoal on this site suggests past intense fire activity. The oldest tree on the site was a 123-year-old Douglas-fir sampled outside the fixed-radius plots. We could not determine whether this tree was part of a post-fire or between-fire cohort. A fire event is suspected at 102 years or slightly earlier. Six lodgepole pines were established 99 to 102 years ago (Figure 1). Continuous recruitment of this species occurred for the next two decades (Figure 1). A few Douglas-fir, paper birch, and trembling aspen (*Populus tremuloides* Michx.) also established during the two decades after this suspected fire event (Figure 1). Douglas-fir, Douglas maple, and willow (*Salix* spp. L.) have established in the decades up to the present.

The present stand is dominated by lodgepole pine, a few Douglas-fir veterans are present, and the understorey has many deciduous stems in addition to Douglas-fir (Figure 2). One veteran ponderosa pine was observed on an inaccessible rock outcrop at the top of the site. Stem breakage or windthrow of living and recently dead lodgepole pine is common on the site. Current regeneration consists of Douglas-fir and trembling aspen (Table 2).

Thirty-one percent of all lodgepole pine (Figures 2 and 3) have died. The probable cause is an interaction between mountain pine beetle and root disease. Pitch tubes from beetles were obvious at this site and root disease was not confirmed in the pine. The greatest mortality, however, occurs in the smaller pine (Figure 3). Since bark beetles prefer larger trees, it is likely that the smaller trees were weakened by root disease. *Armillaria* was confirmed in a few understorey Douglas-fir (< 2 m height) examined between plots, therefore the disease likely exists on lodgepole pine. Mountain pine beetle hazard for this stand (BC Ministry of Forests and Ministry of Environment, Lands and Parks 1995a) is only moderate. Although the lodgepole pine are all older than 80 years and represent 97 percent of the conifer basal area, the low density of live conifers (700 sph) reduces the stand hazard. Lodgepole pine is more vigorous on this site than on nearby mixed stands. The increment for this species has decreased during this decade compared to the previous decade (Figure 4).

Discussion

Disturbances in the Dry ICHdw of the West Arm Demonstration Forest

Fire

The Interior Cedar-Hemlock dry warm (ICHdw) subzone variant is distinguished from the adjacent moist warm Interior Cedar-Hemlock variant (ICHmw2) by the presence of large ponderosa pine veterans on the driest sites and by grand fir on the moister sites (Braumandl and Curran 1992). The ponderosa pine stands in the ICHdw appear to reflect a different fire regime from the rest of the subzone, which lacks ponderosa pine. Fire disturbance studies comparable to the ICHdw stands of the West Arm Demonstration Forest include fire ecology studies from northeastern Washington, northern Idaho, and northwestern Montana. In these studies, comparable stands are described as ponderosa pine type (Brown et al. 1994), driest part of the grand fir series (Agee 1994), and driest western hemlock-western redcedar forests (Arno and Davis 1980). The structure of these forests is created by frequent stand-maintaining fires that occur on average every 25 to 100 years (Arno 1980; Smith and Fischer 1997). These low intensity fires do little damage to the overstorey trees other than fire scarring ponderosa pine and Douglas-fir (Arno and Davis 1980; Brown et al. 1994). These open stands are dominated by ponderosa pine with lesser Douglas-fir, western larch, and lodgepole pine (Arno 1980; Agee 1994). Moister sites in these forests have a greater range of fire intensities including stand-replacement fires (Arno 1980; Brown et al. 1994). Stand-replacement fire can occur because greater fuel buildup results from a longer fire-free period. Fire frequencies on these moister sites have an estimated frequency of 70 to 250 years (Arno 1980). The result is post-fire stands on moist sites that can be dominated by a variety of species including Douglas-fir, grand fir, lodgepole pine, and western larch (Brown et al. 1994; Agee 1994). Lack of fire on these sites that included active fire suppression has resulted in denser stands of Douglas-fir and grand fir during this century.

We found that stand-replacement fires had not occurred on two older stands in the ICHdw of the West Arm Demonstration Forest for at least 244 to 319 years (Table 1). A series of stand-maintaining fires was followed by establishment of ponderosa pine and Douglas-fir (Figure 1). The last two stand-maintaining fires allowed lodgepole pine to establish. These two sites have been fire-free for a century, Douglas-fir is now a more dominant part of these sites (Figure 2), and regeneration on these sites is primarily Douglas-fir (Table 2). The mid-slope site also has grand fir and western redcedar regeneration. A younger stand on a dry, westerly aspect had a fire approximately 102 years ago that caused almost complete stand mortality. This fire was followed by dense lodgepole pine regeneration (Figure 1), a few Douglas-fir established within two decades of the fire, and the regeneration cohort (Table 2) plus the smallest diameter classes of the present stand consist of Douglas-fir and trembling aspen (Figure 2).

Our results for the three stands sampled in the ICHdw are comparable to other studies we have described from Washington, Idaho, and Montana. Prior to a century ago, two of our stands were fire-maintained and were dominated by ponderosa pine with some Douglas-fir. The fire return interval was approximately the same as the estimated minimal frequency for other dry, open stands of ponderosa pine and Douglas-fir (Arno 1980). Western larch is mentioned as a less frequent dominant in these stands. Although not sampled in our plots, we have observed this species at low frequency in other stands in the ICHdw. The stand dominated by lodgepole pine at Busk Creek fits the description of a stand-replacement fire resulting from less frequent disturbance. Lack of fire for approximately a century on all three of our sites is yielding denser stands and a shift toward Douglas-fir and grand fir. Western redcedar, trembling aspen, and paper birch have also established during this fire-free interval. Thus the ICHdw stands with veteran ponderosa pine in the West Arm Demonstration Forest probably had high frequency, low intensity ground fires that created an open stand structure. These fires also created suitable germination sites for ponderosa pine, western larch, Douglas-fir, and lodgepole pine. Moister sites had more intensive fires at longer intervals. This allowed grand fir to be a more dominant species in these stands.

Insects and diseases

We found that in the West Arm Demonstration Forest the main active forest health agents were *Armillaria* root disease, the interaction of this disease with Douglas-fir and mountain pine beetles, and the independent action of these insects. We suspect there is potential for future activity in these forests from defoliating insects, bark beetles, and Douglas-fir mistletoe. The following paragraphs describe the disturbance regimes of these insects and diseases and how their activities have changed with lack of fire, including active fire suppression. In particular, we are looking for silvicultural activities that can control or minimize the effects of these forest health agents on tree growth.

Numerous studies from Washington, Idaho, and Oregon describe how selective timber harvesting and fire exclusion in the Douglas-fir and grand fir series have altered these stands from open-growing western larch and ponderosa pine stands to over-stocked, multi-storied stands dominated by Douglas-fir, grand fir (Mutch et al. 1993; Brennan and Hermann 1994; Hessburg et al. 1994; Campbell and Liegel 1996), and western hemlock (Lehmkuhl et al. 1995). These forested series are similar climatically and vegetatively to the ICHdw variant. Historically, fire was the more prominent disturbance in these forests and an important mechanism for thinning stands (Campbell and Liegel 1996). Insects and diseases had lower impact on these forests, creating patches of structural diversity and wildlife habitat. Fire exclusion has created conditions allowing increased disease (Rogers 1996). In the northwestern states, outbreaks of western spruce budworm (*Choristoneura occidentalis* Freeman) and Douglas-fir tussock moth (*Orgyia pseudotsuga* [McD.]) are now more frequent and severe in fire-suppressed stands (Hessburg et al. 1994). Bark beetle mortality is greater as stands are more stressed from overstocking and defoliation. The changes to forest structure have increased root disease incidence and dwarf mistletoes (Campbell and Liegel 1996).

Historically, western pine bark beetles (*Dendroctonus brevicomis* Leconte) typically remained at endemic levels attacking single ponderosa pine trees that were damaged, stressed, injured, or suppressed (Hessburg et al. 1994; Campbell and Liegel 1996). The resulting snags created diversity and provided abundant food and nesting sites for wildlife species (Hessburg et al. 1994; Lehmkuhl et al. 1994). Ponderosa pine becomes more susceptible to western pine bark beetles as vigour declines (Lehmkuhl et al. 1994). Reduced vigour of this ponderosa pine results from competition, drought, or age. Increased density yields more root contact for root disease spread. Thus the combination of reduced vigour and greater root contact cause greater mortality (Hessburg et al. 1994). On the other hand, western pine bark beetle hazard decreases in some areas as the number of hosts declines below a threshold.

Mountain pine beetles are important mortality agents of lodgepole pine. Historically, outbreaks were not severe in eastern Oregon and Washington due to greater structural diversity across the landscape (Hessburg et al. 1994). In the past, fires in pine stands were more frequent, less widespread, and more random. Due to variable tree age and size distribution across the landscape, beetles threatened only small patches of the landscape at a given time. Also, by thinning from above, the beetles produced an uneven-aged stand that better withstood beetle outbreaks (Haack and Byler 1993). Present-day stands of lodgepole pine are homogeneous over large areas allowing beetle populations to build rapidly and to persist (Hessburg et al. 1994). Outbreaks of mountain pine beetle in areas of eastern Washington and Oregon are now more extensive and damaging than historic levels. Furthermore, hazard is increasing in other stands as the trees approach a susceptible size (Lehmkuhl et al. 1994). Alternately, where seral species are declining and climax species invading, hazard across the landscape is decreasing.

Douglas-fir beetle numbers increase as Douglas-fir trees begin to dominate stands similar to the ICHdw in eastern Washington and Oregon (Hessburg et al. 1994). Mortality caused by Douglas-fir beetle has increased as the proportion of stressed Douglas-fir has increased in stands. Susceptibility increases following repeated defoliator attacks, increased drought, and greater incidence of root disease. Much of the current damage in eastern Oregon and Washington occurs in riparian areas and along ridgetops where trees are large. These areas were missed by fires (Hessburg et al. 1994).

Pine engraver beetles (*Ips pini* Say) attack young, dense pine stands, especially on poor, dry sites. Historically, frequent fires kept stands from getting dense and engraver beetle populations typically remained low (Hessburg et al. 1994). Douglas-fir beetles (*Scolytus unispinosus* Leconte) and fir engraver beetles (*S. ventralis* Leconte) are similar in their role of removing weakened, fire-scorched trees (Hessburg et al. 1994). Fir engraver beetle populations increase during periods of drought, in areas with root disease, and after western spruce budworm defoliation (Hessburg et al. 1994; Campbell and Liegel 1996).

Defoliation by western spruce budworm and Douglas-fir tussock moth in the Douglas-fir and grand fir series has increased in incidence and duration. The causes are increased stocking density and a species shift to Douglas-fir following fire suppression in these stands (Williams et al. 1980; Brennan and Hermann 1994; Mutch 1994). Tree ring analysis showed that historic outbreaks were small in area and short in duration (Hessburg et al. 1994). The outbreaks were limited by discontinuous distribution of host species across the landscape (Hessburg et al. 1994). For example, Douglas-fir occurred in pockets repeatedly missed by fire. Hazard increased with greater host abundance, stand density, canopy layering, and more continuous stands across the landscape (Lehmkuhl et al. 1994). Recent insect outbreaks have been larger, more intense, and more frequent (Campbell and Liegel 1996). Increased defoliation causes greater tree damage and increases the susceptibility to bark beetles, root disease, and other pests. Although defoliator hazard is currently low in the West Arm Demonstration Forest, as determined using the Defoliator Management Guidebook (BC Ministry of Forests and Ministry of Environment, Lands and Parks 1995b), this example provides insight regarding potential impact following disruptions of natural disturbance regimes.

Armillaria root disease can act as a primary or secondary mortality agent varying with habitat type and stand development history (McDonald et al. 1987a). In frequently burned stands, root disease plays a more opportunistic role by killing only trees stressed by competition, fire scorch, lightning, mechanical injury, or declining vigour (Hessburg et al. 1994). The resulting small scattered gaps increase structural diversity of landscapes and stands as various tree, shrub, and herb species invade. These gaps also provide forage and cover for wildlife (Hessburg et al. 1994).

In contrast, areas repeatedly missed by fire, such as wet or rocky areas, develop shade-tolerant understories below the overstorey of seral species. As tree densities increase and species shift to those more susceptible to root disease, such as Douglas-fir and grand fir, visible and persistent root disease centres develop (Byler et al. 1993; Hessburg et al. 1994; Campbell and Liegel 1996). There is concern that fire exclusion combined with past harvesting of more disease-resistant seral species, including western larch and western white pine (*Pinus monticola* Dougl. ex D. Don), is yielding more late seral or climax stands. The result is increased root disease hazard over a large portion of the landscape (Byler et al. 1993; Hessburg et al. 1994; Lehmkuhl et al. 1995). Regeneration within these disease centres continues to be killed (Byler et al. 1993). Root disease has progressed in these stands from endemic to epidemic (Hessburg et al. 1994). *Armillaria* appears to increase in activity with current forest management, including greater fire control (Shaw et al. 1976).

McDonald et al. (1987b) found *Armillaria* to be outside its ecological range in the cold-dry and hot-dry sites in the Northern Rocky Mountains of the United States. Hot-dry habitat types include the warmest and driest stands of Douglas-fir and ponderosa pine. Cold-dry habitats include the coldest and driest stands of subalpine fir (*Abies lasiocarpa* [Hook.] Nutt.). *Armillaria* was found in all cedar-hemlock and most grand fir forest types. Several studies have concluded that *Armillaria* distribution and pathogenicity vary with moisture conditions. In British Columbia, all slope positions in the Interior Cedar Hemlock subzone were found to favour *Armillaria* growth (Cruickshank et al. 1997). In Idaho, wet sites with low water retention capability and dry sites with good water retention capability favour *Armillaria* growth (Williams and Marsden 1982). In greenhouse experiments, greater incidence occurred in coarse soils (sandy loam) versus dense clay loam soils. This may be related to different moisture levels (Blenis and Mugala 1989). Disease incidence is often higher if the site is disturbed by harvesting, road building, or spacing (McDonald et al. 1987a). *Armillaria* is also least pathogenic on more productive sites. In addition to *Armillaria*, researches have noted an increase in S-type annosus root disease (*Heterobasidion annosum* [Fr.] Bref.) of true fir in old, selectively harvested areas. This disease enters the host through stump surfaces (Campbell and Liegel 1996).

Susceptibility to dwarf mistletoe (*Arceuthobium* spp.) is low in stands with reduced tree densities, fewer canopy layers, and high diversity of tree species (Campbell and Liegel 1996). Dwarf mistletoe is generally less prevalent in stands with frequent fire intervals. Moderate- to high-intensity fires destroy infected trees, and this disease slowly invades from perimeter stands as the new stands develop. Dwarf mistletoe distribution and persistence has increased in stands where fire exclusion has allowed Douglas-fir densities to increase. In northern Idaho, Douglas-fir dwarf mistletoe infected 43 percent of the Douglas-fir of one study area (Hessburg et al. 1994; Campbell and Liegel 1996).

Larch dwarf mistletoe (*Arceuthobium laricis* [Piper] St. John) was the most prevalent parasitic species prior to fire exclusion (Hessburg et al. 1994; Campbell and Liegel 1996). The deciduous nature, high crowns, and thick bark of western larch make mature trees very fire resistant. This increases the probability that larch mistletoe plants will survive. Where mistletoe-infected western larch persists after low-intensity fires, the mistletoe rapidly spreads to the new stand (Hessburg et al. 1994). Western larch also has few insect and disease pests and readily persists in fire-maintained stands. Lehmkuhl et al. (1995) found, however, that larch dwarf mistletoe had decreased or remained constant in the forests of eastern Washington and Oregon. The frequency of western larch may have decreased because of past timber harvesting.

This review of insects and diseases currently or potentially affecting the ICHdw forests of the West Arm Demonstration Forest suggests silvicultural practices that should minimize the impacts of these forest-health agents. In general, stands should have lower stocking densities and a variable species composition compared to current stands. Lower densities of Douglas-fir would reduce the susceptibility of this species to Douglas-fir beetle, *Armillaria* root disease, and defoliating insects. Mortality of lodgepole pine could be reduced by creating a landscape mosaic of stands with variable ages and sizes of lodgepole pine interspersed with stands dominated by other tree species. Individual stands of lodgepole pine should also have variable ages and sizes of trees. Western larch, ponderosa pine, and blister rust-resistant western white pine should form a large proportion of these relatively open stands in the West Arm Demonstration Forest. These stands should also have proportionally fewer Douglas-fir and grand fir. Stands dominated by early seral species should be more resistant to *Armellaria* root disease.

Application

Assessment of Natural Disturbance Processes for the ICHdw of the West Arm Demonstration Forest Landscape Plan

We summarize in Table 3 the natural disturbance processes used in the West Arm Demonstration Forest Total Resource Design Workshop (Duff 1994a,b; Diaz and Bell 1997). We updated the process information derived from literature review and stand sampling in the same table. This table also includes literature values and a few calculations that use published fire history values. Notes, observations, and some data from the natural disturbance study are also listed. This includes information on fire, insects, diseases, wind, and other stand disturbance processes. Based on literature review and stand sampling from the present study, the dry stand type in the ICHdw described in the Total Resource Design Workshop for the West Arm Demonstration Forest does not merit modification or subdivision, but some modifications or additions to management activities and structural objectives are recommended.

The range of fire frequencies for surface and stand-replacement fires used in the Total Resource Design Workshop was smaller than the range of values from the published literature (Tables 3 and 4). The workshop values for surface fires were comparable to our fire scar assessment (Table 1). The stand-replacement interval (SRI) value of 118 years for the ICHdw found in a recent study for the Nelson Forest Region (Pollack et al. 1997) was within the workshop range. This SRI represents the average for stand-replacement fires in the ICHdw subzone. The Workshop disturbance values for fire in this unit, while acceptable, could be assigned a greater range that includes more frequent fires. The literature values for ground fires and the values for most frequent mixed-intensity fires are similar (Table 3), which suggests a gradation in terms of fire frequency and type. Thus a wide range of fire type, frequency, and disturbance area likely occurred in the ICHdw, although we have insufficient information available to know whether the full range has occurred within the West Arm Demonstration Forest drainages. Most fire history studies refer to large areas. It is easy to envision individual watersheds within a study area that would have greater or lesser frequency of fires because local climate, topography, and proximity to other forests where fires have started vary. The fire scar analysis of the present study found that the average frequency of stand-maintaining fires for stands in the ICHdw was more frequent than for individual trees. Time between fires at an individual tree may be twice the interval of the stand as a whole. It is better to view stand-maintenance burns as affecting different parts of the stand for each fire. Thus mimicking fire in this variant should consider fire treatments at different intervals for different parts of a stand.

Table 3

A summary of disturbance patterns for the Dry Warm Interior Cedar-Hemlock (ICHdw) subzone in the West Arm Demonstration Forest derived from a Total Resource Design Workshop (January 1994, Nelson, BC) and natural disturbance study.

Fire	Disease/Insects	Wind	Comments
Total Resource Design Workshop[a]			
Frequency - surface fires: 30-50 years. Surface fires start on rock outcrops, coarse soils, and S & W facing slopes. Surface fires produce open stands & less dense fuel loading. Surface + crown fires: 100-150 years.	Lots of insects in this zone. Armillaria root disease indicated by hardwood patches. Eventually replaced by coniferous species (35-40 years). Allow for stand composition to change with seral stages.	Wind funnels up and down Kootenay Lake, occasionally deflects into side creeks. Wind creates small gaps with some trees left in the gaps. Wind activity linked to dead/dying trees.	
Natural Disturbance study			
(i) Type: ground or stand-maintaining fires. Frequency: 18 yr (Arno 1980); 4-43 yr (mean= 11-24 yr) (present study)	Root disease most obvious in moist sites, present throughout the ICHdw and in centres. Hazard increasing where tree densities are increasing and species are shifting to more susceptible species.	Stem breakage or windthrow of living or recently dead lodgepole pine common on some sites, especially dry or shallow soils. Snow storm enhanced wind damage.	Time since last stand-replacing fires for Busk Creek site is 102 years. Time since complete stand-replacement for Kokanee Creek sites is 244-314 years. These sites have had stand-maintenance burns since then. Ground fires affect only parts of stands, reburn of individual trees ranged from 4-43 years.
(ii) Type: mixed intensity fires. Frequency: 33-150 yr (Agee 1994; Arno and Davis 1980)	Douglas-fir beetle incidence and hazard increasing		Fire exclusion has increased hazard for many pests due to shift in tree species

(iii) Type: crown or stand-replacement fires. Frequency: 24-300 yr (McCune 1983; Brown et al. 1995); 118 yr (Pollack et al. 1997).

Estimated rate of stand-replacement fires[b]: 9.1-114.1 ha/yr (0.33-4.17%/yr). Estimated rate of ground fire: 137.6 ha/yr (5.5%).

especially on dry sites with increasing fir component. Douglas-fir dwarf mistletoe and defoliators absent from WADF.

composition and increased tree densities. More Douglas-fir has increased beetle and root disease hazard; more grand fir has further increased root disease hazard. Root disease role may change from passive where low-vigour trees are removed to aggressive where larger areas of vigorous trees are killed.

[a] Compiled from Duff (1994a,b) and Diaz and Bell (1997).

[b] Proportion of forest that burns every year is equal to reciprocal of fire cycle. Therefore area burned for 35-year cycle = 1/35 x area of variant. Area of ICHdw in West Arm Demonstration Forest = 2,739 ha; dry ICHdw (02/01 site series) = 2,477 ha.

Agee (1994) estimates that grand fir forests in northeastern Washington, similar in climate and species composition to the ICHdw, would have had fire disturbance patches of 500 to 1,000 ha for severe fires under extreme weather conditions. The annual rate of burn for stand-replacement and ground fires in the ICHdw was calculated using the range of fire frequencies in the published literature and the area of this variant in the West Arm Demonstration Forest (Table 3). The estimated range for the annual rate of stand-replacement fires in the ICHdw (0.33 to 4.17 percent) would overlap a rate calculated for Douglas-fir/grand fir forests in the Selway-Bitterroot Wilderness of northern Idaho and Montana (Brown et al. 1994). In the Selway-Bitterroot study, a burn averaged 501 ha per year out of 56,379 ha of this type of forest. This is equivalent to 0.89 percent per year. The annual rate of underburning in the ICHdw, using literature values for comparable sites, was 5.5 percent or 152 ha. The fire frequency from Arno (1980) for surface fires is within the range obtained from fire scar analysis in the West Arm Demonstration Forest (Tables 1 and 2). These natural disturbance statistics could be used as annual treatment targets or as fire mimics using combined timber harvesting, prescribed burns, and appropriate stand or patch retention. The ICHdw has greater options for mimicking natural disturbance because both stand replacement and stand maintenance could be considered. A wide range of opening sizes, whether treatment units or retention patches, should be considered in planning and management. Patch sizes could be achieved through single time treatment regimes on larger blocks or through aggregated strategies of smaller patches distributed over time. Ultimately, size of patches and the range of acceptable sizes will be determined by social acceptance of the Total Resource Design plan or by regulations within the Forest Practices Code of British Columbia. For the landscape planner and the silviculturist in the West Arm Demonstration Forest, mimicking natural disturbance processes within social constraints may be more art than science. Many options exist in terms of size and timing of treatments for particular harvesting units.

The disturbance patterns used in the Total Resource Design Workshop noted the role of *Armillaria* root disease and a general high level of insect activity in this variant. As noted earlier, we found that root disease and Douglas-fir beetles are probably having the greatest impact on stand development in the ICHdw of the West Arm Demonstration Forest. Root disease is a major concern in the forests of moist sites where species composition and densities have been altered by fire exclusion. Where seral species once dominated at low densities, more Douglas-fir and grand fir are invading, perhaps creating conditions for increased pathogenicity of root disease. Where pure lodgepole pine stands exist, mountain pine beetle populations

Table 4

**Structural objectives for the driest sites in the Dry Warm Interior
Cedar-Hemlock (ICHdw) subzone derived from a Total Resource Design
Workshop (January 1994, Nelson, BC) and natural disturbance study.**

Description	Structural objectives	Forest health objectives

Total Resource Design Workshop*

Description	Structural objectives	Forest health objectives
Found on S&W aspects and coarse soils; small fires and large trees surviving alone. Predominantly ponderosa pine, Douglas-fir, and some western larch. Hardwood patches present. Fires: surface fires 30-50 years and stand-replacement fires every 100-150 years. Heavily diseased with frequent insect attacks.	Wide spacing; multistoried canopy; some denser, mature patches on wet sites and fire exclusions. Numerous large veterans. Group selection with denser patches in moist pockets; occasional larger clearances with a few large trees retained.	

Natural Disturbance study

Description	Structural objectives	Forest health objectives
Historic stands were open and park-like with numerous large veterans. Found on S&W aspects and coarse soils; low intensity fires and large trees surviving alone. Predominantly ponderosa pine, Douglas-fir, and some western larch. Hardwood patches present. Fires: surface fires every 11-24 years and mixed intensity fires every 33-150 years.	Wide spacing; multistoried canopy; some denser patches on wet sites and fire exclusions. Group selection with denser patches in moist pockets; occasional larger openings with a few large trees retained. Consider commercial thinning, single tree selection from below, and uneven-aged management.	Low insect/disease hazard if low density/open-growing structure maintained. In lodgepole pine stands, create mosaic of age classes across the landscape to reduce impact of beetle outbreaks.

---

* Descriptions and structural objectives compiled from Duff (1994a,b) and Diaz and Bell (1997). The driest sites in the ICHdw include the 02 and 01a site series (Braumandl and Curran 1992).

may increase. A landscape mosaic of age classes creates natural breaks in susceptible stand types and can reduce the impact of mountain pine beetle outbreaks.

The Total Resource Design Workshop disturbance patterns indicated that wind affects individual trees that are dead or dying (Table 3). This was confirmed in 1997 through field observation in the West Arm Demonstration Forest. Wind damage was observed most frequently in lodgepole pine and with greater frequency on shallow soils in proximity to bedrock outcrops or on soils with high contents of coarse fragments. The form of the lodgepole pine is a significant factor. Removing this species from stands, managing for alternative species, or using lower stocking densities should be considered where wind and snow are problems for timber production. The ICHdw has lower snowfall than the subzones that occur at higher elevation in the West Arm Demonstration Forest (Braumandl and Curran 1992), and the process of snow break for enhancing wind damage should be less. Lower snowfall should also result in less snow accumulation and, combined with gentler slopes, less potential for snow creep. Snow press of suppressed trees was common in open growing stands (Kokanee Creek, upper slope). Avalanche activity, in terms of pathways, runout zones, and deposits is rarely observed in the ICHdw.

Assessment of Structural Objectives in the Driest Sites of the ICHdw of the West Arm Demonstration Forest Landscape Plan

The stand types, descriptions, and structural objectives from the Total Resource Design landscape plan have been summarized for the dry ICHdw (Table 4) (Duff 1994a,b; Diaz and Bell 1997). Table 4 also includes input from the present study and objectives for forest health derived from the field sampling and literature review. The objectives for forest health and other aspects of silviculture are not blanket objectives for all areas; they are guidelines for a given management area. For example, if *Armillaria* root disease is a concern in parts of the ICHdw, then it may not be appropriate to manage for large Douglas-fir veterans in this unit. In contrast, if wildlife values at the stand level are considered high for the unit, managing for large veterans of Douglas-fir may be quite appropriate.

The Total Resource Design Workshop recommended wider spacing and an open stand structure for the driest sites of the Interior Cedar Hemlock dry warm subzone (Table 4). We agree with this and would specify that large Douglas-fir, ponderosa pine, and western larch be retained. The levels of retention would have to be compatible with other forest health considerations. The estimated annual rates for stand-replacement and ground fires (Table 3) could be used to set the rate of timber harvesting if this reflects people's values for mimicking past disturbance patterns. Values of 152 ha per year for ground fire treatment or mimicking and 9 to 114 ha per year for stand-replacement treatment or mimicking seem compatible with present management practices. Historically, the driest sites in the ICHdw probably had more ground fires and fewer stand-replacement fires because less fuel was available. Therefore the emphasis should be on treatments that mimic ground fire in the driest sites of the ICHdw. On these sites, even the most drastic treatment should consider some level of retention of large ponderosa pine and Douglas-fir. Ponderosa pine is readily established after fire or other disturbance and grows well in the drier parts of this subzone.

The most important forest health objective for the driest sites of the ICHdw is maintaining or creating low-density open-growing stands through commercial thinning or selection harvesting. Densities in pure lodgepole pine stands could be reduced by beetle-proofing. A mosaic of patches with different age classes should also be considered for lodgepole pine stands. Stand densities should vary with moisture in this stand type and drier sites should be the least dense. If this low-density stand structure is maintained, overall insect and disease hazard should be low. Root contacts important for the spread of root disease would be minimal. Drier stands should be dominated by ponderosa pine and Douglas-fir; more moist sites should have mixed species. On the more moist sites, retaining more early seral species will also ensure minimal root disease hazard. Lower stand densities will likely reduce bark beetle hazard.

Summary and Conclusions

A landscape plan was developed for the West Arm Demonstration Forest at a workshop in January 1994. The workshop did not use stand sampling to support or develop the Total Resource Design. As part of an adaptive management strategy as reported in this paper, we used literature review and stand sampling to revise and update information on the natural disturbances, stand types, and structural objectives of the West Arm Demonstration Forest landscape plan. We expanded structural objectives to include greater detail on forest health objectives.

Information on natural disturbance processes and the resulting patterns used in the West Arm Demonstration Forest Total Resource Design should be modified. The range of fire frequencies used in the Total Resource Design should be increased to include more frequent fires. A statistic based on historic annual rate of burns is suggested for possible use in determining rate of treatments in the biogeoclimatic ecosystem classification variants of the West Arm Demonstration Forest. The ICHdw variant, the driest ecological unit in the West Arm Demonstration Forest, has increased root disease activity due to greater stand density and a greater proportion of more susceptible tree species. Bark beetles are quite active on dry sites in this variant. Snow storm activity during recent winters was observed to be enhancing wind damage of trees in most of the West Arm Demonstration Forest.

The following suggestions are made for the dry ICHdw stand type and its structural objectives in the Total Resource Design plan. Forest health objectives in this stand type should include stands with greater western larch and blister-rust-resistant western white pine combined with less Douglas-fir. This should address the endemic root disease problems. In general, management in this stand type can best address forest health problems by reducing stocking densities and selecting more tolerant seral species. This stand type should be managed for open stands of ponderosa pine, western larch, and possibly Douglas-fir.

We recommend that future Total Resource Design workshops for developing landscape plans include more detailed literature review of natural disturbance ecology and stand sampling as part of the preliminary inventory and data collection. This would provide a better basis and probably a more flexible framework for developing the landscape plan. We also recommend that the results of this study be applied to a landscape modeling exercise in the West Arm Demonstration Forest. A proper landscape analysis is needed to assess the impact of our recommendations on landscape processes such as mountain pine beetle dynamics.

Finally, historic landscape patterns are interesting and help us to understand landscape and stand level processes. Studying historic patterns confirms the high variability of natural systems and processes. Ultimately, understanding the ecological processes will enable people to understand the risks and choose the appropriate management systems that optimize social values. The Total Resource Design of the West Arm Demonstration Forest provides an excellent framework for ecosystem-based management.

References

Agee, J.K. 1993. Fire ecology of Pacific Northwest forests. Island Press, Washington, DC and Covelo, CA.

Agee, J.K. 1994. Fire and weather disturbances in terrestrial ecosystems of the eastern Cascades. General Technical Report PNW-GTR-320, US Department of Agriculture (USDA) Forest Service, Portland, OR.

Arno, S.F. 1980. Forest fire history in the northern Rockies. Journal of Forestry 78:460-465.

Arno, S.F., and D.H. Davis. 1980. Fire history of western redcedar/hemlock forests in northern Idaho. Pp. 21-26 *in* Proceedings of the fire history workshop, 20-24 October 1980, US Department of Agriculture (USDA) Forest Service, Fort Collins, CO.

Atkins, M.D., and L.H. McMullen. 1960. On certain factors affecting Douglas-fir beetle populations. Pp. 857-859 *in* Proceedings of the Fifth World Forestry Congress, Seattle, WA.

Backhouse, F. (compiler). 1993. Wildlife tree management in British Columbia. Workers' Compensation Board, British Columbia Silviculture Branch, and Canada-British Columbia Partnership Agreement on Forest Resource Development: Forest Resource Development (FRDA) II.

BC Forest Service. 1995a. Biodiversity guidebook: Forest Practices Code of British Columbia. BC Ministry of Forests, Victoria, BC.

BC Ministry of Forests and Ministry of Environment, Lands and Parks. 1995a. Bark Beetle Management Guidebook: Forest Practices Code of British Columbia. BC Ministry of Forests and Ministry of Environment, Lands and Parks, Victoria, BC.

BC Ministry of Forests and Ministry of Environment, Lands and Parks. 1995b. Defoliator Management Guidebook: Forest Practices Code of British Columbia. BC Ministry of Forests and Ministry of Environment, Lands and Parks, Victoria, BC.

Blenis, P.V., and M.S. Mugala. 1989. Soil affects *Armillaria* root rot of lodgepole pine. Canadian Journal of Forest Research 19:1638-1641.

Braumandl, T.F. and M.P. Curran (compilers and eds.). 1992. A field guide for site identification and interpretation for the Nelson Forest Region. Land Management Handbook No. 20. British Columbia Ministry of Forests Research Branch, Victoria, BC.

Brennan, L.A. and S.M. Hermann. 1994. Prescribed fire and forest pests: Solutions for today and tomorrow. Journal of Forestry 92:34-37.

Brown, J.K., S.F. Arno, S.W. Barrett, and J.P. Menakis. 1994. Comparing the prescribed natural fire program with presettlement fires in the Selway-Bitterroot Wilderness. International Journal of Wildland Fire 4:157-168.

Byler, J.W., R.G. Krebill, S.K. Hagle, and S.J. Kegley. 1993. Health of the cedar-western white pine forests of Idaho. *In* D.M. Baumgartner, J.E. Lotan, and J.R. Tonn (eds.). Interior cedar-hemlock-white pine forests: Ecology and management. Symposium proceedings, Cooperative extension, Washington State University, Pullman, WA.

Byler, J.W., M.A. Marsden, and S.K. Hagle. 1987. Pp. 52-56 *in* S. Cooley (ed.). Proceedings of the 1986 Western International Forest Disease Work Conference. US Department of Agriculture (USDA) Forest Service, Portland, OR.

Campbell, S., L. Liegel (tech. coords.). 1996. Disturbance and forest health in Oregon and Washington. General Technical Report PNW-GTR-381, US Department of Agriculture (USDA) Forest Service, Portland, OR.

Cruickshank, M.G., D.J. Morrison, and Z.K. Punja. 1997. Incidence of *Armillaria* species in precommercial thinning stumps and spread of *Armillaria ostoyae* to adjacent Douglas-fir trees. Canadian Journal of Forest Research 27: 481-490.

Diaz, N., and D. Apostol. 1992. Forest landscape analysis and design. General Technical Report PNW R6 ECO-TP-043-92, US Department of Agriculture (USDA) Forest Service, Portland, OR.

Diaz, N., and S. Bell. 1997. Landscape analysis and design. Pp. 255-269 *in* K.A. Kohm and J.F. Franklin (eds.). Creating a forestry for the 21st century: The science of ecosystem management. Island Press, Washington, DC.

Duff, J. 1994a. Total Resource Design. Application to the West Arm Demonstration Forest. Proceedings of a workshop, January 1994. British Columbia Ministry of Forests, Nelson, BC.

Duff, J. 1994b. Total Resource Design: Documentation of a method and a discussion of its potential for application in British Columbia. MSc thesis, University of British Columbia, Vancouver, BC.

Everett, R., C. Oliver, J. Saveland, P. Hessburg, N. Diaz, and L. Irwin. 1994. Adaptive ecosystem management. *In* M.E. Jensen and P.S. Bourgeron (eds.). Vol. 2, Ecosystem management: Principles and applications. General Technical Report PNW-GTR-318, US Department of Agriculture (USDA) Forest Service, Portland, OR.

Forman, R.T.T., and M. Godron. 1986. Landscape ecology. John Wiley and Sons, New York, NY.

Furniss, M.M. 1965. Susceptibility of fire-injured Douglas-fir to bark beetle attack in southern Idaho. Journal of Forestry 63:8-11.

Furniss, M.M., M.D. McGregor, M.W. Foiles, and A.D. Partridge. 1979. Chronology and characteristics of a Douglas-fir beetle outbreak in northern Idaho. General Technical Report INT-59, US Department of Agriculture (USDA) Forest Service, Ogden, UT.

Haack, R.A., and J.W. Byler. 1993. Insects and pathogens: Regulators of forest ecosystems. Journal of Forestry 91:32-37.

Hagle, S.K. 1985. Monitoring root disease mortality. Report Number 85-27, US Department of Agriculture (USDA) Forest Service, Missoula, MT.

Haney, A., and R.L. Power. 1996. Adaptive management for sound ecosystem management. Environmental Management 20(6):879-886.

Hessburg, P.F., R.G. Mitchell, and G.M. Filip. 1994. Historical and current roles of insects and pathogens in eastern Oregon and Washington forested landscapes. General

Technical Report PNW-GTR-327, US Department of Agriculture (USDA) Forest Service, Portland, OR.

James, R.L., and D.J. Goheen. 1981. Conifer mortality associated with root disease and insects in Colorado. Plant Disease 65:506-507.

Lehmkuhl, J.F., P.F. Hessburg, R.L. Everett, M.H. Huff, and R.D. Ottmar. 1994. Historical and current forest landscapes of eastern Oregon and Washington. Part I: Vegetation pattern and insect and disease hazards. General Technical Report PNW-GTR-328, US Department of Agriculture (USDA) Forest Service, Portland, OR.

Lehmkuhl, J.F., P.F. Hessburg, R.D. Ottmar, M.H. Huff, R.L. Everett, E. Alvarado, and R.E. Vihnanek. 1995. Assessment of terrestrial ecosystems in eastern Oregon and Washington: The eastside forest ecosystem health assessment. *In* R.L. Everett and D.M.Baumgartner (compilers). Ecosystem management in western interior forests, Symposium proceedings. Cooperative extension, Washington State University, Pullman, WA.

Lejeune, R.R., L.H. McMullen, and M.D. Atkins. 1961. The influence of logging on Douglas-fir beetle populations. Forestry Chronicle 37:308-314.

Lotan, J.E., and W.B. Critchfield. 1990. Lodgepole pine. *In* R.M. Burns and B.H. Honkala (Tech. Coords.). Silvics of North America. Vol. 1, Conifers. Agriculture Handbook 654. US Department of Agriculture (USDA) Forest Service, Washington, DC.

McDonald, G.E., N.E. Martin, and A.E. Harvey. 1987a. Occurrence of *Armillaria* spp. in forests of the Northern Rocky Mountains. Research Paper INT-381, US Department of Agriculture (USDA) Forest Service, Ogden, UT.

McDonald, G.E., N.E. Martin, and A.E. Harvey. 1987b. *Armillaria* in the Northern Rockies: Pathogenicity and host susceptibility on pristine and disturbed sites. Research Note INT-371, US Department of Agriculture (USDA) Forest Service, Ogden, UT.

Mutch, R.W. 1994. Fighting fire with prescribed fire: A return to ecosystem health. Journal of Forestry 92:31-33.

Mutch, R.W., S.F. Arno, J.K. Brown, C.E. Carlson, R.D. Ottmar, and J.L. Peterson. 1993. Forest health in the Blue Mountains: A management strategy for fire-adapted ecosystems. *In* T.M. Quigley (ed.). Forest health in the Blue Mountains: Science perspectives. General Technical Report PNW-GTR-310, US Department of Agriculture (USDA) Forest Service, Portland, OR.

Pollack, J.C., H. Quesnel, C. Hauk, and H. MacLean. 1997. A quantitative evaluation of natural age class distributions and stand replacement intervals in the Nelson Forest Region, Vol. 1. Technical Report TR-015, British Columbia Ministry of Forests, Nelson, BC.

Rogers, P. 1996. Disturbance ecology and forest management: A review of the literature. General Technical Report INT-GTR-336, US Department of Agriculture (USDA) Forest Service, Ogden, UT.

Rudinsky, J.A. 1966. Host selection and invasion by the Douglas-fir beetle, *Dendroctonus pseudotsugae* Hopkins, in coastal Douglas-fir forests. Canadian Entomology 98:98-111.

Shaw, C.G., L.F. Roth, L. Rolph, and J. Hunt. 1976. Dynamics of fir and pathogen as they relate to damage to a forest attacked by *Armillaria*. Plant Disease Reporter 60:214-218.

Shugart, H.H. 1998. Terrestrial ecosystems in changing environments. Cambridge University Press, Cambridge, UK.

Simon, J., S. Christy, and J. Vessels. 1994. Clover-Mist fire recovery: A forest management response. Journal of Forestry 92:41-44.

Smith, J.K., and W.C. Fischer. 1997. Fire ecology of the forest habitat types of northern Idaho. General Technical Report INT-GTR-363, US Department of Agriculture (USDA) Forest Service, Ogden, UT.

Stewart, A.J. 1994. Forest insect and disease conditions in the West Arm Demonstration Forest. Forest Insect and Disease Series Pest Report 94-25, Canadian Forest Service, Victoria, BC.

Stewart, G., and E. Todd (eds.). 1995. Wildlife/danger tree assessor's course workbook. Canada-British Columbia Partnership Agreement on Forest Resource Agreement. Forest Resource Development Agreement (FRDA II), 3rd edition, British Columbia Ministry of Forests, Victoria, BC.

Williams, J.T., R.E. Martin, and S.G. Pickford. 1980. Silvicultural and fire implications from a timber type evaluation of tussock moth outbreak areas. Pp. 191-196 *in* Proceedings of the Sixth Conference on Fire and Forest Meteorology. Society of American Forester, Seattle, WA.

Williams, R.E., and M.A. Marsden. 1982. Modelling probability of root disease center occurrence in northern Idaho forests. Canadian Journal of Forest Research 12:876-882.

Forest Practice Certification: Options and Implications

Chris Ridley-Thomas and David Bebb

Abstract

Forest practice certification is in its infancy both within Canada and worldwide. Within Canada there are currently three major certification processes to choose from: the International Organization for Standardization's (ISO) Environmental Management System Standard (ISO 14001), the Canadian Standards Association's (CSA; now called CSA International) Sustainable Forest Management System Standard (Z808 and Z809), and the Forest Stewardship Council's (FSC) Principles and Criteria for Forest Stewardship. While the ISO and CSA standards provide a framework for managing environmental impacts, the FSC process focuses on achieving specific regional performance standards (where they exist and have been approved by the FSC Secretariat). This paper examines some of the implications of the approaches to certifying forests, the synergies that exist between the major certification processes, and how these may influence strategies for achieving certification.

The Case for Certification

To maintain market share in today's marketplace, forest product retailers must be able to demonstrate to consumers that the wood products they are buying come from well-managed forests. As a result, forest practice certification processes have been developed that allow forest companies to validate their performance to their retailers and consumers through a credible independent audit process.

In the past, declarations of compliance with self-declared codes of practice (such as the Forest Alliance's Principles of Sustainability), a good record of compliance with local legislation, and periodic tours of woodlands operations by retailers often provided adequate assurance of the quality of forest management. With the intensification of debate among the forest industry, environmental groups, and government over what constitutes good forestry and sustainable forest practices, the arguments have become technical in nature. These technical debates do not help retailers or customers answer the question, Is the company managing the forest well? At the same time, environmental groups are putting considerable pressure on retailers to buy wood products from certain sources only (Greenpeace Canada 1998). While these campaigns have generally been targeted against selected companies or regions, it has become clear that, to justify the source of their wood products, retailers need some form of independent verification of the quality of forest management practices. Such verification, if accepted by consumers, would alleviate much of the potential for negative publicity that these retailers risk in the marketplace. Many Canadian forest companies who supply retailers in markets where the debate over sustainability and forest practices is intense expect to achieve distinct advantages through obtaining independent verifications (Price 1998).

A company should consider a number of factors when determining whether certification of its forest practices is an appropriate step to take. These include the following:

(1) the level of pressure from customers and buyers' groups
(2) the level of pressure from environmental non-governmental organizations, both direct and through customers
(3) market access concerns (e.g., lack of certification may create a barrier to access in some markets)
(4) potential market discounting of prices on uncertified wood
(5) certification of competitors
(6) the ability to demonstrate due diligence
(7) the cost of different certification options
(8) the time it will take to implement certification options.

When assessing the potential costs and benefits of certification, it is important to consider what may happen during the lag time between deciding to pursue certification and actually achieving certification. If it will take 12 months to achieve certification, companies should consider how customers may react if, as will likely be the case, certified wood products are available on the market from another source prior to the company achieving certification.

Forest Practice Certification Options

There are currently three main certification options available to Canadian forest companies. These are:

(1) the International Organization for Standardization (ISO) Environmental Management System (EMS) Standard (generally referred to as ISO 14001)
(2) the Canadian Standards Association (CSA) (now called CSA International) Sustainable Forest Management (SFM) System Standard

Table 1

Main features of the International Organization for Standardization (ISO), Canadian Standards Association (CSA), and Forest Stewardship Council (FSC) standards.

Feature	Certification standard (applicability)		
	CSA (Canada)	FSC (global)	ISO 14001 (global)
Status in BC as of November, 1998	No woodlands operations certified	No woodlands operations certified	No woodlands operations certified
Nature of standard	Management system framework with some performance elements	Performance standards	Management system framework
Object of certification	Management system	Product	Management system
Product label	No	Yes	No
Basis of standard	Based on CCFM criteria and indicators for sustainable forest management	Based on FSC principles and criteria, supplemented by regional standards (where they exist)	Based on ISO EMS standard, supplemented by any additional forest management commitments included in policy statements
Public involvement	Mandatory requirement in forest management system development	Requirement to consult with persons and groups directly affected by forest management operations	Requirement to consider the views of interested parties in developing environmental objectives and targets

	No	No	Yes
Applicable to associated mills	No	No	Yes
Forecasting of future forest conditions	Explicitly required by the standard	Not required	Not explicitly required but may be necessary to set objectives and targets
Forest management performance requirements	Regulatory compliance is the minimum; others to be developed through a public process	Included in FSC principles and criteria (e.g., Principle #9: maintenance of natural forests)	Regulatory compliance is the minimum; others derived from commitments in environmental policy
Management plan	Required	Required	Requires the development of an environmental management program
Documented operational controls	Required for all aspects of the SFM system	Required for a specific list of forest practices	Required for all activities associated with significant aspects
Performance monitoring	Required for all indicators	Required for selected indicators	Required for system performance and all objectives and targets
Internal audits	Required	Not required	Required
Continuous improvement	Explicitly required by the standard, which sets an adaptive management framework	Implied by requirement to incorporate monitoring results in management plan revisions	Explicitly required by the standard, which sets an adaptive management framework

(3) the Forest Stewardship Council (FSC) Principles and Criteria for Forest Stewardship.

At the present time few woodlands operations in Canada are registered under any of these standards, although a considerable number are working towards achieving them. Table 1 shows the main features of each of these standards.

All of the standards in Table 1 have requirements that go beyond any of the regulatory frameworks currently in place in Canada. Significant differences exist, however, in both the time required for implementation and the cost to the company. For example, under the ISO 14001 standard, much of the required system development work occurs within the company (e.g., no externally set performance measures, no requirements to develop objectives through a public process), making the standard relatively quick and simple to implement. In addition, the development of the required management system elements may be relatively straightforward for larger forest companies, as they usually have many of the components of a woodlands Environment Management Standard already in place. As a result, ISO 14001 can provide obvious benefits for those with short-term certification goals and those wishing to maintain a simple, flexible system that is adaptable to management needs (International Organization for Standardization 1996).

In contrast, the Canadian Standards Association (CSA) and Forest Stewardship Council (FSC) standards, being more prescriptive, are more complex and expensive to implement. From our experience, the CSA standard is more challenging to implement because of the requirement to involve the public in developing the management system and performance measures, although this varies from province to province depending on how well the public participation process required under the CSA standard fits with existing regulatory requirements for each province. British Columbia's companies in particular are at a disadvantage in this respect because of the differences between existing legislated public processes and those required under the CSA standard. These companies are put in the position of having to "fill in the gaps" between what has been accomplished through existing public processes and the additional requirements of the CSA standard. The most critical difference among the standards identified in Table 1, however, is the fact that the ISO and CSA standards are management system standards, whilst the Forest Stewardship Council standard is a performance standard.

Management system standards are designed to foster an adaptive management approach intended to achieve continual improvement in forest management practices over time (Canadian Standards Association and CSA International 1996a,b). There are no preset levels of performance that a forest manager must meet beyond compliance with regulatory requirements.

This has led to criticism of both the ISO and CSA standards by some environmental organizations who are concerned that, in the absence of predefined performance measures, conformance with these standards may not provide sufficient assurance of good forest stewardship (Evans 1995).

While it is true that the ISO 14001 and CSA standards do not prescribe uniform performance requirements, the level of forest stewardship under these standards is still expected to exceed current regulatory requirements. The CSA standard specifically incorporates principles and criteria for sustainable forest management (SFM) into the design of the management system. Also, forest companies opting for ISO 14001 certification can voluntarily commit to SFM in their environmental policy, thereby making SFM a performance requirement of the Environmental Management System (Canadian Standards Association and CSA International 1996a,b; International Organization for Standardization 1996). Both standards also require that specific objectives and targets be developed as a component of the management system (by the company in the case of the ISO standard and through a public participation process in the case of the CSA standard), and these may exceed regulatory requirements in many instances. In addition, the standards require a commitment to continual improvement that is expected to lead to a gradual improvement in forest practices, even where regulatory compliance is already being achieved.

In contrast, the performance standard approach taken by the Forest Stewardship Council requires certain preset levels of performance to be achieved. These are developed on a regional basis by Forest Stewardship Council working groups and are intended to incorporate the input of a broad range of stakeholders. Where regional standards do not yet exist, the certification body (the firm doing the certification audit) is required to develop standards based on the Forest Stewardship Council's general principles and criteria (Wenban-Smith et al. 1998). These tailored standards are then made available for public review and comment prior to their adoption for the certification audit.

The Forest Stewardship Council's process is preferred by many environmental organizations, who are uncomfortable with standards that allow for the local development of performance requirements either by forest companies (in the case of ISO 14001) or public stakeholders (in the case of the CSA standard). Because the externally developed Forest Stewardship Council performance requirements tend to be fairly prescriptive, however, companies who obtain Forest Stewardship Council certification for their woodlands may find that this leads to constraints on their management flexibility.

Under both the ISO and CSA standards, the management system itself is certified (under ISO 14001 the management system may apply to all or part of a company's operations, while under the CSA standard the management system applies to a defined forest area). In contrast, both forest managers and forest products (where chain of custody can be verified) may be certified by Forest Stewardship Council's accredited certification bodies (Evans 1995). The Forest Stewardship Council's ability to certify specific products is a considerable benefit as this allows retailers to more easily demonstrate to consumers that their products come from well-managed forests.

Potential Synergies Between Certification Processes

Each of the certification options discussed has perceived strengths and weaknesses. Both the CSA and ISO standards are viewed as being strong on process and controls but relatively weak in setting uniform standards of forest practice performance. At the same time, the Forest Stewardship Council's standard, while prescribing performance standards that all certified operations are required to achieve, is relatively weak in providing the management framework necessary to ensure that the performance standards are achieved on the ground. In addition, while there is provision within the Forest Stewardship Council standard to incorporate monitoring results into subsequent management plans (Forest Stewardship Council 1999), the Forest Stewardship Council standard has little focus on continuous improvement of forest practices.

To overcome the perceived weaknesses of the various certification standards that are available, a number of forest companies are choosing to pursue certifications that combine two or more standards and expect to achieve certain synergies as a result. One approach that is gaining popularity involves developing an ISO 14001 Environmental Management System that incorporates Forest Stewardship Council performance standards. For example, AssiDomän, a large Swedish-based forest company, intended to certify all of their company woodlands under both ISO 14001 and Forest Stewardship Council by the end of 1998 (AssiDomän 1997). Similarly, because both ISO 14001 and the CSA standard are based on a management system framework, some Canadian companies see an ISO 14001 Environmental Management System registration as an effective interim step on the way to developing a system that complies with the more prescriptive CSA standard.

Not all combinations of certification standards are as compatible as the examples given above. While Forest Stewardship Council performance criteria are set externally, the CSA standard requires that they be set through a public process. As a result, to make Forest Stewardship Council and CSA standards compatible, the participants in the public process would have to agree that the performance targets prescribed by the Forest Stewardship Council are appropriate for the landbase to which both certifications would apply.

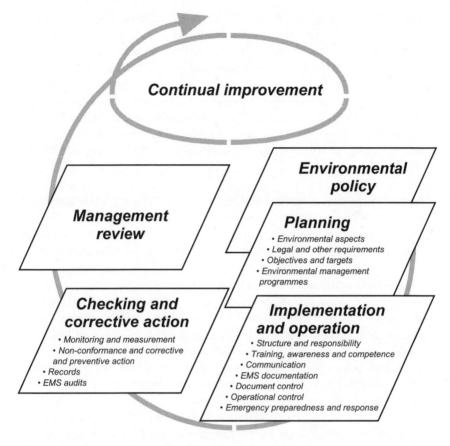

Figure 1. Key features of an environmental management system (from International Organization for Standardization 1996)

Forest Practice Certification and Ecosystem Management

As mentioned previously, both the ISO and CSA standards employ management system frameworks as outlined in Figure 1. Various components of the management system, particularly those that relate to checking and corrective action and management review, are designed to incorporate adaptive management, leading to the continuous improvement of forest practices. As more is learned about the responses of forest ecosystems to various management approaches, the related processes, indicators, and targets can be expected to be continually refined under both standards. This adaptive approach to management mirrors the concepts upon which ecosystem management is based, making both the ISO and CSA standards particularly compatible with the ecosystem management approach.

Forest Stewardship Council certification requirements, however, are generally more prescriptive than either ISO 14001 or the CSA Sustainable Forest Management System standard, particularly where regional standards exist. As a result, where regional Forest Stewardship Council standards limit the nature of the forest practices that may be undertaken (such as prescribing specific silvicultural systems or treatments that must be used under certain stand conditions, or prohibiting forest harvesting in certain stands or forest types), the ability to practice adaptive management (and hence ecosystem management) on Forest Stewardship Council certified woodlands will be constrained.

Factors Influencing the Credibility of Forest Practice Certification

A number of factors will determine the marketplace credibility of each standard as more companies become certified. Some of the more significant determinants of marketplace credibility are discussed in the following sections.

The Requirements of the Standard Selected

As noted previously, the various certification standards specify different requirements for certification. Forest Stewardship Council performance standards are set externally and provide a high degree of consistency within forest regions. At the same time, both ISO 14001 and the CSA standard set explicit requirements for management system components, resulting in a high degree of consistency in Environmental Management System structure among companies. Performance requirements under ISO 14001 and the CSA standard may vary among forest companies depending on the nature of the forest management commitments in company environmental policies (in the case of ISO) and the nature of local values, goals, indicators, and objectives identified during the public process (in the case of CSA). The approach that looks best to retailers and consumers will obviously influence which standards prevail.

Marketplace Acceptance

When selecting an appropriate standard, cost, flexibility, and simplicity of implementation have to be balanced against the level of public acceptance that each standard has in the marketplace. Little will be gained by obtaining certification under a particular standard if it is not well received by a company's customers.

Experience has shown that the public tends to more readily accept certification schemes that are endorsed by environmental organizations (Palmer 1995). The Forest Stewardship Council standard, which has been well marketed by various environmental organizations, currently enjoys a high acceptance level in some regions, particularly Europe. The ISO 14001 standard also has a high degree of name recognition, but has not been strongly marketed to date. The CSA standard, although it arguably may set the highest forest management standard of all, currently lags behind both the Forest Stewardship Council and ISO 14001 standards in public acceptance, primarily because it is a Canadian rather than an international standard.

Registrar/Certification Body Reputation

Maintaining the credibility of any forest practice certification involves a partnership between the firm carrying out the certification audit and the forest company involved. Both parties have a vested interest in ensuring that the certification is seen to be credible, as do all other audit firms and forest companies who subscribe to the standard. Where an audit firm is shown to have granted certification based on dubious or inadequate audit evidence, the reputation of all parties will be affected. Where such problems are found to be widespread, the value of the standard will be undermined and the considerable time and expense required to become certified to the standard will become harder for forest companies to justify.

The Nature and Extent of Verification Procedures Carried Out to Confirm Compliance with the Standard Selected

In British Columbia, many forest practice certifications will likely be subject to extensive scrutiny and debate, irrespective of the standard chosen. To maintain marketplace credibility, the procedures used to verify conformance to the chosen standard(s) will have to be defensible or the marketplace credibility of the standard(s) and the companies involved will suffer. British Columbia forest companies will need to factor in such scrutiny when selecting an audit firm to complete the certification.

All three certification standards outline, in varying degrees, the nature and extent of audit evidence that must be collected to support a certification decision. The level of audit guidance provided may not be sufficient to ensure a consistent approach. As a result, interpretations of required audit evidence varies among audit firms, leading to variations in the quality of the field verification work.

Already the quality of some of the audits performed to support certification decisions have been questioned (Armson 1998; Kiekens 1998). The absence of strong audit guidance for certifiers creates a risk that similar questions will be raised as more companies become certified. This has the potential to seriously reduce the credibility of the standards concerned.

To be defensible, a field audit of forest practices must be carried out to support the certification decision. This audit must be based on sufficient and appropriate evidence. More than just staff interviews and a review of documentation, it must assess an appropriate sample of field sites to determine conformance with the standard being assessed. Audits must look not only at the achievement of regulatory and internal performance requirements (i.e., a simple compliance audit), but extend to assessing of the effectiveness of implementation of operational controls, the adequacy and accuracy of monitoring efforts, and the adequacy and timeliness of actions taken to correct any performance problems identified. Only audit processes with this kind of rigour will be sufficient to withstand external scrutiny.

Conclusions

From our examination of certification standards, we can draw a number of conclusions.

- The pressure to certify forest practices continues to grow, particularly in markets where consumers of forest products have a negative perception of the quality of forest practices used in a particular forest company or region.
- Canadian companies wishing to pursue forest practice certification have three main options: ISO 14001, the CSA standard, or the Forest Stewardship Council. Which standards are selected should be based on each company's particular circumstances.
- Each of the three available certification standards has perceived strengths and weaknesses. Potential synergies may be available to those companies who pursue certification under some combination of the existing standards.
- The combination of the management system framework and adaptive management approach upon which both ISO 14001 and the CSA standard are based suggest that both standards are particularly well suited to the application of ecosystem management concepts. The Forest Stewardship Council standard, which is more prescriptive, may pose greater challenges to those forest companies who might wish to practice ecosystem management.
- A number of factors combine to determine the marketplace acceptance of a certification standard. Chief among these are the consumer's perception of the requirements of the standards selected, the effectiveness of the process used to market the standards, the reputation of the audit firm selected to perform certification work, and the adequacy of the audit procedures carried out in support of the certification decision.

References

Armson, K.A. 1998. Forest certification. The Forestry Chronicle 74(3):284.

AssiDomän AB. 1998. 1997 Environmental Report. Stockholm, Sweden.

Canadian Standards Association and CSA International. 1996a. CAN/CSA-Z808-96. A sustainable forest management system: Guidance document. Rexdale, ON.

Canadian Standards Association and CSA International. 1996b. CAN/CSA-Z809-96. A sustainable forest management system: Specifications document. Rexdale, ON.

Evans, B. 1995. Technical and scientific elements of forest management certification programs. Pp. 279-303 *in* Proceedings of the 1996 University of British Columbia-Universiti Pertanian Malaysia Conference on the Ecological, Social and Political Issues of the Certification of Forest Management. Putrajaya, Selangor, Malaysia.

Forest Stewardship Council. 1999. Forest Stewardship Council principles and criteria. Online: <http://www.fscoax.org/principal.htm>.

Greenpeace Canada. 1998. Greenpeace takes rainforest protection message to the U.S. Media release 20 October 1998, Toronto, ON.

. International Organization for Standardization. 1996. ISO 14001. Environmental management systems: Specifications with guidance for use. Geneva, Switzerland.

Kiekans, J.P. 1998. Is an FSC certificate really worth the paper it is written on? Malaysian Timber Bulletin 4(6):12-13.

Palmer, J. 1995. Monitoring forest practices. Pp. 91-117 *in* Proceedings of the 1996 University of British Columbia-Universiti Pertanina Malaysia Conference on the Ecological, Social and Political Issues of the Certification of Forest Management. Putrajaya, Selangor, Malaysia.

Price, W. 1998. Which standard will stand? Timber and the Environment: A Supplement to Timber and Wood Products International 384(6286):10-13.

Wenban-Smith, M.G., T.J. Synnott, and J.R. Palmer. 1998. FSC accreditation manual. Forest Stewardship Council, Oaxaca, Mexico.

Ecosystem Management in the Muskwa-Kechika Management Area: Implementing Strategic Land-Use Decisions in Northern British Columbia

Ron Rutledge

Abstract

The Northern Rocky Mountain region, known as the Muskwa-Kechika, is a vast wilderness area in northeastern British Columbia. It has millions of hectares of rugged mountains, pristine valleys, wild watersheds, and significant resource development values including natural gas, forestry, and mineral resources. The region also offers economic opportunities for tourism and backcountry recreation.

Recently approved land use plans for the region reached consensus on strategic management objectives for 4.4 million hectares including both Protected Areas and Special Management Zones. A major goal for the management of the Muskwa-Kechika is that the entire area be managed as an ecological unit while allowing resource development and use.

The legislation establishing the management objectives for the Special Management Zones requires the development of local strategic plans before approval of certain operational activities related to forestry, oil and gas development, and commercial recreation. Each plan for these activities must address the protection of ecological values and prescribe management actions or conditions of tenure that maintain the ecological integrity of the area.

Landscape Unit Objectives are developed to conserve biodiversity and will initially focus on old growth retention, stand structure, and patch size. Pre-Tenure Plans for oil and gas include assessing environmental values with an emphasis on fish and wildlife habitats. The Recreation Management Plan is developed to minimize human-induced changes to ecosystem components susceptible to impacts from wilderness recreation use. By considering ecological values in these plans, we have the potential to assist in meeting the desired future conditions legislated for the Muskwa-Kechika Management Area.

Introduction

The Rocky Mountain region, known as the Muskwa-Kechika, is a vast wilderness area in northeastern British Columbia and is one of the few remaining large, intact, and almost unroaded areas south of the 60th parallel. "It supports a diverse number of large mammals including moose, elk, mule deer, whitetail deer, caribou, plains bison, mountain sheep, mountain goat, wolves, black bears and grizzly bears in population densities of global significance" (Province of British Columbia 1997b). The area encompasses the eastern foothills of the Muskwa range, the Kechika ranges of the Cassiar Mountains, and the northern portion of the Rocky Mountain Trench.

The Muskwa-Kechika's millions of hectares of rugged mountains, pristine valleys, and wild watersheds contain significant resource development values including well-defined oil and gas resources in the eastern portion, a variety of metallic and non-metallic resources in the central and western portion, and high timber values in other portions of the area. The region also offers economic opportunities in tourism and backcountry recreation.

Recently approved land-use plans for the region reached consensus on strategic management objectives or desired future conditions for over 14 million hectares. Five broad categories of Resource Management Zones delineated for the region include Enhanced and General Management Zones (9.2 million hectares), Agricultural/Settlement Zones (0.6 million hectares), and over 4.4 million hectares of Special Management Zones (3.3 million hectares) and Protected Areas (1.1 million hectares) known as the Muskwa-Kechika Management Area (MKMA).

A major goal for managing the Muskwa-Kechika Management Area is that the entire 4.4 million hectares be managed as an ecological unit. While resource development is permitted outside of Protected Areas, operational plans must consider and address all significant values identified such as fish and wildlife habitat, wilderness and backcountry recreation, tourism, visual quality, and major river corridors (Province of British Columbia 1998a).

Strategic Land-Use Planning

The approach to strategic land-use planning in British Columbia changed significantly in the early 1990s. At the regional level, planning was initiated by the Commission on Resources and the Environment. A number of plans were developed and approved for four regions in British Columbia: Vancouver Island, Cariboo-Chilcotin, West Kootenay-Boundary and East Kootenay (Province of British Columbia 1998a). Strategic land-use planning was also begun at the sub-regional level through the Land and Resource Management Planning (LRMP) process.

In northeastern British Columbia, two working groups were initiated for the Fort Nelson and Fort St. John communities. Representatives in the groups came from government, industry, conservation, recreation, First Nations, and other interested stakeholder groups. After years of work, the Fort Nelson and Fort St. John Land and Resource Management Plans were approved by the provincial government in 1997.

Muskwa-Kechika Management Area Act

The Fort Nelson and Fort St. John Land and Resource Management Plans recommended, and government subsequently approved, the establishment of specific management objectives for a combination of Special Management Zones (SMZs) and Protected Areas (PAs). The Special Management Zones, Protected Areas, and their objectives were given legal status directly through 1998 legislation – the Muskwa-Kechika Management Area Act. The legislation requires the development of local strategic plans before approval of certain operational activities related to forestry, oil and gas, and commercial recreation tenures. There is one exception to this requirement. Designated officials and district managers may approve an operational activity in the absence of a local strategic plan if they all agree that the local strategic plan would not affect the operational activity.

Before approval of a Forest Development Plan, a Landscape Unit Objective must be established under Section 4 of The Forest Practices Code of British Columbia Act (Province of British Columbia 1995). Dispositions of Crown land for oil and gas development and any subsequent operational activities must be preceded by a Pre-Tenure Plan developed in accordance with the Memorandum of Understanding Respecting Operational Land Use Planning for Oil and Gas Activities in Northeast British Columbia. A Recreation Management Plan is required before Crown land can be used for commercial recreation.

Not only must these plans be consistent with the Special Management Zones and Protected Area management objectives developed by the Fort Nelson and Fort St. John Land and Resource Management Plans, they also must be consistent with direction given in the Muskwa-Kechika Management Plan adopted by government in 1997. To achieve this consistency, each of the local strategic plans must address the protection of ecological values and must prescribe management actions or conditions of tenure that maintain the ecological integrity of the area. The following is a summary of the ecological considerations that will be addressed in these plans to achieve the desired future conditions for the Muskwa-Kechika Management Area and apply an ecosystem-based management approach to an area larger than Denmark or Switzerland.

Landscape Unit Objectives

Slocombe (1998) has identified a number of substantive ecosystem-based management goals. He states, "Substantive goals refer to desired states or characteristics of the ecosystem being managed." Examples of substantive goals related to the biophysical environment include biodiversity, ecological linkages and flows and sustainability. The first of these, conserving biological diversity, is one of the primary objectives of the Forest Practices Code of British Columbia (BC Forest Service 1995).

As Fenger (1996:71) points out, "The FPC Act [Forest Practices Code of British Columbia] establishes landscape units as one of the frameworks for conserving biodiversity." Typically, landscape units range from 75,000 to 100,000 hectares and are defined by ecological boundaries such as watersheds. Section 5 of the Strategic Planning Regulation of the Forest Practices Code of British Columbia (BC Forest Service and BC Environment 1995) sets out the six elements of biodiversity to be addressed by Landscape Unit Objectives. Following is a summary of these six elements.

Retention of Old Growth Forest

Since old growth forests usually contain an abundance of plant and animal species that are sometimes at high risk from timber harvesting, guidelines for biodiversity conservation specify old-growth targets by natural disturbance types. The central component of old-growth retention is the establishment of spatially fixed old-growth management areas where the size and location of old forest patches is an important determinant for operational planning.

Stand Structure

Forest stand structure is the second element in the biodiversity conservation strategy. The Biodiversity Guidebook of the Forest Practices Code of British Columbia (BC Forest Service 1995) " ... describes the retention of forest stand structure in terms of wildlife trees and coarse woody debris ... that provide valuable habitat for conservation or enhancement of wildlife" (Fenger 1996:74). Landscape Unit Objectives for stand structure will primarily focus on maintaining stand structure attributes in the form of coarse woody debris, standing live trees, and standing dead trees.

Temporal and Spatial Distribution (Patch Size)

Patch-size targets can be beneficial to biodiversity by allowing a variety of openings more reflective of the natural patterns of distribution and by reducing fragmentation. Targets will be determined consistent with the natural disturbance type.

Seral Stage Distribution

The intent of the seral stage distribution is to provide a variety of different aged forests (from early seral to old growth) across a landscape. Targets in the Biodiversity Guidebook of the Forest Practices Code of British Columbia have been set for early and mature seral stages as well as for old growth as described earlier.

Landscape Connectivity

Since forest harvesting can result in fragmentation affecting biodiversity conservation, the Biodiversity Guidebook of the Forest Practices Code of British Columbia gives recommendations for connectivity based on the type and degree of connectivity found in each natural disturbance type. The goal of connectivity is "to soften the matrix of harvested and developed areas, and provide for suitable habitat across landscape units and between retention areas" (Fenger 1996:74).

Species Composition

The Biodiversity Guidebook of the Forest Practices Code of British Columbia gives recommendations for species composition that focus on maintaining a significant portion of each landscape unit (by natural disturbance type) in communities with plant species composition similar to those communities that have developed through natural succession. For most disturbance types, an important component of maintaining biodiversity is to maintain a variety of species from early seral communities to climax communities.

In the initial stages of implementation of the Forest Practices Code of British Columbia, priorities for landscape unit objectives will focus on old-growth retention, stand structure through wildlife tree patches, and patch size.

Pre-Tenure Plans

In July 1996 a number of provincial and federal natural resource ministries and energy industry representatives approved a Memorandum of Understanding Respecting Operational Land Use Planning for Oil and Gas Activity in Northeast British Columbia. This agreement applies to the exploration, production, and transportation of oil and gas with a particular emphasis on environmentally sensitive areas such as those in the Muskwa-Kechika Management Area. Parties to the agreement include the British Columbia Ministry of Environment, Lands and Parks; the British Columbia Ministry of Forests; the British Columbia Ministry of Energy and Mines; Westcoast Energy; the Canadian Association of Petroleum Producers; the Habitat and Enhancement Branch of the Federal Department of Fisheries and Oceans; and the National Energy Board.

The primary objective of the agreement is to establish a clear framework for operational-level planning for oil and gas development, production, and transportation that incorporates the assessment of environmental and other values and protection of a sustainable environment. More specifically, plans are developed to ensure that impacts from oil and gas activities are minimized on sensitive fish and wildlife and their habitats, timber, range, recreation, cultural/heritage, and archeological values. The following is a summary of the description of ecosystem characteristics/sensitivity and operational guidelines in the pre-tenure plan.

Ecosystem Descriptions and Interpretations

A number of information sources are used to ensure that a full range of ecological values are considered in developing conditions of tenure for oil and gas developments. Baseline data, collected on ecosystem maps using Terrestrial Ecosystem Mapping (TEM), follow the British Columbia Provincial Government's Resources Inventory Committee mapping standards of 1:50,000 scale for the planning area. "Ecosystem mapping is the stratification of a landscape into map units, according to a combination of ecological features, primarily climate, physiography, surficial material, bedrock geology, soil and vegetation" (Province of British Columbia 1998b).

Ecosystem maps provide basic information on the distribution of ecosystems from which interpretations can be developed and the sensitivity of ecosystem components can be determined. Interpretations most useful for pre-tenure planning for oil and gas developments include critical fish and wildlife habitat and habitat capability and suitability, site sensitivity, site index and site productivity, soil sensitivity, surface erosion potential, flooding and high water, and ecosystem rarity.

These interpretations are then used to develop management strategies that eliminate or reduce impacts from oil and gas development on the components of the ecosystem identified through Terrestrial Ecosystem Mapping. Specific practices are also developed to, for example, minimize vegetation impacts on wildlife species that use a particular habitat type. Practices could include recommending the distribution of coarse woody debris on disturbed sites that are being reclaimed, or they could include managing the shape of clearings to minimize natural forest openings and to increase edge effects.

Operational Guidelines

The purpose of having operational guidelines is to provide proponents and/ or operators with guidelines that will be employed at all levels, from geophysical survey operations through initial well drilling to abandonment and reclamation of development. While specific operational techniques are identified in the plan, they are by no means the only procedures permitted.

Proponents and operators are encouraged to use and develop techniques that will reduce environmental impacts (BC Ministry of Environment, Lands and Parks and BC Ministry of Energy and Mines 1995).

General guidelines address the potential for mitigation measures if they are needed to protect sensitive ecological values from development activities such as well drilling, road construction, pipeline and utility rights-of-way, and general infrastructure. These guidelines also set out the level of detail expected in the habitat impact assessments that are to be completed by proponents. Specific guidelines are also developed that describe restrictions related to topsoil salvaging; timing and seasonal restrictions; fisheries, water, and water crossings; geophysical operations; roads (access/closures); exploratory wells and drilling; well pads; pipelines; waste, facilities, and power lines; and personnel and camp management.

The last set of specific guidelines covers reclamation activities. The primary objectives for reclamation are "to return the site to a condition where self-sustaining natural vegetation provides: (1) wildlife habitat acceptability equal to or greater than initial conditions; and (2) erosion control equal to or greater than conditions found on adjacent undisturbed sites" (BC Ministry of Environment, Lands and Parks and BC Ministry of Energy and Mines 1995:33). Guidelines are set out regarding recontouring, scarification, topsoil spreading, seeding, maintenance, and monitoring the effectiveness of reclamation activities.

Recreation Management Plans

The local strategic plan required by the Muskwa-Kechika Management Area Act dealing with Crown land use for commercial recreation is the Recreation Management Plan. As was stated earlier, the Muskwa-Kechika Management Area has significant wilderness recreation and high potential for increased economic opportunities in tourism and backcountry recreation. Both the Fort Nelson and Fort St. John Land and Resource Management Plans acknowledged this potential and developed a number of strategic management objectives to protect current wilderness recreation values and to promote public and commercial recreational use of the Muskwa-Kechika Management Area.

The Land and Resource Management Plan working groups also recognized the importance of the "setting" in which wildland recreation occurs. As Hammitt and Cole (1987:3) point out, "In *wildland* recreation the importance of the environment or setting for activities is greater than in *developed* recreation situations. Moreover, these wildland settings are largely natural, and management strives to maintain a natural appearance."

The dependence of wilderness recreation activities on a natural setting is clearly reflected in the following examples of management objectives and recommended strategies developed for outdoor recreation and tourism values in the Fort Nelson Land and Resource Management Plan's General Management Direction (Province of British Columbia 1997a).

Management Objective	*Recommended Strategy*
Maintain or enhance ecological integrity in areas subject to impacts from recreation use.	More detailed plans will address the impacts of recreation activities on ecological integrity, for example wildlife disruption, damage to plant communities, and water quality.
Ensure the continued existence of quality experiences in areas used for commercial tourism.	Manage levels of recreation use to maintain the quality of the experience and the natural environment.

The recreation management plan is the vehicle stipulated in the Muskwa-Kechika Management Area Act to accomplish these types of objectives. Following are some major ecosystem components identified by Hammitt and Cole (1987) that can be impacted by wilderness recreation use and that will be addressed through specific management prescriptions and monitoring activities developed in the Recreation Management Plan.

Soil Conditions
The majority of negative impacts to soil conditions from recreation use begin with destruction of organic matter in the soil surface and compaction of soils and snow, primarily from trampling by people, horses, and all-terrain vehicles. Changes to soil characteristics such as aeration, temperature, moisture, nutrition, and soil organisms affect the soil's capability to support existing vegetation and establish new growth. In addition, compaction increases the potential for erosion and the diversion of natural watercourses.

Vegetation Conditions
Vegetation impacts start with damages to soil conditions and are exacerbated by the effect of trampling, most commonly resulting in decreases in vegetation cover leading to changes in species composition. More tolerant (i.e., resistant or resilient) species begin to outnumber those species less resistant to damage (e.g., tree seedlings, low-lying shrubs, and lichens). Growth rates are also affected which causes changes in age and structure of vegetation species. "Loss of vegetation cover exacerbates such impacts as loss of organic matter and increased erosion" (Hammitt and Cole 1987:13).

Wildlife
Potential adverse effects on wildlife species from recreation use include direct killing (hunting and trapping) and harassment in addition to habitat alterations from the vegetation impacts described above. Alterations to

population structures and species composition can be caused by decreases in food sources and distribution of home sites (e.g., snags for cavity-nesting birds).

Water
Recreation use can directly and indirectly reduce water quality in wilderness settings. Direct impacts include the introduction of pathogens resulting from improper disposal of human waste, or pollution such as oil and gasoline residue from boat motors or discarded fuel drums. Indirect sources include increased sedimentation from erosion associated with soil and vegetation impacts. Together, these impacts can cause changes in aquatic plant growth and reduced amounts of dissolved oxygen leading to disturbances to aquatic fauna.

While the areal extent of ecological impacts from public and commercial recreation use in the Muskwa-Kechika Management Area is probably not extensive relative to the size of the area, the potential nonetheless exists for human-induced changes to natural ecosystems and preferred natural settings. One of the primary purposes of the Recreation Management Plan for the Muskwa-Kechika Management Area is to assess the ecological condition of the natural settings where current wilderness recreation takes place. Recommendations will then be developed that:

(1) identify specific management objectives regarding desired future conditions
(2) select measurable impact indicators reflecting the degree of change in ecological conditions
(3) set impact indicator standards that define the maximum acceptable level of change in the impact indicators and thus the ecological conditions they reflect.

Research in recreation ecology has shown that changes to ecological conditions in the Muskwa-Kechika Management Area are inevitable, due to both natural and human factors. The goal of natural resource managers who will implement the Recreation Management Plan is to keep within acceptable levels the character and rate of change caused by human factors. Only then can the desired future conditions for both public and commercial recreation resources set out in the strategic land-use decisions developed for the Muskwa-Kechika Management Area be achieved.

Conclusion
Successful implementation of strategic land-use decisions in northern British Columbia is, in part, possible only if an ecosystem management approach is followed in future plans and management activities. For the

Muskwa-Kechika Management Area, the challenge is formidable due to the vastness of such an ecologically diverse area, the complexity of issues, increasing industrial uses, and the high standards for desired future conditions developed in the consensus-based planning approach. The ecological values in the legally required local strategic plans described have the potential to assist in meeting this challenge, resolving the issues, and attaining the desired future conditions for the Muskwa-Kechika Management Area.

References

BC Forest Service. 1995. Biodiversity guidebook: Forest Practices Code of British Columbia. British Columbia Ministry of Forests, Victoria, BC.

BC Forest Service and BC Environment. 1995. Strategic planning regulations: Forest Practices Code of British Columbia Act. British Columbia Ministry of Forests and Ministry of Environment, Lands and Parks, Victoria, BC.

BC Ministry of Environment, Lands and Parks and BC Ministry of Energy and Mines. 1995. Upper Sikanni management plan. BC Ministry of Environment, Lands and Parks and BC Ministry of Energy and Mines, Victoria, BC.

Fenger, M. 1996. Implementing biodiversity conservation through the British Columbia Forest Practices Code. Forest Ecology and Management 85:67-77.

Hammitt, W.E., and D.N. Cole. 1987. Pp. 3-25 *in* Wildland recreation ecology and management. John Wiley and Sons, New York, NY.

Province of British Columbia. 1995. Forest Practices Code of British Columbia Act, Revised Statutes of British Columbia, Chapter 159.

Province of British Columbia. 1997a. Fort Nelson land and resource management plan. Inter-Agency Management Committee, Prince George, BC.

Province of British Columbia. 1997b. Muskwa-Kechika management plan. Order in Council Number 1367, Schedule 3. Queen's Printer, Victoria, BC.

Province of British Columbia. 1998a. Fort Nelson land and resource management plan summary. Inter-Agency Management Committee, Prince George, BC.

Province of British Columbia. 1998b. Standards for terrestrial ecosystem mapping. Resource Inventory Committee, BC Environment, Victoria, BC.

Slocombe, D.S. 1998. Defining goals and criteria for ecosystem-based management. Environmental Management 22(4):483-493.

Do We Really Want to Be Stunned by Visions?

Stephen M. Smith

Abstract

In common with other healthy movements, environmentalists produce books and articles at a furious pace. Other environmentalists then review these works, writing admiring comments like "stunningly visionary." The thesis of this paper is that being "stunned by visions" is the wrong frame of mind for resource managers who are trying to maintain biodiversity. This paper gives some examples of popular visions that either do not match nature's reality or don't make sense. I explore the (sometimes stunning) consequences of acting on these visions. I examine the ideas that humans are apart from nature and that commercially profitable resource use is bad. I suggest some principles for managing biodiversity and provide some criteria for measuring success.

Introduction

One day I saw the phrase "stunningly visionary" used in a glowing review by one environmentalist of another environmentalist's book. My thesis in this paper is that being "stunned by visions" is exactly the wrong frame of mind for those approaching the management of natural resources.

Visions are an important part of human life. Visions are empowering when they are part of a call to adventure or to a vocation. For example, Mahatma Ghandi's vision of a united India free of British rule was an empowering vision. But visions can also stun or disempower us. In the year 1294 AD, Saint Thomas Aquinas, author of the vast "Summa Theologica," experienced a sort of thunderclap from aloft – a vision of God. After that he neither wrote nor dictated anything more. In our time, Brian Wilson of the Beach Boys singing group became stunned from too many LSD trips. After this, he required almost constant attention and lost his great ability to write songs.

If we become stunned by visions, we lose competence and are unable to function in the day to day world. After the stunning vision we may indeed achieve a state of grace, but we also become disengaged from the earthly world of action and events. Resource management cannot be disengaged from the world of action and events.

This paper is an assembly of ideas expressed by other people and does not represent original thoughts of my own. The paper contains verbatim extracts from and summaries of existing books, research papers, and magazine articles. Any errors in assembling the ideas expressed by others are my own and do not belong to the original authors. The paper contains three main sections: Visions of Doom, Visions of Nature, and Visions of Commerce. These are followed by a concluding discussion.

Visions of Doom

The discussion of doom draws heavily from "Environmental Scares," *The Economist*, 20 December 1997.

In 1798 Thomas Robert Malthus inaugurated a grand tradition of environmentalism with his best selling pamphlet on population. Malthus argued that since population tended to increase geometrically (2, 4, 6, 8) and food supply to increase arithmetically (1, 2, 3, 4), the starvation of Britain was inevitable and imminent. Almost everybody thought he was right. He was wrong.

In 1865, an influential book by Stanley Jevons argued that Britain would run out of coal in a few short years. In 1914, the United States Bureau of Mines predicted that American oil reserves would last ten years. In 1939 and again in 1951, the United States Department of the Interior said American oil would last 13 years. Wrong, wrong, and wrong.

Visions of ecological doom, including recent ones, have a terrible track record. Doom enthusiasts appear to think that a long history of incompetence and wrong predictions somehow makes them more likely to be right in the future.

In 1972, the Club of Rome published a highly influential report called Limits to Growth. The report said total global oil reserves amounted to 550 billion barrels. We could use all of the proven reserves of oil in the entire world by the end of the next decade, said President Jimmy Carter shortly afterwards. Sure enough, between 1970 and 1990 the world used 600 billion barrels of oil. So, according to the Club of Rome, reserves should have been overdrawn by 50 billion barrels by 1990. In fact, by 1990 unexploited reserves amounted to 900 billion barrels – not counting the tar shales of which a single deposit in Alberta contains more than 550 billion barrels.

The Club of Rome made similarly wrong predictions about natural gas, silver, tin, uranium, aluminum, copper, lead, and zinc. In every case, it said finite reserves of these minerals were approaching exhaustion and prices would rise steeply. In every case except tin, known reserves have actually grown since the Club's report. In some cases, they have quadrupled.

The Club of Rome's mistakes have not tarnished its confidence. More recently it tried again with *Beyond the Limits*, a book that essentially said that although we were too pessimistic about the future before, we remain equally pessimistic about the future today.

In 1980 the environmentalist Paul Ehrlich made a bet with an economist called Julian Simon. Dr. Ehrlich chose five minerals: tungsten, nickel, copper, chrome, and tin. He and Simon agreed how much of these metals US$1,000 would buy in 1980. They further agreed to recalculate how much that amount of metal would cost ten years later (still in 1980 dollars). Dr. Ehrlich agreed to pay the difference if the price fell, Dr. Simon to pay if the price rose. Ten years later, Ehrlich sent Simon a cheque for US$570. Dr. Simon won easily. Indeed, he would have won even if they had not adjusted prices for inflation. He would have won if Dr. Ehrlich had chosen virtually any mineral. Of 35 minerals, 33 fell in price during the 1980s. Only manganese and zinc were exceptions.

The 1983 edition of a British high school textbook said zinc reserves would last ten years and natural gas 30 years. By the 1993 edition the author had wisely removed references to zinc (rather than explain why it had not run out), and he mocked his own earlier forecast for natural gas by giving it 50 years.

So much for minerals. The record of mispredicted food supplies is even worse. Consider two quotations from Paul Ehrlich's best selling books in the 1970s.

> Agricultural experts state that a tripling of the food supply of the world will be necessary in the next 30 years or so, if the 6 or 7 billion people who may be alive in the year 2000 are to be adequately fed. Theoretically such an increase might be possible, but it is becoming increasingly clear that it is totally impossible in practice.

> The battle to feed humanity is over. In the 1970s the world will undergo famines – hundreds of millions of people are going to starve to death.

Ehrlich was not alone. Lester Brown of the Worldwatch Institute began predicting in 1973 that population would soon outstrip food production. He still does so every time a temporary increase in wheat prices occurs. In 1994, after 21 years of being wrong, he said, "After 40 years of record food production gains, output per person has reversed with unanticipated abruptness." Two bumper harvests then followed and the price of wheat fell to record lows. Yet Mr. Brown's pessimism remains as impregnable to facts as his views are popular with newspapers.

The facts on world food production are truly startling for those who have heard only the doomsayers' views. Since 1961, the population of the world has almost doubled, but food production has more than doubled. As a result, food production per person has risen by 20 percent since 1961. This improvement is not confined to rich countries. According to the Food and Agricultural Organisation (FAO), calories consumed per capita per day are 27 percent higher in the third world than they were in 1963. Deaths from famine, starvation, and malnutrition are fewer than ever before.

Global 2000 was a report to the United States President written in 1980. The report predicted that world population would increase faster than world food production so that food prices would rise by between 35 percent and 115 percent by the year 2000. So far, the world food commodity index has fallen by 50 percent.

In the early 1980s acid rain became the favourite cause of doom. Lurid reports appeared of widespread forest decline in Germany where half the trees were said to be in trouble. By 1986 the United Nations reported that 23 percent of all trees in Europe were moderately or severely damaged by acid rain. What happened? They recovered. The biomass stock of European forests actually increased during the 1980s. The damage all but disappeared. Forests did not decline, they thrived. A similar gap between perception and reality occurred in the United States. A ten-year, $700 million official study concluded, "There is no evidence of a general or unusual decline of forests in the US or Canada due to acid rain."

Today the mother of all environmental scares is global warming. Here the jury is still out, though not according to President Clinton. Before you rush to join Mr. Clinton's consensus, compare two quotations:

> Meteorologists disagree about the cause and extent of the cooling trend ... But, they are almost unanimous in the view that the trend will reduce agricultural productivity for the rest of the century. (*Newsweek* 1975)

> Scientists concluded – almost unanimously – that global warming is real and the time to act is now. (United States Vice President Al Gore 1992)

Only one environmental scare in the past 30 years bears out the most alarmist predictions made at the time: the effect of DDT on birds of prey, otters, and some other predatory animals. Every other environmental scare has been either wrong or badly exaggerated. Will you believe the next one?

Environmental scare stories now follow such a predictable line that we can chart their course.

Year 1: The year of the scientist who discovers some potential threat.

Year 2: The year of the journalist who oversimplifies and exaggerates it.

Year 3: The year of the environmentalists who join the bandwagon and polarise the issue. Either you agree that the world is about to end and are fired by righteous indignation, or you are a paid lackey of big business.

Year 4:   The year of the bureaucrat. A conference is organised keeping public officials well supplied with airline tickets and limelight. This diverts the argument from science to regulation. A totemic target is the key feature.

Year 5:   The year of the villain. Pick one and gang up on him.

Year 6:   The year of the sceptic who says the scare is exaggerated.

Year 7:   The year of the quiet retreat. Without fanfare, the official consensus on the size of the problem shrinks.

Examples are world population estimates and greenhouse warming. Human population went from an "explosion," to an asymptotic rise to 15 billion, to an asymptote of 12 billion, to an asymptote of less than 10 billion. This last figure infers that population will never double again since the Earth already supports more than five billion people. Greenhouse warming went from "uncontrolled' to 2.5 to 4 degrees per 100 years, to 1.5 to 3 degrees per 100 years (according to the United Nations).

What Is Wrong with Doom?

Isn't it good to exaggerate the potential ecological problems the world faces so that people will take action? Most definitely not. Consider the consequences.

The exaggeration of the population explosion leads to a form of misanthropy that comes dangerously close to fascism. Dr. Ehrlich is an unashamed believer in the need for coerced family planning. Another eco-guru, Garrett Hardin, has said that "freedom to breed is intolerable." If you think population is "out of control" you might be tempted to agree to such drastic curtailments of liberty. If you know that the graph is flattening, you may be more tolerant of your fellow human beings.

Exaggeration of potential problems distorts research priorities and may weaken its integrity. According to the book *Are We Scaring Ourselves To Death?* (Cohl 1997), Dr. Kenneth Olden, Director of the United States National Institute of Environmental Health Sciences, admitted to the Los Angeles Times (13 September 1994) that "often scientists hype and build up their own findings [to get] more funding." A doomsday scenario reported in the media is a sure-fire ticket to research support. Dr. Olden also refers to climatologist Stephen Scheider's explanation in *Discover* magazine (October 1989) that scientists have to "capture the public's imagination" to get funding. One consequence of doomsaying may well be a misallocation of funds and mistaken priorities for research and improving the environment. For example, many lives of Third World people are made miserable or shortened by pervasive water and air pollution. Yet in 1992, the Earth Summit in Rio ignored this and proclaimed that global warming is a horror. To make Rio a fashionably correct event about Western guilt-tripping, a troubling but speculative concern was cynically and cruelly put above real and preventable loss of lives from Third World water and smoke pollution.

Scaring ourselves to death is itself a scary proposition. Not only can our fears be far worse than the things we are afraid of, but the fears themselves can harm us more. Who loses in the exaggeration of potential ecological problems? You do, the ordinary citizen. Who wins? The lawyers, the media, politicians, regulators, and special interest groups.

You lose by putting time, energy, worry, and money into risks that are not necessarily serious. You lose by giving attention to negligible (but dramatic) risks at the expense of real risks. For example, our political system invests billions of dollars seeking unobtainable safety while failing to provide resources necessary to protect our future.

Lawyers win. The Superfund project in the United States provided $13.5 billion for cleaning up toxic waste sites. One quarter of this, almost $3.5 billion, went to lawyers. Were the Superfund sites cleaned up? No. In the end only 15 percent of the sites were cleaned.

Regulators win from a climate of fear. Over-regulation greatly reduces the quality of our lives but provides lucrative employment for the regulators.

Politicians win by regularly playing on our fears. Governor Pete Wilson of California campaigned on the crime issue, building people's fears of carjacking and drive-by shootings. Suburbanites who lived in areas where crime was lowest were the voters who elected him.

Special interest groups are also winners when people become scared. Environmentalists are quick to accuse their opponents in business of having vested interests. But their own incomes, their advancement, their fame, and their very existence can depend on supporting the most alarming versions of every environmental scare. H.L. Mencken reminded us that "the whole aim of practical politics is to keep the populace alarmed – and hence clamorous to be led to safety – by menacing it with an endless series of hobgoblins, all of them imaginary."

Visions of Nature
The discussions of nature draw heavily from *A Moment on Earth* by Gregg Easterbrook and *The Masks of God* by Joseph Campbell.

What is nature? The idea of the earth as both a bearing and nourishing mother was and is prominent in primitive religions. In hunter societies, it is from her womb that the game animals come. Their timeless archetypes exist in the underworld, or dancing ground, and the flocks on earth are but temporal manifestations sent for the nourishment of man. According to planter or crop-growing societies, it is in the mother's body that the grain is sown. The ploughing of the earth is a begetting and the growth of the grain is a birth.

Yet the divine mother who gives birth to life also brings death. For example, the Hindu goddess Kali is a cannibal-mother who personifies all-consuming time. She is the terrible one of many names whose stomach is a void and so can never be filled and whose womb is giving birth forever to all things. A river of blood has been pouring continuously for millennia, from beheaded offerings, into Kali's void to be returned, still living, from her womb.

Another version of the earth as mother is the Gaia metaphor. This is suscribed to by some environmentalists and is linked to the ideas of deep ecology. In Greek mythology Gaia was the goddess of the earth. She conspired with Hades, the lord of the underworld, to seduce Persephone, the daughter of Zeus and Demeter. Persephone, we are told, was playing in a meadow when she spied a glorious plant with a hundred fragrant blossoms. This plant had been sent up expressly to seduce her by the goddess Earth (Gaia), at the behest of Hades. When Persephone hurried to pluck the fragrant blossoms, the earth gaped and a great god appeared in a golden chariot to carry her down into his abyss.

More recently, the goddess Earth has appeared in the Gaia Hypothesis, constructed by the British chemist James E. Lovelock and the American biologist Lynn Margulis. The Gaia hypothesis describes the idea of complementary evolution of life and environment. The hypothesis is that the earth's living matter, air, oceans, and land surface form a complex system that can be seen as a single organism and that has the capacity to keep our planet a fit place for life. The Gaia hypothesis intimates that individual species might sacrifice themselves for the benefit of all living things.

Deep ecology, a tributary but not mainstream part of environmentalism, is one expression of the Gaia hypothesis. In deep ecology humans are just another creature, equal in moral value to all others. Some deep ecologists extend equal moral value to rocks, plains, and water. Hence the mantra "let the rivers live." Deep ecology was invented by Arne Naess, a Norwegian philosopher. Regarding universally equal moral value, Naess maintained that "it is against my intuition of unity to say I can kill you because I am more valuable." But it was not against that intuition to say "I will kill you because I am hungry." Would Naess object if a poor man who was hungry killed Naess to take his wallet? For those modern-day deep ecologists who endorse "ecocentrism," which grants rocks and plains the same ethical significance as living things, some equally awkward questions arise. For example, if it's fine for a tiger to savage an antelope and it's fine for a hungry robber to gun down a passer-by, was it fine for all those people in Honduras to drown in the rivers and mudflows during Hurricane Mitch? After all, according to deep ecology, Mitch was only expressing its right to blow, and the rivers were only expressing their right to flow.

Whatever one's personal view of nature, the fact is that nature has destroyed virtually everything it has ever made with the exception of life itself. Fashioning new species and habitats, nature acts like a Renaissance artist, painting the most recent work over the older one already on canvas. In nature's case, "painting over" has happened countless times. For example, in British Columbia the formation of the original Coast Mountains ended about 45 million years ago. For the next 40 million years they eroded to perhaps a low chain of hills. Even by ten million years ago, the Coast "Mountains" had become so low there was no rainshadow behind them. The British Columbia Interior was cloaked in lush vegetation. Then, beginning about five million years ago, a subterranean mass of molten rock flowed directly beneath the eroded hills. The heat of this subduction zone not only created volcanoes, but also heated the thick crust along the British Columbia coast. As the crust warmed up, it expanded. Geologists calculate that this thermal expansion alone would result in a two-kilometre uplift of the Coast Mountains. They have, in fact, risen about this much in the last five million years, and they continue to rise today.

Where do humans fit in nature? Are we part of nature or separate from it? Some environmentalists say we are part of nature but are somehow simultaneously separate from and in conflict with nature. According to some, we are a cancer upon the earth.

Many people feel guilty about "interfering" with nature. Millions of people who live in "urban green land" appear to feel so guilty about every part of their behaviour that they suffer from a kind of environmental neurosis.

At least some of this angst appears to be rooted in the traditional religions of Western culture. For example, a key part of several religions, like Judaism and Christianity, is a serious falling out between humans and God. In short, we were booted out of the Garden of Eden. Consequently, for many Christians, a key part of daily religious practice is a lifelong attempt to come back into communion with God. These humans have become separated from God and Nature and must always strive for a reunion. For other Christians, God and humans were reunited by the self-sacrifice of Jesus. These Christians are always in communion with God. In other religions like Hinduism or Buddhism a tragic and defining split between humans and God never occurred. Here humans never have been apart from nature.

For those of us who are not so fretful the picture is clear. Earth is our home and we depend upon it for our food, shelter, organization of our societies, and creation of our wealth. The Earth provides the foundation and context for the personal adventure that builds a life. The Earth is beautiful and familiar, not ugly and alien. Of course we are part of nature. Humans live in most parts of the world and are absent from some. Places empty of humans are no more natural than places teeming with our kind. They are just different.

What does all this have to do with biodiversity? For one thing, the knowledge that nature loses everything it gains – except life itself – is central to understanding nature and its biodiversity. The reality is that ecosystems change all the time with or without human help. Historically speaking nearly all species have proven expendable. Humanity has evolved during a period obviously favourable to us. In particular, the past 14,000 years or so have been good times. During this this period the Earth's climate has been temperate, rainfall has been plentiful, volcanic activity has been low, few killer rocks have fallen from space, the ozone layer has been stout, and, so far as we know, no cosmic radiation storm front has moved through the solar system. Any of these conditions could change.

Should the conditions end that now favour humans, life itself will go merrily on. Nature will simply raise up new creatures. Nature has no favourite species or ecosystems. Only humans have these. Take the spotted owl, for example. Part of the spotted owl saga rests on the plausible idea that small isolated populations are more prone to extinction than large populations spread over a range of habitats. If the spotted owl of Washington and Oregon is a lonely, isolated breed then the odds of peril rise. But if these birds belong, as they in fact do, to a large genetic family existing in a wide range of habitats from British Columbia to Mexico and from the Pacific Coast to West Texas, then any local population downtrend is unlikely to lead to an unstoppable species descent. If all spotted owls are one species, as the Audubon Society says they are, then there are thousands and thousands of these birds, and the Northern Spotted Owl is not a species. Even so, the United States Endangered Species Act treats the Northern Spotted Owl as a separate species, citing locally "distinct" populations. This treatment supports the theory that local variations in climate and diet convert creatures into different species.

Consider one consequence. According to this theory, a black man who lives in Vancouver, gets rained on, and eats salmon would be a different species than a white man who lives in stifling humidity in Louisiana and dines on gumbo. By this Northern Spotted Owl theory the human race contains hundreds of entirely distinct species.

Consider another consequence of the spotted owl fiasco. Thousands of honest people lost their livelihoods. Affluent environmentalists with white collar sinecure should ask themselves whether they destroyed thousands of skilled labour jobs to satisfy an ideology and boost returns on fund-raising drives.

What about other species? Since nature began, most species that have lived on Earth have long since become extinct. In fact, species extinction is nature's norm, with or without human assistance.

Today we have no clear estimate of how many species exist. The following estimates for the total number of species existing on Earth have been made during the last 240 years.

1758	Linnaeus	9,000
1987	Peter Raven	2,200,000
1987	Paul Ehrlich	4,000,000
1994	Paul Ehrlich	100,000,000
1995	Catalogued species	1,400,000

These widely differing estimates, sometimes by the same person, undermine the credibility of all estimates of species numbers.

Roughly since the 1970s, ecologists have claimed a rising number of species loss caused by human activity. In the same period, researchers have supposed the natural world to contain far more species than once believed. For example, Thomas Lovejoy of the Smithsonian Institute has called human-caused species loss "a potential biological transformation unequalled since perhaps the disappearance of the dinosaur." Russell Train, head of the World Wildlife Fund, has declared that owing to species loss "the future of the world could be altered drastically." Edward Wilson says human-caused species loss may eventually equal the damage that occurred 65 million years ago when the dinosaurs vanished.

Environmentalists seem to thrill to the vision of a dreadful people-caused species loss. At the same time they also endorse the evidence that the known number of species is increasing. To correct for this inconvenience, as estimates of total species have increased, the loss estimates have also increased. Some examples of increasing estimates of species loss largely due to humans are:

1979	N. Myers	100 species per year
1980	"Global 2000"	274 species per day between 1980 and 2000
1995	N. Myers	30,000 species per year
1995	E.O. Wilson	137 species per day

What is the natural rate of extinction? Nobody really knows, but one species per year globally is a common estimate. A recent book, *Life In the Balance* by Niles Eldredge, provides data from which to estimate a general extinction rate. An appendix to the book lists the animal species thought to have become extinct since about 1600 AD. The list includes corals, molluscs, crustaceans, insects, fishes, amphibians, reptiles, birds, and mammals. In total, 489 species are listed, an average total extinction rate of about 1.2 species per year. According to the list, about 200 of these extinctions were caused or contributed to by human interference. This would leave the average natural rate of extinction as 0.7 animal species per year and that from humans at about 0.5 species per year, averaged over the past 400 years.

If extinctions truly have gone from one per year to 274 per day, the consequence would be stunning. According to "Global 2000," since the year 1980 about 1.8 million species should have become extinct. This number is actually larger than the 1.4 million living species catalogued globally by 1995!

Where are the missing 100,000 annual extinctions of "Global 2000," or the missing 50,000 annual extinctions of E.O. Wilson? Where are the horrors of human-caused mass extinction? The species corpses should be piled high. Why can't we find them?

According to Gregg Easterbrook in *A Moment on the Earth*, the United States Endangered Species Act in 1973 first listed 67 species. Of that list, 44 are now stable or improving, 20 are in decline, and seven are gone. According to E.O. Wilson's estimate of 137 per day, 1.1 million species should have gone extinct globally since 1973. The United States has about 6 percent of the world's land mass, so about 66,000 species should have gone extinct in the United States since 1973. The reality is that only seven did. This is less than the expected natural rate of one species per year.

Easterbrook also comments that in the United States Pacific Northwest, according to David Wilcove of the Environmental Defense Fund, no vertebrates are known to have gone extinct in those forests since World War II. Likewise, the Nature Conservancy reports no known extinctions of vascular plants in the Pacific Northwest during this period.

Again, using the E.O. Wilson estimate and prorating, 75,000 species should have gone extinct in the Pacific Northwest since the Second World War. In the past, Wilson asserted that Paleo Indians in North America caused mass extinction about 11,000 years ago when they learned to hunt. Wilson also asserts that after such a major extinction ten million years are required for biological diversity to recover naturally. Astoundingly, the same Dr. Wilson also insists that biological diversity today is the highest ever seen in the Earth's history! These three assertions by Wilson cannot all be true. Together they make no sense.

Besides having favourite species, environmentalists have favourite ecosystems. Top of the pops are wetlands. Biologists like wetlands because they are active ecospheres where competition is ongoing and biodiversity is high. Wet, saturated soils have only one drawback. Easterbrook provides an estimate that, owing to natural decomposition of organic matter, global wetlands emit about 115 million tons per year of methane, a potent greenhouse gas. The world's cattle are the source of about 60 million tons per year of the same gas, methane. That is, wetlands produce nearly twice as much methane as cattle. Environmentalists like Jeremy Riffkind often say that methane emissions from livestock are an outrage. They say this while simultaneously crying out that the world needs more wetlands. Either methane emissions from cattle are an outrage or the world needs more wetlands. The two things cannot both be true at the same time.

Visions of Commerce

This section draws heavily from articles in *The Economist* newspaper, *Adam Smith in his Time and Ours: Designing the Decent Society* by Jerry Z. Muller, and *Costing the Earth* by Frances Cairncross.

A great debate took place in eighteenth century Europe on the role of commerce in society. In the early 1700s, European society and government were structured to impede the free movement of labour, capital, and goods. Property rights and laws favoured a narrow elite who benefited from legal restrictions to sell commodities or labour. Legal monopolies were common. Similarly, guilds and related organizations had a right to limit the supply of labour in specific occupations. Many different interest groups were able to promote their own interests at the expense of public interest. Not surprisingly, also existing at the same time was a large and miserable class of working poor.

By the 1770s, Britain was beyond the stage at which the margin between hunger and the level of agricultural production was so narrow that the survival of a large part of the society depended on the harvest. In earlier times, the slimness of the margin had justified government regulation of everything from how the land was worked to when and where grain could be sold.

As the nation became wealthier, not just its elite but its working people as well, expectations were changing. For perhaps the first time in history, acquiring a basic minimum of food, shelter, and clothing was a nearly universal expectation. Contemporary observers were struck by the relative ease with which an ordinary labourer could support himself and his family. Wage rates increased gradually for most of the century. As wages rose and the costs of production fell in agriculture and the manufacturing of basic necessities like textiles for clothing, the standard of living rose. What were once luxuries became decencies. Then decencies became necessities. Tea, a luxury beverage of the upper classes at the start of the 18th century, was the daily drink of road workers by mid-century. Objects long reserved for the rich came within the reach of a large part of society. Blankets, linens, pillows, rugs, curtains, pewter, glass, china, brass, copper, and ironware flowed into English homes.

But powerful segments of society were upset at the improved lives of ordinary people. Elite writers responded in consternation to the rising living standards of the labouring classes. Economic arguments were added to the moralistic denunciation of luxury as something that promoted sin and undermined civic virtue. Higher wages, it was said, would undermine the will to work because workers would work only long enough to meet their traditional requirements. After that, they would choose leisure over income. Supporting this opposition to the betterment of the working classes were two important and powerful traditions, the old Christian and classical Civic Republican traditions.

The following passage describes a vision of commerce from the old Christian tradition expressed by Father Thomassin in his "Traité de Negoce et de l'usure" published in 1697.

> Those who accumulate possessions without end and without measure, those who are constantly adding new fields and new houses to their heritage; those who hoard huge quantities of wheat in order to sell at what to them is the opportune moment; those who lend at interest to poor and rich alike, think they are doing nothing against reason, against equity, and finally against divine law, because, as they imagine they do no harm to anyone and indeed benefit those who would otherwise fall into great necessity ... [yet] if no one acquired or possessed more than he needed for his maintenance and that of his family, there would be no destitute in the world at all. It is thus this urge to acquire more and more which brings so many poor people to penury. Can this immense greed for acquisition be innocent, or only slightly criminal?

The following example expresses the position and vision of Civic Republicanism, which derived from classical Greek origins. It is from the "Essay Upon the Probable Methods of Making a People Gainers in the Balance of Trade" by the English political economist Charles Davenent in 1699.

> Trade, without doubt, is in its nature a pernicious thing; it brings in that wealth which introduces luxury; it gives rise to fraud and avarice; and extinguishes virtue and simplicity of manners; it depraves a people and makes way for that corruption which never fails to end in slavery, foreign or domestic. Lycurgus, in the most perfect model of government that was ever formed, did banish it from his commonwealth.

Both the old Christian and Civic Republican traditions were suspicious of commerce, regarding it as a barrier to the pursuit of virtue. In the ideal political regime, the best citizens ruled, and in their rule they rewarded virtue. Those who engaged in trade would play no political role. Most classical writers saw no justification for the merchants' deriving income from buying and selling. They assumed that the material wealth of humanity was more or less fixed, which meant that the gain of some was a loss to others (in modern terms, trade was a zero sum game). Profits from trade were morally illegitimate.

Along with the disparagement of wealth was a denigration of merchants and the pursuit of profit. The 18th century church fathers also preserved the classical assumption that since the material wealth of humanity was more or less fixed, the gain of some could only come at the loss of others.

Civic Republicanism assumed that political participation was the highest form of activity. The needs of the community took precedence over other moral demands. Only a shared vision of the public good would hold society together. The Civic Republican tradition stressed collective liberty in the sense of freedom from foreign domination, and individual liberty in the sense of the freedom of citizens to participate in political activity. The spectre haunting those steeped in this tradition was corruption and self-interest. Corruption arose when those who should devote themselves to public virtue instead pursued private material interests.

The traditions of the 18th century church fathers and of Civic Republicanism are with us today. One consequence of the church fathers' vision of commerce is that the pursuit of profit endangers the soul. The search for gain in this world is likely to lead to loss of the next. There was an additional consequence of Civic Republicanism. Possession of property was a prerequisite for citizenship. By freeing men from the need to engage in productive activity, Civic Republicanism allowed them to devote themselves to the fate of the commonwealth. From Aristotle through to 19th century defenders of slavery in the United States South, the Civic Republican tradition assumed that citizenship would be limited to those who did not need to work for a living.

In the 18th century, views contradictory to the old Christian tradition and to Civic Republicanism were expressed. One of these views held that commerce was part of a providential design to permit humans to enjoy the widely scattered fruits of the earth. People supporting this view were optimistic about the opportunities for general betterment. They believed that improvements in the standard of living, not only of the rich but also of the working poor, could be gained by extending greater market freedom to all parts of the economy. Adam Smith's ideas of enlightened self-interest articulated this side of the debate. Other writers of a similar mind endorsed a culture that valued sociability and that looked for guidance to philosophy, science, literature, and the arts. One of their aims was to legitimize the pursuit of material prosperity as a worthy political goal. A major premise of *The Wealth of Nations* was that material wealth can be created and increased by human endeavour. Moreover, trade creates wealth in a way that benefits all parties and not by making one person's gain another person's loss. Many of the proponents of economic growth argued that material prosperity was the prerequisite for civilization. Society could not and need not be governed by some faith or purpose that dictated the proper distribution of property. Included in these arguments was the idea of civil jurisprudence. This stressed the rule of law and regarded the protection of property from arbitrary confiscation by government as a central freedom. Rather than valuing the liberty to participate in government, it valued freedom from government, a freedom guaranteed by law.

Today's environmental debate in British Columbia has close parallels with the eighteenth century debate on commerce. Much environmentalism is a new version of the old hatred of business by would-be utopians. Many of the policies of greens are more concerned with stopping companies from making money than with making the world a cleaner, greener place. Look at the environmentalists' enthusiastic and gleeful attempts to organise boycotts of British Columbia wood products. Have environmentalists assumed the mantles of the old Christian and Civic Republican traditions?

Many environmentalists wince at the very mention of industry. As they rightly see it, industrial activity is the immediate cause of most environmental damage. The environmentalists' solution is to banish industry. This is a foolish attitude. Perhaps in the Middle Ages human activity had little lasting impact on nature's balance. That day is gone for good. Today 5.8 billion people live on earth. Even in the poorest countries, where the lives of millions of people are barely touched by industry as the developed world knows it, the human impact on the environment is immense.

It is also utopian and hypocritical for the citizens of urban green land to think that the poor will not want the same trappings of Western wealth enjoyed by their own green selves. Top of the shopping list for every liberated East European is a family car. Many products that harm the environment have made life better for people. Above all, disposable products have frequently improved life at the cost of extra pressure on the environment. However much the environmentalists bemoan the throw-away society, powerful human forces will continue to encourage it.

In British Columbia history all governments have pursued policies of development. Presumably their constituents supported these policies. The aim has been to have more and bigger villages, towns, and cities with more and better services and facilities. To date, most of this development has been paid for by creating wealth from resource industries. It is important for environmentalists to understand these forces. Rather than yearning for a world that can never be created, they need to help develop incentives for industry to support human needs in the most benign way. British Columbia environmentalists are failing to do this. Instead, environmental groups expertly twang public heart strings. Supposedly educated and intelligent people mouth meaningless slogans. The form of intolerance called political correctness condones shouting down and suppressing other viewpoints. At times it seems we are entering a second dark age where knowledge, reason, and common sense are exchanged for whimsy, intolerance, and new-age fuzzy wuzzy.

In the western world, capitalism and the economic growth it has fostered have freed countless people from poverty and created the means to help others. No other economic system comes even close to matching its achievements in any aspect of economic or social progress. The main varieties of capitalism, American, European, and East Asian are different, but all have four essentials that need to be preserved if liberal economics is to go on to further success.

First and foremost, capitalism separates, to a high degree, the realms of politics and economics. Decisions about what goods and services are provided, by whom, to whom, and for how much, are made for the most part in markets by willing buyers and sellers. Governments in capitalist countries participate in markets often in a big way, as buyers or sellers, or as regulators. They do not usurp the price system altogether. In capitalist countries, the extent of government intervention is a matter of politics, but the way government intervenes is mostly a matter of economics.

If government intervenes by designating resource prices without respect to the market or by setting tariffs or quotas that favour or discriminate against selected industries, disaster results. Among other things, this leads to bigger government. Next, more interventions are needed to sort out the mess surprisingly caused by the first interventions, and so on.

Second, private ownership has usually been a feature of capitalist economies. It is a natural outcome of separating politics from economics. It is not a necessary outcome because, in achieving that separation, control matters more than ownership and ownership does not guarantee control. For example, in southern China, enterprise managers were given increasing freedom to run their businesses themselves. Even without private property, a separation of politics and economics was achieved, and the price system began to direct the allocation of resources. India has private ownership, but until the reforms of the 1990s it also had a system of state control that rivaled that of the Soviet Union. A factory making bicycles needed government permission to increase its output, or to reduce it, or to start making a new kind of bicycle. This "Licence Raj" was so pervasive and intrusive that, in effect, it unified the realms of politics and economics despite the existence of private property. In general, private control of productive assets brings wealth to society. Government control of assets causes loss of wealth to everyone, making people poorer than they should be.

The third essential of capitalism is that decisions about allocating resources are highly decentralized. Instead of an explicit organizing intelligence, there is a spontaneous and unwitting coordination. Instead of planned cooperation, there is competition. This competition extends far beyond static competition among existing producers and their products. It encompasses competition among new, would-be producers; among ideas for products yet to be invented; among alternative means of production and among different modes of industrial organization.

The fourth essential is innovation. Because capitalism is decentralized and competitive, it is especially good at conducting experiments. Experiments can be conducted on a small scale and at correspondingly small expense in resources. Successful experiments reap big rewards. That, of course, is the incentive to undertake the experiment in the first place. The resulting profits will also be the signal for others to follow, so successful innovations (of product, service, production method, or mode of organization) are quickly taken up by others. Equally important, experiments that fail – as most do – can usually be abandoned with comparatively little pain and at no cost to the politically powerful. These conditions offer maximum encouragement for efficient innovation. It is no surprise, therefore, that western capitalism has been relentlessly innovative.

The past 150 years of material advance in the West has shown beyond doubt that social change is tightly linked to economic growth. In fact, they are nearly, if not quite, the same thing. Yet throughout the period, people have feared change as much as they have wanted growth.

This ambivalence is more acute than ever before. Expectations of continually rising living standards are now so deeply rooted that annual growth rates below two percent are regarded as deplorable. They prompt talk of stagnation, or even a crisis of capitalism. Yet, for almost all of the period that humans have lived, the average rate of economic growth each year has been, to the nearest round number, zero. Growth of two percent per year is enough to double output every 35 years. Western electorates demand not merely growth, but rapid growth. At the same time, they are increasingly suspicious of change – growth by its other name.

A growing preference for stability (or less change) may help to explain diminishing support in the West for open markets and free trade. These fears are bound to be increased by prospects of faster growth in Eastern Europe and the third world. Ironically, the spread of capitalism in the poorest part of the world may undermine support for market economics in the countries where it has already worked well.

With greater wealth comes greater demand for public goods that unassisted capitalism may fail to supply. Among the most important of these goods is a clean environment. The most ardent preservationists explicitly reject capitalism. They argue that economic growth and the protection of the environment are incompatible goals. They are strangely unimpressed by the fact that rich economies are generally proportionally less polluted than poor ones. It seems unlikely that this is due to existing regulation. To some extent, markets reward clean processes. Capitalism and a clean environment may not be entirely incompatible.

Nevertheless, many environmental resources are neither owned nor priced in any functioning market. If incomes continue to rise, governments are likely to increase their regulatory demands, and sometimes rightly so. By itself this need not be much of a setback for liberal economies. Increased regulation will probably happen alongside other moves to advance the political realm at the expense of the price system. For example, environmentalists and protectors of declining industry have already formed an influential alliance in the United States.

In addition to these trends is the tendency for public spending to claim an ever-growing share of national income. Yet the perception of most voters is the reverse. They think spending has been cut and public services diminished. In one way the voters are right, because governments have been getting less in return for their spending. One reason for this is that prices for many government services have been rising faster than prices in general. Therefore, if governments are to provide a broadly constant level of social services in the future, their spending (and the taxes needed to finance it) will keep rising as a share of national income.

Against the pressures threatening to undermine capitalism in the coming years, the strongest countervailing force is likely to be technology. In many industries technological progress has reduced the fixed costs of production making it easier for small firms to compete with larger ones. Technology has also developed new products and services that broaden the possibilities of competition. To deal with this, governments will try to devise new systems of regulation. Fortunately, this is difficult and it is likely that technology will continue to move faster than government.

As these opposing forces work themselves out, governments will have two choices. One is to give way to the pressures that will impede the market system – i.e., to favour more trade protection; to favour more help for declining industries; to expand the welfare state; and to increase the regulation of cross-border trade. These actions operate most powerfully against change and hence against growth. The alternative choice is to extend the scope of the market. This means freer trade; policies to protect workers unlucky enough to be in declining industries rather than policies to save jobs; and a welfare state that helps the poor, not the middle class as it does now.

Conclusions

This paper has assembled several very old and contradictory traditions of thinking about the future of our planet, nature, and commerce. All of these traditions are vigorously alive today and lead directly to contradictory views about ecosystem management and the maintenance of biodiversity. As always, there is a common ground, there are differences, and there are principles for management that can be established.

Common Ground

First, we all agree that the earth is a wonderful place. To some the divine is everywhere. To others the divine is everything. To the rest it's just a really neat place in which to be alive.

Second, our species has flourished over the past 14,000 years or so, largely because of a benign climate and stable geological conditions. Over the far reaches of time we may not be so lucky.

Third, no such thing as a static, ideal state of nature exists. In the past British Columbia has been hotter, colder, flatter, and lower than it is today. Before that it didn't exist. Most of the animals and plants we see today are descendants of immigrants that colonized British Columbia after the retreat of the ice ages only ten thousand years (about 330 human generations or 100 tree generations) ago. Since then the climate in every region of the province has changed through many cycles. Along with the changes in climate have been changes in the animals and the forests. For example, in the far northwestern part of British Columbia it is thought that the scrub tundra that prevailed immediately following the retreat of the glaciers quickly gave way to spruce *(Picea spp.)* woodland. Then there was a general increase in warmth and moisture until about 6,000 years ago. Following a further trend to a drier climate about 4,000 years ago, subalpine fir *(Abies lascioparca)* and then lodgepole pine *(Pinus contorta)* joined the spruce forest about 3,000 years ago. The ecosystems we see today in northern British Columbia have probably been in place for about 3,000 years (100 human generations, or 30 tree generations).

Forests come and go, with or without humans. As climates have changed, forests have come and gone. Even when land is stripped to bedrock by glaciers or blanketed with ash by volcanoes, the forest will return if the climate is right. This is equally true of forests cut by humans. Whether they are left alone or helped by foresters, the forests will return after logging and can eventually be as beautiful in their own right as the forests they replace.

Fourth, losing species is a natural phenomenon, albeit one that saddens humans. Furthermore, when species that we admire are lost or endangered by human actions, anger embitters our sadness.

Fifth, we all like to see contiguous areas of big, old trees.

Sixth, barring sudden climate changes or catastrophic geological events, we all want to use natural resources in ways that allow the use to continue profitably for a good number of human generations; that will support a stable set of animal and plant species; and that will provide plenty of pretty views. In other words, we all like the idea of sustainable resource management.

If there are six points on which we all agree, there are others on which there is no agreement.

Uncommon Ground

In addition to the common ground on which people generally agree, equally important ideas are the subject of deep disagreements.

First is the difference between those who live on "Planet Plenty" and those who live on "Planet Doom." Some of us see abundance, some see scarcity. Some see improvements in the opportunities for each of us to build a decent, satisfying life. Others see an unlimited need for legions of government-paid social workers and grief counselors.

Another difference lies between the "hairspray ecosystem" crowd that wants to set nature in some static and never-changing state and those people who believe life and nature are continually energized by change.

A third difference exists between the "commerce is good" and the "commerce is evil" believers. These differences extend into the social and political realms.

A fourth major difference is between those who see humans as apart from nature and those who see humans as part of nature and interdependent with it. If humans are apart from nature, then human activity is unnatural. If humans are part of nature and interdependent with it, then all human acts are natural.

Suggestions for Ecosystem Management

The first requirement of ecosystem management is to discriminate between good sense and nonsense. Visions that stun people into incompetence are to be avoided. Ideas that have ridiculous consequences should be exposed. Statements that require two or more conflicting realities to be simultaneously true must be identified.

To the author, the following premises make sense:

- Humans are part of nature and are interdependent with it.
- The earth is far closer to being "Planet Plenty" than to being "Planet Doom."
- Nature is dynamic and will never stop changing until life itself ceases to exist.
- Trade is what makes the ship go. Trade and commerce bring economic freedom without which political freedom is not possible.

These four premises plus the six areas of common ground identified earlier form a basis for sensible and sustainable management of ecosystems and biodiversity.

A central teaching of ecology is that humans are part of nature and inter-dependent with it. All our acts are natural in this sense. It is hard to see how the word "unnatural" can be used as a definition of all human activity and "natural" be applied to all non-human activity. Human intervention com-pared with non-human intervention is surely no different in nature. Nature does not play favourite among earth's species. Catastrophic events like ma-jor earthquakes, volcanic eruptions, hurricanes, tornadoes, tidal waves, fires, and avalanches all rival human capabilities. Diseases, insects, and grazing by herbivores have all had profound effects on forested ecosystems. But 5.8 billion humans are bound to affect their neighbourhood and are bound to have an impact on most ecosystems.

As for forests, the only way to stop them from growing back after fire, death by disease or blowdown, or logging is by purposefully interfering with the process of forest renewal. Common ways of doing this are by plough-ing it every year and planting crops, by grazing livestock so that they eat every tree and seedling that grows, or by covering land with cement and buildings.

It is not surprising that many people associate deforestation and destruc-tion of forests with logging. After all, the first stage of deforestation is the removal of trees. But deforestation is a two-stage process. The second stage involves human activity directed at preventing the forest from growing back where it is cut. It is the second stage, the permanent conversion of land to another use, that results in deforestation. Logging followed by reforestation is not deforestation. Deforestation is logging followed by conversion to ag-riculture, human settlement, and industrial development.

Today we see a strong interest in the ecology of forests as a whole. The interest is expressed in two different ways. First, a strong movement to pro-tect and preserve remaining first growth forests as wilderness areas is evi-dent. So long as these areas are large enough, it is possible to provide relatively intact ecosystems in which all species dependent on old forest cover can flourish. The extent of the area required to sustain the larger predators, such as grizzly bears and big cats, is still being debated.

Second, where forests are managed for wood production, pressure is grow-ing to protect as many features of the original forest as possible. This is best accomplished by planting native tree species and by retaining and enhanc-ing critical elements of the forest such as standing dead trees required by cavity nesting birds. This approach to forestry has been called "new for-estry," "messy forestry," and "ecologically based forestry." The challenge is to provide habitat for as many species as possible in managed forests while, at the same time, maintaining a profitable forest industry.

The most common term associated with improved forest management is *sustainable forestry*. Sustainable forestry means adding the social and eco-nomic issues to the environmental issues and balancing all three. On one level, sustainability is an ideal state in which the actions of today's generation have no adverse impact on the opportunities of future generations. On another level, it is a practical, rational approach to conserving natural resources so future generations have more choices than if we squandered them. No perfect state of sustainability exists. It is a relative state.

Sustainable forestry, at its simplest, means a commitment to retain trees on a site. But which species of trees? It may be possible to cut down a native coniferous forest and replace it with non-native hardwoods such as maple and oak and still have a sustainable forest. But some of the other species that lived in a coniferous forest may not survive in a hardwood forest. In that sense, they would not be sustained even though the hardwoods themselves could be sustained over a hundred generations. Some other plant and animal species would prefer the hardwoods over the original conifers and hence could be more sustainable in a new environment created by human management.

Sustainable forestry is not about what happened to forests in the past but about what will happen to the land in the future. Therefore the most important factors underlying sustainability are the institutional arrangements that ensure the forest land will be properly tended and not be converted irrationally to other uses such as agriculture or urban development. Such things as ownership, control, land tenure, management agreements, and forest laws are the factors that bear most heavily on sustainability. If the timber-producing land base is reduced to nothing by preservation, conversion to real estate, or agriculture, there cannot be any sustainable forestry. No amount of details about forest practices, environmental protection, or public participation can produce sustainable forestry if we have nowhere left to cut trees and nowhere left to grow them.

To what future time frame does sustainability apply? No state of affairs is sustainable eternally. Climate will change, ice ages will return, and eventually the sun itself will burn out. We need not think in these geological time frames. For example, our new-age friends might choose 12 years as a sustainability horizon because this is how long the planet Jupiter requires for a circuit of the zodiac and a return to its moment of retrograde motion in the sign of Cancer. Then again, we might notice that at the moment of the spring equinox, March 21, the heavens are never quite in the position they were in the year before. Each year there is a very slight lag of about 50 seconds. This amounts to one degree in 72 years, 30 degrees in 2,160 years (one sign of the zodiac), and completes one cycle of the zodiac, returning to the original alignment in 25,920 years. Here we have three more alternatives (72 or 2,160 or 25,920 years) for our sustainability horizon. Or maybe we should use the 21,000-year cycle in which the timing of the Earth's closest approach to the sun moves through the year (right now the Earth is closest to the sun in January). Cosmic numbers aside (and believe me there are plenty more), horizons for sustainable forestry are properly obtained by considering the length of tree and human generations. One human generation, the time taken to grow to maturity and produce a family, is often taken as 30 years. Different tree species have different life spans, but 100 years is a time in which all tree species have grown to maturity and have achieved full powers of reproduction. Some species, such as alder *(Alnus spp.)* and birch *(Betula spp.)*, live for only 60 to 80 years. If we say that three human generations and one tree generation are approximately encompassed by 100 years, then 100 years seems a reasonable time frame for sustainability. This time frame will be continuously rolled over by each successive human generation.

A sustainability horizon of 100 years does not mean that all trees older than 100 years will be cut down. Nor does it mean that no trees younger than 100 years will be cut down. It just means that forecasts about the future state of all our forested ecosystems should extend about 100 years from now.

We have many issues to address under the general heading of sustainable forestry. Some of them are road building standards, old growth protection, wildlife preservation, fisheries, biological diversity, and forest renewal. Sustainable forestry must take into account the full range of values found on the land. These include spiritual and cultural values as well as material concerns. There are special places in forests where natural features combine to encourage contemplation and wonder. There are places of deep cultural and historical significance where commercial activity must be curtailed. And there are many other forest areas in British Columbia where a skillful and never-ending cycle of forest harvesting and renewal will release an abundant and sustainable flow of material wealth.

In sustainable management the fate of old-growth forests is second only to clearcutting as a flashpoint for environmental campaigns to preserve wilderness. Already in British Columbia, extensive old-growth preserves have been created by national parks, provincial parks, ecological reserves, old-growth reserves or management areas, riparian management reserves, critical wildlife habitat reserves, environmentally sensitive areas, and all the other areas that are set aside from the timber harvesting land base. Many millions of hectares of old-growth forest have already been reserved by these means across a wide range of ecosystems.

In addition to these large reserves, other areas of commercially valuable old growth are not protected by the long list of mechanisms given above. These areas have been designated for timber production and will eventually be logged and converted to new forests. These are the areas that appear on television and on high school classroom walls juxtaposed with ugly pictures of recently clearcut and/or burned land. The new forest that comes in after logging to form beautiful stands of vigorous young trees never appears on the television screen or the classroom wall. The old-growth forest has been idealised by images of the most photogenic trees contrasted with the worst possible portrayal of charred clearcut devastation. Logging is portrayed as total destruction with no chance of recovery. The old-growth trees, apparently, will never change despite the fact it is at least 10,000 years since the last ice age and most members of the main coniferous species in British Columbia die and fall over between 300 and 400 years of age. That means many of the first-growth forests in British Columbia in reality are thirtieth-growth forests.

Directional and Cyclical Forest Change

To advance forest ecosystem management and the maintenance of biodiversity it is vital for foresters to continually provide a coherent account of forest ecosystems. In particular, we need to increase our understanding of the forces that determine forest structure and that cause changes in forests over time.

Forests undergo continual change that can be classed as directional change and cyclical change. Directional change induced by an environmental change or by intrinsic properties of the tree species is called forest succession. Cyclical change is usually caused by a reaction between the forest and its habitat, which produces repeated cycling through the same vegetational phases.

Directional change conceives plants occupying bare ground and progressing toward one or more stable forms via the successive replacement of one community by another of different growth form. Interior British Columbia stands of lodgepole pine established after fire, which then develop a spruce understory and eventually become spruce stands, are an example of directional change. Cyclical change has several phases that are repeated in cycles on the same piece of ground. Cyclical change was documented by A.C. Poole in 1937 for the mixed temperate rain forest in New Zealand. On any fairly large area undergoing cyclical change the vegetation usually occurs as a mosaic of patches, each one of which is in a different phase. A good example of this was described by A.S. Watt in 1947 for the regeneration complex that occurs on raised bogs. Repeated cycles of logging and subsequent forest renewal are, by definition, cyclical forest change.

These two types of change, directional and cyclical, found in natural plant communities, provide a model for ecosystem management in British Columbia. Areas set aside or managed to provide old-growth values should be recognized as areas where directional change is the dominant management objective. The forests are moving toward one or more stable forms. Timber producing areas should be recognised as areas where cyclical change is the objective. Important wildlife areas should also be managed for a component of cyclical change using prescribed burning. For example, young aspen *(Populus tremuloides)* stands provide important habitat for hare *(Lepus spp.)*, black and grizzly bear *(Ursus americanus and U. horribilis)*, deer *(Odocoileus spp.)*, elk *(Cervus canadensis)*, moose *(Alces alces)*, grouse *(Bonasa umbellus)*, and a number of smaller birds and animals. If directional change occurs, incoming spruce trees eventually destroy the aspen habitat values. Prescribed burning returns the cycle to open range land, which is then invaded by young aspen stands, and so on. Areas managed for directional change will tend to have large areas in the same stage of development. Areas managed for cyclical change will exhibit a mosaic of patches, each of which is in a different development phase. Patch size distribution patterns will be different in directional change areas than in cyclical change areas.

It is possible to model vegetational change using well-established techniques so that the probable consequences of different directional change and cyclical change management strategies can be predicted. Further, if we understood better the relationships between forest ecosystem structure and wildlife habitat, we could also predict future quantities of available habitat under different management strategies.

To repeat the point: directional change for parks and old-growth preserves; cyclical change for the timber harvesting landbase and for components of wildlife producing areas.

People

Where do people fit into ecosystem management? First of all, people who need to work for a living are obviously more interdependent with nature than those who are independently wealthy. Second, working people must be given the respect they deserve. They should not be treated as pieces in a board game by environmentalists, industrialists, or labour unions. This does not mean that they should be sheltered from the consequences of natural and economic laws. It means that they should be treated fairly and allowed to choose their own lifestyles and means of earning a living.

Wealth creation benefits all of society and provides the best opportunities to create a decent and free society. For growth to continue in a sustainable way, government has to support private enterprise as a facilitator and not as a manager. It should promote enterprise mainly in the sense that government does not go out of its way to undermine it. When government fosters a pro-business culture in this way, an area of the economy is gradually carved out as business territory. Relations between government and business come to be governed by stable rules. Enterprise is secure enough to invest and grow.

In British Columbia, the clash of different interest groups (stakeholders) in a vicious zero-sum game, where each win equals someone else's loss, has removed our ability to be sensible. In fact, we have become ridiculous. In a province with such abundant natural resources and so many opportunities to create wealth from tending the forests, we are also being very stupid.

References

Bhagwati, J. 1988. Protectionism. MIT Press, Cambridge, MA.

Cairncross, F. 1992. Costing the earth. Harvard Business School Press, Boston, MA.

Campbell, J. 1956. The hero with a thousand faces. Meridian Books, New York, NY.

Campbell, J. 1959, 1962, 1964, 1968. The masks of God (four volumes). Viking Press, New York, NY.

Cannings, R., and S. Cannings. 1996. British Columbia: A natural history. Greystone Books, Vancouver, BC.

Cohl, H.A. 1997. Are we scaring ourselves to death? St. Martin's Griffin, New York, NY.

Easterbrook, G. 1995. A moment on the earth. Penguin Books, New York, NY.

The Economist: December 1997, Environmental Scares; September 1993, The Future of Capitalism; September 1989, Poor Man's Burden, A Survey of the Third World; April 1993, An Editor's Farewell.

Eldredge, N. 1998. Life in the balance. Princeton University Press, Princeton NJ.

Frazer, J.G. 1922. The golden bough. MacMillan, London, UK.

Gwartney, J., R. Lawson, and W. Block. 1996. Economic freedom of the world 1975-1995. Fraser Institute, Vancouver, BC.

Moore, P. 1995. Pacific spirit. Terra Bella Publishers, Vancouver, BC.

Muller, J.Z. 1993. Adam Smith in his time and ours: Designing the decent society. Free Press, New York, NY.

Poole, A.C. 1937. A brief ecological survey of the Pukekura State Forest, South Westland. New Zealand Journal of Forestry, Vol. 4.

Smith, A. 1776, 1783. An inquiry into the nature and causes of the wealth of nations. Fifth edition (1819), three volumes.

Watt, A.S. 1947. Pattern and process in the plant community. Journal of Ecology 35:1-22.

Wilson, E.O. 1998. Consilience. Alfred A. Knopf, New York, NY.

Worrying Issues about Forest Landscape Management Plans in British Columbia

Gordon F. Weetman

Abstract

Some questions are posed about forest landscape planning in British Columbia. What is their scientific basis? Who will prepare them? What are the attributes of a sustainable forestry plan? How will they be done? What are the roles of Registered Professional Foresters, the provincial government, the forest industry, and the Land and Resource Management Plan (LRMP) processes? Why is Alberta so different? Each question is addressed briefly.

Introduction

Landscape level planning in British Columbia is still in its early development phases. The province has a feature that makes it almost unique in the world: it has about 50 million ha of forested landscape, owned by the province, about one half of it being old natural conifer forests.

Figure 1 shows the age class distribution of several Timber Supply Areas in British Columbia. The presence of old natural forests across the wet climate landscape makes British Columbia quite different from the drier Alberta climate. In fact, most of the boreal forest in Canada, in contrast to British Columbia's forests, has a frequent fire history that results in age classes that are younger and fragmented. The preponderance in British Columbia of Natural Disturbance Type One (NDT1 as defined by BC Forest Service 1995), ecosystems with rare stand-initiating events, differs from Alberta, which is mainly NDT3, ecosystems with frequent stand-initiating events. It is obvious that, in many parts of British Columbia, only old natural forests will be harvested for several more decades.

A Science Basis

The science basis for landscape planning in British Columbia is currently governed by ideas, notions, and concepts from conservation biology. The Biodiversity Guidebook of the Forest Practice Code of British Columbia (BC

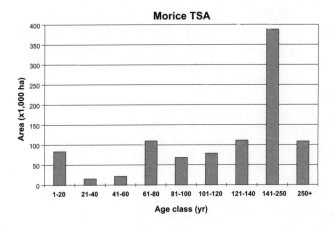

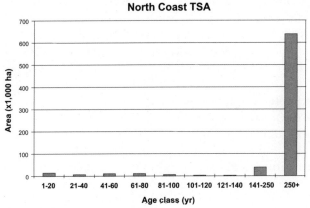

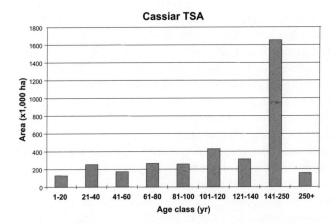

Figure 1. Age class distributions of three Timber Supply Areas (TSA) in Prince Rupert Forest Region, British Columbia, illustrating the preponderance of old natural forests.

Forest Service 1995) stipulates landscape units and seral stage requirements for different Natural Disturbance Types, plus green-up and adjacency constraints. The science basis for these requirements has been questioned. The Alberta Forest Management Science Council (1997,1999) has identified the elements of sustainable forest management as:

(1) ecological integrity and inherent disturbance
(2) a desired future landscape
(3) social and economic values and public involvement
(4) spatial and temporal scales
(5) adaptive management.

British Columbia has many levels of planning of forested landscapes that involve higher level plans with visions, strategies, and zoning into Resource Management Zones. The 1998 "Guide to writing strategic resource management plans" in British Columbia points out that the province has no forest management plans presently, although there is a logical place for them in the hierarchy of plans.

What Is a Forest Management Plan?

According to G. Baskerville, former Chair, Department of Forest Resource Management, University of British Columbia (pers. comm. 1994), for a plan to be credible, implementable, and auditable, and for it to conform with continuous learning and improvement through adaptive management, the plan should have the following nine components.

(1) define a specific forest area for management (order of 200,000 ha)
(2) designate a manager (an actual person) responsible for the design and implementation of management of the defined forest
(3) create a general description of values sought from the forest (this should be a consensus process much as in a present LRMP)
(4) list values to be obtained from the specific forest (targets)
    • What level for each value?
    • What spatial distribution?
    • What temporal distribution?
    • Makes rules for tradeoffs in event of conflict amongst values (this should be a consensus process).
(5) simulate 100-year forest development without management intervention. Show 100-year availability of values (other than timber) by decade and by broad geographic location (for credibility this must be a rigorous process).
(6) simulate 100-year forest development with management interventions (three or four alternative strategies).

Show 100-year availability of timber and other values by decade and by broad geographic location (for credibility this must be a rigorously technical process).

(7) build an implementation guide for the strategy of choice
- a harvest schedule: what will be cut, where, when, with first 20-year detail, 100-year broad picture, all of it stand-type specific, age-class specific, location specific
- a treatment schedule: stand conditions to be treated, timing of a treatment, location specific, and related to harvest schedule
- availability schedule: show values (timber/non timber) that will be available in amount/quality, at what locations, with detail for the first 20 years (for credibility of the plan, this must be a rigorously technical process).

(8) implementation over time
- follow harvest schedule or record/justify deviations
- follow treatment schedule or record/justify deviations
- use a Forest Practices Code to ensure quality of implementation
- assess actual availability of all values and compare to forecast (for credibility of the plan, this must be a rigorously technical process).

(9) periodic review: compare values actually achieved to those forecast, determine basis for difference, review/revise goals, review/revise plan as outlined above (this should be a consensus process).

A key requirement in this list is the spatially explicit planning of harvest locations across the landscape in a way that meets the owners' desired future forest or landscape condition. The concept is that such a desired future forest condition can be identified and that this future condition then becomes the objective of all the protection, harvesting, and renewal actions on the landscape for the next few decades. Audits judge whether the actions are really going to achieve the condition, not whether the Code has been obeyed.

Dr. Baskerville has pointed out that if it is not technically possible to achieve these future conditions as described, it is unprofessional (and perhaps dishonest) to suggest you are managing the forest. Only once the steps suggested are in place does the Code have an appropriate context for quality control of local actions, and in this context the Code should be an excellent tool.

Need for Separate Planning
I have never seen a plan that meets these requirements in British Columbia. The reasons are several:

(1) We have not had a way to "grow" all those thousands of inventory polygons through space and time and produce maps that show what

the future landscapes would look like. This is where the McGregor Model Forest, located near Prince George, British Columbia, has led the way in the development of the "ECHO" planning system. The ECHO planning system uses the McGregor Model Forest spatial analysis techniques that grow every inventory polygon and allow for the production of scenarios with maps of what forested landscapes will look like in the future, based on assumptions and constraints.

(2) Because such planning is not required by law for Timber Supply Areas in British Columbia, money and expertise to do it have not been provided.

(3) Current planning involves operating plans for the next five years on Timber Supply Areas (i.e., a plan that shows where quota will be cut).

(4) It is very difficult to identify a future landscape condition that is the objective of management. Such a future landscape condition should identify all the values the British Columbia public wants to see maintained on the land. From attending two years of Land and Resource Management Plan meetings, I am conscious of how difficult it is to agree on zoning and planning a forested landscape when the table members have little appreciation of the dynamics of the landscape and the future consequences of decisions made today. Again, this is where the future landscape maps produced by the McGregor Model Forest have been very helpful to Land and Resource Management Plans.

The next round of Annual Allowable Cut determination in British Columbia will have to use spatially based computer modeling so that the public can actually see, when land use decisions are made, what will happen to the forests and what the Annual Allowable Cut trade-offs are.

Who will prepare landscape level plans for sustainable forestry in British Columbia? It is clear that the task is complex, requires large data sets, and requires spatial simulation of landscapes. The 1999 Forest Practices Code Landscape Unit Planning Guide is a daunting document (BC Forest Service 1999a). The British Columbia Chief Forester's staff have neither the budget nor the capability to tackle plans of such complexity for 25 million ha of commercial forest land in British Columbia.

Table 1 shows an example of resource information required by the McGregor Model Forest Association for scenario planning. Such a list clearly indicates that forest landscape level planning for Crown lands in British Columbia, nearly all of which are under license to forest companies, requires cooperation among all resource agencies. All parties involved need help running scenarios of feasible options. It should be noted that allowable cut is not "calculated" and thereby is excluded from the scenario that provides for a sustainable landscape. The features of the scenarios are outlined by Weetman (elsewhere in this volume). This approach has already been proposed by the quota holders in the Morice and Lakes Timber Supply

Table 1

A sample list of resource information required for sustainable forestry landscape level planning using scenarios as used by McGregor Model Forest Association (1998).

SCENARIO PLANNING PROJECT TECHNICAL WORK PLAN

We will maintain an up-to-date library containing the full compendium of forest related legislation and Forest Practices Code guidebooks and will ensure all legislative requirements not covered within higher level plans are also accounted for within this list. Based on this review, resource information, which may be required, includes the following.

- Forest inventory planning files (FIP)
- Major licensee silviculture information database (MLSIS)
- Integrated silviculture information database (ISIS)
- In-house databases
- Digital forest cover mapping (FCI)
- Visual landscape mapping (VQOs)
- Forest ecosystem mapping (FENs)
- Special management zone mapping
- Riparian management area mapping (RMAs)
- Riparian classification/inventory maps
- Steams, lakes and wetlands inventories
- Fish inventories (salmonoid spawning, habitat, and rearing)
- Total chance plan maps (TCPs)
- Watershed boundaries
- Landscape unit boundaries
- Compartment boundaries
- Biogeoclimatic zone maps
- Natural Disturbance Types
- Silviculture treatment unit mapping
- Terrain stability mapping
- Existing and proposed timber development maps
- Heritage and cultural resource maps
- Wildlife habitat maps
- Terrestrial ecosystem Maps
- Forest health/pest risk/pest hazard maps
- Recreation Management Classification
- Recreation Opportunity Spectrum
- Recreation Emphasis Areas
- Identified Ungulate Winter Range (caribou medium habitat)
- Caribou High Habitat
- Caribou Corridors
- Grizzly Bear Habitat
- Bull Trout
- Songbird Inventory

Areas and the Robson Valley District (Northwood Inc. 1998; BC Forest Service 1999b). The Land and Resource Management Plan process can and does provide a zoning of forested lands. These zones can be inspected in the scenario planning.

The Foresters Act in British Columbia provides Registered Professional Foresters (RPFs) with the exclusive right to title and practice. Practice includes the preparation of forest management plans. When plans were mainly timber based for "sustained yield" forestry, they were not so complex. For "sustainable forestry," landscape plans require close cooperation of many resource specialists such as Registered Professional Biologists (right to title only), landscape architects, sociologists, and economists. If Registered Professional Foresters are to sign the plans and be accountable for them, they in turn must understand all the issues, analyses, and resource information. This clearly 'ups the ante' for "due diligence" for Registered Professional Foresters whose association does not yet have a requirement for continuing education.

Conclusion

Landscape level planning in British Columbia requires large and complex sets of resource information, a technically rigorous process, sets of analytical steps similar to those proposed by G. Baskerville, and close cooperation between resource professionals, the government, and tenure holders. It is clear that the days of tenure holders submitting management plans to various government agencies for approval is better replaced by initial cooperative planning by all agencies to develop feasible scenarios for sustainable forest management that will meet with the Chief Forester's and the public's approval. A relaxation of "command and control" and "obey the code" attitudes by government is needed, to be replaced by cooperative planning involving all parties and using the new spatially based tools of landscape scenario planning.

References

Alberta Forest Management Science Council. 1997. Sustainable forest management and its major elements. Alberta Department of Environment Protection, Edmonton, AB.

Alberta Forest Management Science Council. 1999. A desired future forest for Alberta: A model. Alberta Department of Environmental Protection, Edmonton, AB.

BC Forest Service. 1995. Biodiversity guidebook: Forest Practices Code of British Columbia. British Columbia Ministry of Forests, Victoria, BC.

BC Forest Service. 1999a. Landscape unit planing guide. British Columbia Ministry of Forests, Victoria, BC.

BC Forest Service. 1999b. Enhanced Forest Management Pilot Project: Robson Valley District. British Columbia Ministry of Forests, McBride, BC.

McGregor Model Forest Association. 1998. Scenario planning project within the McGregor approach to sustainable forest management. Technical Work Plan. Prince George, BC.

Northwood Inc. 1998. Innovative Forest Practices Agreement. Morice and Lakes Timber Supply Area's Proposal. Northwood Inc., Prince George, BC.

A Process, a Science Basis, and Scenario Planning for Forested Landscapes

Gordon F. Weetman

Abstract

This paper briefly reviews the following three examples of recent issues and technologies for forested landscapes in Western Canada.

(1) a public stakeholder Crown Land use zoning process in British Columbia
(2) an example of a science council's advice to the Alberta government on the science basis for sustainable forestry in Alberta
(3) a federally financed non-government initiative to provide an approach to sustainable forests using scenario planning with a spatially based landscape simulation technology for a Tree Farm Licence in northern British Columbia.

Land and Resource Management Plan (LRMP)

These ongoing Land and Resource Management Plan processes have the following features for Timber Supply Area Landscape Zoning:

- Resource Management Zones (RMZs)

- Protected Areas (Pas)
- Special Management Zones (SPZ)
- General Management Zones (GMZ)
- Enhanced Management Zones (EMZ)

- Issues
- Values
- Goals
- Objectives
- Strategies
- Indicators

The zoning is done by citizen representatives who recognize:

- all resource inventory maps
- landscape units
- allowable cut scenarios

- Agriculture Land Reserve
- Forest Practices Code
- visual quality objectives
- Timber Harvesting Land Base
- socio and economic and environmental indicators
- First Nations concerns.

It should be noted that: "As a Cabinet-approved plan the LRMP provides *direction* for *local level* operational plans" on Crown Lands for the next 10 years. All land use zoning is frozen until the Land and Resource Management Plan reaches a consensus. Land and Resource Management Plans are not landscape-level forest management plans (Province of British Columbia 1993; Land Use Coordination Office 1999).

Some Positive Features of the Land and Resource Management Plan (LRMP) Process of Landscape Zoning

- democratic, citizen decision-making for Crown land
- no politicians involved
- facilitated monthly meetings with codes of behaviour and speech
- over a two- to three-year period, members learn to present and defend their interests and reach a consensus in a courteous and civilized manner
- lots of expert presentations on all resource values and interests
- members develop a good understanding of Timber Supply Area (TSA) resources and features and learn to appreciate each other's interests
- meetings are open to the public; no hidden information
- no lobbying, influence peddling, or politicking allowed; non-table attempts are rejected by the Table
- land-use decision-making is frozen while the Land and Resource Management Plan is in process
- GIS-based maps of resources are very revealing
- all people at the Table are volunteers.

Some Problems with the Land and Resource Management Plan Landscape Zoning Process

- First Nations boycotts
- Annual Allowable Cut impacts of Timber Harvesting Land Base reductions are often not known
- Annual Allowable Cut increases due to "enhanced forestry" are often not known
- absence of any long-term harvest schedule and cut block locations

- no dynamic forecast of what the future landscape will look like 50 to 100 years from now
- some Table members feel that the Forest Practice Code of British Columbia (Province of British Columbia 1996) will somehow ensure forest landscape sustainability
- misunderstandings of the sensitivity analyses in the British Columbia Chief Forester's Annual Allowable Cut scenarios
- difficulties in resolving Table interests and demands in mapped zoning
- some Table members think that Land and Resource Management Plan zoning is "management"; actually it is a first step in a management process.

The Alberta Forest Management Science Council

The Alberta Forest Management Council of nine academics was set up in 1997 to advise the Alberta Land and Forest and Natural Resources Sciences on the science required for better sustainable forest management in the province of Alberta. The highlights of the council report are (from Alberta Forest Management Council 1997):

A. Definition of Sustainable Forest Management

The Alberta Forest Management Science Council defines Sustainable Forest Management as: *"The maintenance of the ecological integrity of the forest ecosystem while providing for social and economic values such as ecosystem services, economic, social and cultural opportunities for the benefit of present and future generations."*

B. Elements of Sustainable Forest Management

The Council regards the maintenance of ecosystem integrity as an essential goal for the sound and sustainable management of Alberta's forests. Knowledge about ecosystem processes, including disturbance ecology and resilience to disturbance, provides the necessary context to identify a Desired Future Forest based on a range of possible scenarios. Social and economic values may be identified, quantified, and addressed through public involvement processes. Analysis and planning to reconcile social and economic values with ecological realities must occur at a variety of temporal and spatial scales. Our ability to achieve the desired future forest is constrained by inherent uncertainties within ecological and human systems as well as our limited understanding; therefore, an adaptive management approach is required for the sustainable future of Alberta's forests.

1. Ecological Integrity and Inherent Disturbance

The conservation of ecological integrity of the forest is a necessary condition for the sound and sustainable management of the forest.

Ecological management of the forest develops and applies understanding of how forest ecosystems sustain themselves over long periods of time. It involves examination of growth, development, and the inherent disturbances that underlie the ecological integrity, dynamics, biological diversity and resilience of forest ecosystems. The knowledge enables managers to develop approaches that work with, rather than against, the processes that underlie forest ecosystem sustainability ...

2. Desired Future Forest Landscape

    Defining a vision of a desired future forest in Alberta is a necessary step in implementing a more sustainable forest management program.

    The characteristics of a desired future forest are forecast from the existing forest (including use commitments) and an understanding of the processes under which it has evolved. The forecast includes trends in human use and changes in other elements of disturbance such as wildfire and predicts outcomes in terms of forest structure, composition, ecosystem flows and benefits. The forecast includes transition flows and states from the existing to the desired future forest ...

3. Social and Economic Values and Public Involvement

    Social and economic values are integral to the selection and attainment of the desired future forest.

    Timber and non-timber values, activity levels and existing commitments of the land base should be quantified and integrated into the projection of the future forest condition and flows using scientific methods to determine market and non-market conditions, trends and the evolution of these trends ...

4. Scales – Spatial and Temporal

    The temporal and spatial scales used to manage the forest must be consistent with the scales of disturbances and processes inherent to the forest and social scales relevant to forest resource use ...

5. Adaptive Management

    Adaptive management monitors progress towards the desired future forest, continually improves the knowledge-base and adjusts actions to correct for deviations.

    Adaptive management is a process of hypothesis testing at the scale of whole systems. It continually evaluates and adjusts management relative to predicted responses, objectives and predetermined thresholds of acceptable change.

    Adaptive management includes improvement of the data and analyses on which forest management predictions are based and testing of the assumptions underlying the management practices carried out on forest lands.

The Council report served as a basis for "The Alberta forest legacy: Implementation framework for sustainable forest management" (Alberta Environmental Protection 1998).

The Mcgregor Model Forest Approach to Sustainable Forest Management

The McGregor Model Forest was established with federal model forest funding on Northwood Inc. Tree Farm Licence 30 (181,000 ha) near Prince George, British Columbia. The five million dollar funding was given not to a government or a company but to an association composed of Canadians tasked with finding ways to better manage forested landscapes. The association's technical steering committee struggled for over a year to find an approach to use before committing to a program and budget. Over the last five years an approach has evolved that is a framework to build objective-driven, results-oriented, sustainable forest management (SFM) plans that are practical, efficient, and meaningful to a wide range of stakeholders.

It is comprised of three interrelated initiatives:

* scenario planning
* strategic and operational planning support
* indicators and adaptive management.

It is complemented by enhanced knowledge resulting from research into ecological processes, forest practices, socio-economic values, and inventories.

Objectives

The objectives are to:

* address social forest concerns by helping decide on the values and conditions for which we must plan and manage
* build confidence in forest management by determining what form of future forest is desired
* create ecologically based Sustainable Forest Management plans with explicit objectives that identify which management activities need to be applied, and where and when to apply them
* develop practical indicator monitoring methods linked to the plan objectives
* evaluate and interpret the monitoring, learning and adapting the plan accordingly.

Steps

The steps involved are as follows (Figure 1):

(1) Assess Opportunities

- Participants define the scope of the management problem, synthesize existing knowledge about the system, and explore the potential outcomes of alternative management actions.
- Explicit forecasts are made about outcomes to assess which actions are most likely to meet management objectives.
- During this exploration and forecasting process, key gaps in understanding the system (i.e., gaps that limit the ability to predict outcomes) are identified.

(2) Design Strategy

- Design a management plan and monitoring program that will provide reliable feedback about the effectiveness of chosen actions.
- Ideally, the plan should also be designed to yield information that will fill the key gaps in understanding that were identified in the first step.
- It is useful to evaluate one or more proposed plans or designs on the basis of costs, risks, informativeness, and ability to meet management objectives.

(3) Implement Strategy

- The plan is put into practice by licensee and agencies.

(4) Monitor

- Indicators are monitored to determine how effective the practices are in meeting the plan objectives.
- Indicators also test the hypothesized relationships that formed the basis for the forecasts.

(5) Evaluate

- Compare the actual outcomes to the forecasts and interpret the reasons for any differences.

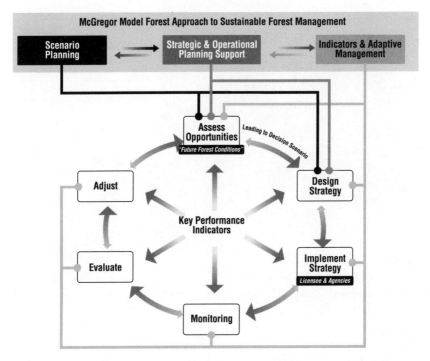

Figure 1. The McGregor Model Forest (Prince George, British Columbia) approach to sustainable forest management.

(6) Adjust

- Practices, objectives, and the models used to make forecasts are adjusted to reflect new understanding.
- Understanding gained through each of these six steps may lead to reassessment of the problem, new questions, and new options to try in a continual cycle of improvement.

Scenario Planning

The first of the three interrelated initiatives, scenario planning, is a facilitated process to assess management opportunities and design sustainable forest management strategies. Diverse participants work cooperatively to define desired future forest conditions (objectives) and explore the management strategies or paths that offer the best hope of achieving the objectives as conditions change over time. Scenario planning is also a method to consider "what's possible" in an uncertain future and to better use the

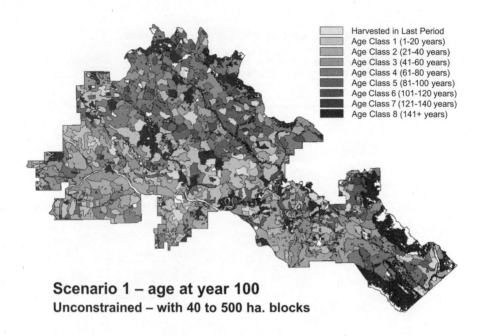

Scenario 1 – age at year 100
Unconstrained – with 40 to 500 ha. blocks

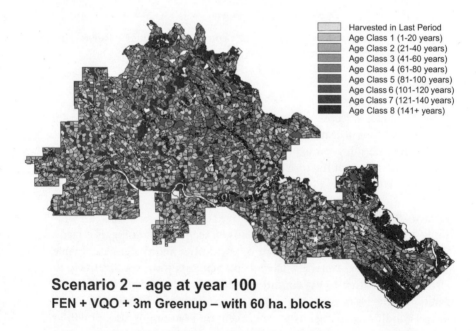

Scenario 2 – age at year 100
FEN + VQO + 3m Greenup – with 60 ha. blocks

Figure 2. An example of modeling management alternatives over huge landscapes within Tree Farm License 30, McGregor Model Forest, Prince George, British Columbia. FEN = Forest Ecosystem Network; VQO = Visual Quality Objectives; 3m Greenup is a cut block adjacency rule.

knowledge and skills of people in designing mutually acceptable and workable management strategies.

Strategic and Operational Planning Support

The second initiative, strategic and operational planning support, involves applying the McGregor Model Forest's suite of modeling, forecasting, and visualization tools to assess the cumulative impacts of how different management strategies may shape the future forest landscape and the possible effects on timber and non-timber values. The suite of tools includes sophisticated blocking, patch, road network, and scheduling models. The models are capable of efficiently handling the spatial and temporal aspects of huge landscapes and help bridge strategic objectives to operational activities.

Indicators and Adaptive Management

The third initiative involves defining, monitoring, evaluating, and reporting on indicators of the forest values that need to be sustained. Linked to plan objectives, indicators act to describe, measure, and track the state of the forest, the values associated with it, and the success of practices in meeting the objectives. Indicators are considered the key to adaptive management – supporting the ongoing adjustment of practices to achieve objectives and highlighting areas that need improvement. The McGregor Model Forest approach used indicators for both forecasting (testing the plan assumptions) and implementing the plan.

Where Is the Approach Being Used?

The approach is being piloted on Northwood Inc.'s Tree Farm Licence 30, an intensively managed area of 181,000 ha near Prince George, British Columbia (Figures 2 and 3). The pilot project addresses mainly company and managing agency issues but will also implement management strategies that are consistent with local Land and Resource Management Plan (LRMP) objectives, particularly biodiversity. The spatial and temporal models employed have been tested on landscapes across British Columbia. The approach is designed to be transferable to other jurisdictions and potential applications are being developed.

Conclusion

Concerns over the sustainability of very large landscapes of natural public forests in Western Canada have raised questions about the scientific basis for sustainability, who will zone the land, and how planning for the future can be done in a spatially explicit and visual way. The three examples given show how these questions are being addressed.

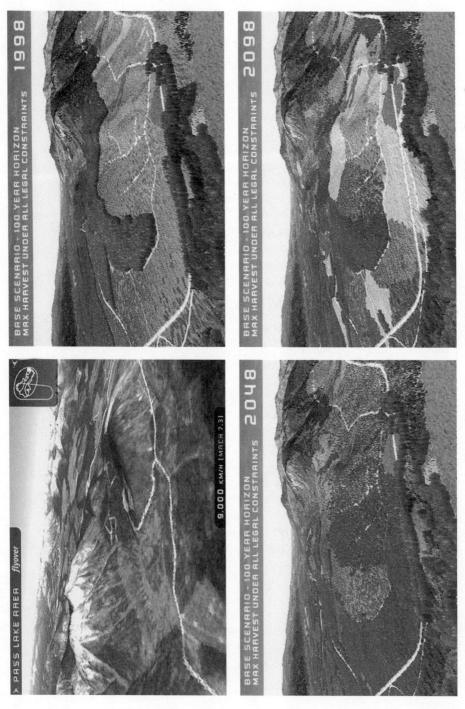

Figure 3. Visualizing potential landscapes by "growing" the landscape in time and space. McGregor Model Forest, Prince George, British Columbia.

References

Alberta Environmental Protection. 1998. The Alberta forest legacy: Implementation for sustainable forest management. Alberta Department of Environmental Protection, Edmonton, AB.

Alberta Forest Management Science Council. 1997. Sustainable forest management and its major elements. Alberta Department of Environmental Protection, Edmonton, AB.

Land Use Coordination Office. 1999. Status report on Land and Resource Management Plans. British Columbia Ministry of Environment, Victoria, BC. On-Line: <http://www.luco.gov.bc/lrmp>.

Province of British Columbia. 1993. Land and resource management planning: A statement of principles and processes, Victoria, BC.

Province of British Columbia. 1996. Forest Practices Code of British Columbia Act, Revised Statues of British Columbia, Chapter 159.

Contributors

Paul Alaback
Professor of Forest Ecology
School of Forestry
University of Montana
Missoula, MT 59812
USA
Telephone: 406-243-5522
Fax: 406-243-6656
E-mail: palaback@forestry.umt.edu

David W. Andison
Consulting Forest Ecologist
Bandaloop Landscape-Ecosystem
  Services
30 Debra Ann Road
Coal Creek Canyon, CO 80403
USA
Telephone: 303-642-1670
Fax: 303-642-7501
E-mail: andisond@aol.com

Dean Apostol
Landscape Architect
23850 SE Borges Road
Gresham, OR 97080
USA
Telephone/Fax: 503-661-6152
E-mail: wordland@mail.aracnet.com

Reiner Augustin
Forest Operations Manager
Kalesnikoff Lumber Company
  Limited
PO Box 3000
Thrums, BC V1N 3L8
Canada
Telephone: 250-399-4211
Fax: 250-399-4170
E-mail: reinera@netidea.com

David Bebb
Consulting Forester
KPMG Consulting
Forestry Specialist Group
9th Floor, 777 Dunsmuir Street
PO Box 10426, Pacific Centre
Vancouver, BC V7Y 1K3
Canada
Telephone: 604-691-3451
Fax: 604-691-3033
E-mail: dbebb@kpmg.ca

Simon Bell
Landscape Architect
British Forestry Commission
231 Corstorphine Road
Edinburgh EH12 7AT
United Kingdom
Telephone: 0131-334-0303
Fax: 0131-334-4473
E-mail: sbell@easynet.co.uk

Patrick W. Daigle
Landscape Ecologist
British Columbia Ministry of Forests,
  Research Branch
PO Box 9519, Stn Prov Govt
Victoria, BC V8W 9C2
Canada
Telephone 250-387-3542
Fax 250-387-0046
E-mail:
  Patrick.Daigle@gems3.gov.bc.ca

Craig DeLong
Landscape Ecologist
Ministry of Forests
1011 4th Avenue
Prince George, BC V2L 3H9
Canada
Telephone: 250-565-6202
Fax: 250-565-4349
E-mail: craig.delong@gems1.gov.bc.ca

Robert G. D'Eon
Consulting Wildlife and Forest
  Ecologist
Kokanee Forests Consulting Ltd.
Box 36
Slocan, BC V0G 2C0
Canada
Telephone: 250-355-2480
Fax: 250-355-2428
E-mail: rdeon@kgis.com

and

Centre for Applied Conservation
  Biology
Faculty of Forestry
University of British Columbia
3rd floor, 2424 Main Mall
Vancouver, BC V6T 1Z4
Canada
Telephone: 604-822-5841
Fax: 604-822-5410

David J. Downing
Manager of Biometrics and Ecology
Timberline Forest Inventory
  Consultants Ltd.
Suite 315, 10357-109 Street
Edmonton, AB T5J 1N3
Canada
Telephone: 780-425-8826
Fax: 780-428-6782
E-mail: djd@timberline.ca

Frederik Doyon
Co-Director
Institut Québécois d'aménagement
  de la forêt feuillue
7 rue St-André
St-André-Avelin, QC J0V 1W0
Canada
E-mail: fdoyon@iqaff.qc.ca

Peter N. Duinker
Professor and Director,
School for Resource and
  Environmental Studies
Dalhousie University
1312 Robie Street
Halifax, NS B3H 3E2
Canada
Telephone: 902-494-7100
Fax: 902-494-3728
E-mail: pduinker@is.dal.ca

B.G. Dunsworth
Forest Renewal and Biodiversity
  Program Leader
Corporate Forestry
MacMillan Bloedel Limited
65 Front Street
Nanaimo, BC V9R 5H9
Canada
Telephone: 250-755-3425
Fax: 250-755-3550
E-mail: bg.dunsworth@mbltd.com

E. Alex Ferguson
Woodlands Manager
Slocan Forest Products Ltd.
705 Delaney Avenue
Slocan, BC V0G 2C0
Canada
Telephone: 250-355-2100
Fax: 250-355-2168
E-mail: eaf@netidea.com

Susan M. Glenn
Associate Professor of Conservation
  Biology and Landscape Ecoloogy
Centre for Applied Conservation
  Biology
Department of Forest Science
University of British Columbia
2424 Main Mall
Vancouver, BC V6T 1Z4
Canada
Telephone: 604-822-4131
Fax: 604-822-9102
E-mail: sglenn@interchange.ubc.ca

Jonathan B. Haufler
Boise Cascade Corporation
PO Box 50
Boise ID 83728
USA
Telephone: 208-384-6013
Fax: 208-384-7699
E-mail: Jon_Haufler@bc.com

Daryll Hebert
Consulting Wildlife and Forest
  Ecologist
Encompass Strategic Resources
RR 1, Site 60, Box 40
Creston, BC V0B 1G0
Canada
Telephone: 250-428-3092
Fax: 250-428-7073

Dan Jepsen
Manager of Aboriginal Affairs and
  Environment
Western Forest Products Limited
2300-1111 West Georgia Street
Vancouver, BC V6E 4M3
Canada
Telephone: 604-665-6200
Fax: 604-665-6268

Bart Johnson
Associate Professor of Landscape
  Architecture
227 Lawrence Hall
University of Oregon
Eugene OR 97404-5234
USA
Telephone: 541-346-3688
Fax: 541-346-2168
E-mail: bartj@darkwing.uoregon.edu

Keith Jones
Principal, R. Keith Jones and
  Associates
2554 Bowker Avenue
Victoria, BC V8R 2G1
Canada
Telephone: 250-598-2635
Fax: 250-598-2630
E-mail: kjones@tnet.net

J.P. (Hamish) Kimmins
Professor of Forest Ecology
Faculty of Forestry
Department of Forest Sciences
University of British Columbia
2424 Main Mall
Vancouver, BC V6T 1Z4
Canada
Telephone: 604-822-3549
Fax: 604-822-9133
E-mail: kimmins@interchg.ubc.ca

Michael Krebs
School of Forestry
University of Montana
Missoula, MT 59812
USA

Peter Landres
Research Ecologist
Aldo Leopold Wilderness Research
  Institute
Rocky Mountain Research Station
US Forest Service
PO Box 8089
Missoula, MT 59807
USA
Telephone: 406-542-4190
E-mail: plandres/
  rmrs_missoula@fs.fed.us

Marcelo Levy
Coordinator
Forest Stewardship Council - Canada
49 Myrtle Avenue
Toronto, ON M4M 2A4
Canada
Telephone: 416-778-5568
Fax: 416-778-0044
E-mail: fscca@web.net

G. Lui
Hugh Hamilton Limited
850 W. 15th Street,
North Vancouver, BC V7P 1M6
Canada
Telephone: 604-980-5061
Fax: 604-986-0361

Dan Mack
Consulting GIS Specialist
Cave Creek Systems
Box 641
Nelson, BC V1L 4B6
Canada
Telephone: 250-352-3085
E-mail: dan@cavecreeksystems.com

Carolyn A. Mehl
Wildlife and Ecosystem Management
  Associates
5 Yellowpine Drive
Boise, ID 83716
USA

Penny Morgan
Associate Professor
Department of Forest Resources
University of Idaho
Moscow, ID 83844-1133
USA
Telephone: 208-885-7507
Fax: 208-885-6226
E-mail: pmorgan@uidaho.edu

Jan Mulder
Alberta Research Council
3rd Floor, 6815-8th Street NE
Calgary, AB T2E 7H7
Canada
Telephone: 403-297-7570
Fax: 403-297-2339
E-mail: mulder@arc.ab.ca

John Nelson
Associate Professor of Forest
  Management
Faculty of Forestry
University of British Columbia
2nd floor, 2424 Main Mall
Vancouver, BC V6T 1Z4
Canada
Telephone: 604-822-3902
Fax: 604-822-9106
E-mail: nelson@interchange.ubc.ca

J.C. Niziolomski
Consulting Forester
Hugh Hamilton Limited
850 W. 15th Street,
North Vancouver, BC V7P 1M6
Canada
Telephone: 604-980-5061
Fax: 604-986-0361
E-mail: chris_niz@hugh-
  hamilton.com

Russ Parsons
Research Assistant
Department of Forest Resources
University of Idaho
Moscow, ID 83844-1133
USA
Telephone: 208-885-7952
Fax: 208-885-6226
E-mail: pars8342@uidaho.edu

Heather Pinnell
Consulting Forester
H. Pinnell Forestry Consulting
RR 3
Nelson, BC V1L 5P6
Canada

Harry Quesnel
Consulting Forest Ecologist
Ecotessera Consultants Ltd.
RR 1
Nelson, BC V1L 5P4
Canada
Telephone: 250-825-4204
Email: hquesnel@netidea.com

Chris Ridley-Thomas
Consulting Forester
KPMG Consulting
Forestry Specialist Group
9th Floor, 777 Dunsmuir Street
PO Box 10426, Pacific Centre
Vancouver, BC V7Y 1K3
Canada
Telephone: 604-691-3088
Fax: 604-691-3033
E-mail: cridley-thomas@kpmg.ca

Gary J. Roloff
Boise Cascade Corporation
1203 Wolf Court
East Lansing, MI 48823
USA

Paul Rosen
School of Forestry
University of Montana
Missoula, MT 59812
USA

Greg Rowe
Consulting Forester
Rowe Forest Management
497 Parkway Road
Campbell River, BC V9W 6C3
Canada
Telephone: 250-923-7093
Fax: 250-923-3903
E-mail: rowe@island.net

Ron Rutledge
Acting Program Manager
Muskwa-Kechika Management Area
Suite 150, 10003-110 Avenue
Fort St. John, BC V1J 6M7
Canada
Telephone: 250-787-3534
Fax: 250-787-3490
E-mail: ron.rutledge@gems9.
 gov.bc.ca

Marcia Sinclair
Public Involvement Consultant
23850 SE Borges Road
Gresham, OR 97080
USA
Telephone/Fax: 503-661-6152

Stephen M. Smith
Consulting Forester
Sterling Wood Group Inc.
264-2950 Douglas Street
Victoria, BC V8T 4N4
Canada
Telephone: 250-384-7161
Fax: 250-384-0321
E-mail: ssmith@sterlingwood.com

C. Wardman
Hugh Hamilton Limited
850 W. 15th Street,
North Vancouver, BC V7P 1M6
Canada
Telephone: 604-980-5061
Fax: 604-986-0361
E-mail:
   cwardman@hugh-hamilton.com

Gordon F. Weetman
Professor Emeritus of Forest Sciences
Department of Forest Sciences
University of British Columbia
3041-2424 Main Mall
Vancouver, BC V6T 1Z4
Canada
Telephone 604-822-2504
Fax 604-822-9102
E-mail: gweetman@interchg.ubc.ca

Ralph Wells
Research Scientist
Centre for Applied Conservation
   Biology
Faculty of Forestry
University of British Columbia
3rd floor, 2424 Main Mall
Vancouver, BC V6T 1Z4
Canada
Telephone: 604-822-8288
Fax: 604-822-5410
E-mail: rwells@interchange.ubc.ca